Design of Materials for Bone Tissue Scaffolds

Design of Materials for Bone Tissue Scaffolds

Editor

Antonio Boccaccio

MDPI • Basel • Beijing • Wuhan • Barcelona • Belgrade • Manchester • Tokyo • Cluj • Tianjin

Editor
Antonio Boccaccio
Department of Mechanics,
Mathematics and Management
Politecnico di Bari
Bari
Italy

Editorial Office
MDPI
St. Alban-Anlage 66
4052 Basel, Switzerland

This is a reprint of articles from the Special Issue published online in the open access journal *Materials* (ISSN 1996-1944) (available at: www.mdpi.com/journal/materials/special_issues/bone_tissue_scaffolds).

For citation purposes, cite each article independently as indicated on the article page online and as indicated below:

LastName, A.A.; LastName, B.B.; LastName, C.C. Article Title. *Journal Name* **Year**, *Volume Number*, Page Range.

ISBN 978-3-0365-3954-6 (Hbk)
ISBN 978-3-0365-3953-9 (PDF)

© 2022 by the authors. Articles in this book are Open Access and distributed under the Creative Commons Attribution (CC BY) license, which allows users to download, copy and build upon published articles, as long as the author and publisher are properly credited, which ensures maximum dissemination and a wider impact of our publications.

The book as a whole is distributed by MDPI under the terms and conditions of the Creative Commons license CC BY-NC-ND.

Contents

About the Editor vii

Preface to "Design of Materials for Bone Tissue Scaffolds" ix

Antonio Boccaccio
Design of Materials for Bone Tissue Scaffolds
Reprinted from: *Materials* **2021**, *14*, 5985, doi:10.3390/ma14205985 1

Óscar Libardo Rodríguez-Montaño, Carlos Julio Cortés-Rodríguez, Antonio Emmanuele Uva, Michele Fiorentino, Michele Gattullo and Vito Modesto Manghisi et al.
An Algorithm to Optimize the Micro-Geometrical Dimensions of Scaffolds with Spherical Pores
Reprinted from: *Materials* **2020**, *13*, 4062, doi:10.3390/ma13184062 5

Maria Contaldo, Alfredo De Rosa, Ludovica Nucci, Andrea Ballini, Davide Malacrinò and Marcella La Noce et al.
Titanium Functionalized with Polylysine Homopolymers: In Vitro Enhancement of Cells Growth
Reprinted from: *Materials* **2021**, *14*, 3735, doi:10.3390/ma14133735 23

Adrian Ionut Nicoara, Alexandra Elena Stoica, Denisa-Ionela Ene, Bogdan Stefan Vasile, Alina Maria Holban and Ionela Andreea Neacsu
In Situ and Ex Situ Designed Hydroxyapatite: Bacterial Cellulose Materials with Biomedical Applications
Reprinted from: *Materials* **2020**, *13*, 4793, doi:10.3390/ma13214793 37

Francesca Posa, Adriana Di Benedetto, Giampietro Ravagnan, Elisabetta Ada Cavalcanti-Adam, Lorenzo Lo Muzio and Gianluca Percoco et al.
Bioengineering Bone Tissue with 3D Printed Scaffolds in the Presence of Oligostilbenes
Reprinted from: *Materials* **2020**, *13*, 4471, doi:10.3390/ma13204471 55

Katarzyna Krukiewicz, David Putzer, Nicole Stuendl, Birgit Lohberger and Firas Awaja
Enhanced Osteogenic Differentiation of Human Primary Mesenchymal Stem and Progenitor Cultures on Graphene Oxide/Poly(methyl methacrylate) Composite Scaffolds
Reprinted from: *Materials* **2020**, *13*, 2991, doi:10.3390/ma13132991 67

Ana R. Bastos, Lucília P. da Silva, F. Raquel Maia, Sandra Pina, Tânia Rodrigues and Filipa Sousa et al.
Lactoferrin-Hydroxyapatite Containing Spongy-Like Hydrogels for Bone Tissue Engineering
Reprinted from: *Materials* **2019**, *12*, 2074, doi:10.3390/ma12132074 79

Haruka Ishida, Hisao Haniu, Akari Takeuchi, Katsuya Ueda, Mahoko Sano and Manabu Tanaka et al.
In Vitro and In Vivo Evaluation of Starfish Bone-Derived β-Tricalcium Phosphate as a Bone Substitute Material
Reprinted from: *Materials* **2019**, *12*, 1881, doi:10.3390/ma12111881 95

Gianluca Percoco, Antonio Emmanuele Uva, Michele Fiorentino, Michele Gattullo, Vito Modesto Manghisi and Antonio Boccaccio
Mechanobiological Approach to Design and Optimize Bone Tissue Scaffolds 3D Printed with Fused Deposition Modeling: A Feasibility Study
Reprinted from: *Materials* **2020**, *13*, 648, doi:10.3390/ma13030648 111

Daniel Martinez-Marquez, Ylva Delmar, Shoujin Sun and Rodney A. Stewart
Exploring Macroporosity of Additively Manufactured Titanium Metamaterials for Bone Regeneration with Quality by Design: A Systematic Literature Review
Reprinted from: *Materials* **2020**, *13*, 4794, doi:10.3390/ma13214794 . . . . **131**

Nikola Stokovic, Natalia Ivanjko, Drazen Maticic, Frank P. Luyten and Slobodan Vukicevic
Bone Morphogenetic Proteins, Carriers, and Animal Models in the Development of Novel Bone Regenerative Therapies
Reprinted from: *Materials* **2021**, *14*, 3513, doi:10.3390/ma14133513 . . . . **175**

Felice Roberto Grassi, Roberta Grassi, Leonardo Vivarelli, Dante Dallari, Marco Govoni and Gianna Maria Nardi et al.
Design Techniques to Optimize the Scaffold Performance: Freeze-dried Bone Custom-made Allografts for Maxillary Alveolar Horizontal Ridge Augmentation
Reprinted from: *Materials* **2020**, *13*, 1393, doi:10.3390/ma13061393 . . . . **199**

Iris Jasmin Santos German, Karina Torres Pomini, Ana Carolina Cestari Bighetti, Jesus Carlos Andreo, Carlos Henrique Bertoni Reis and André Luis Shinohara et al.
Evaluation of the Use of an Inorganic Bone Matrix in the Repair of Bone Defects in Rats Submitted to Experimental Alcoholism
Reprinted from: *Materials* **2020**, *13*, 695, doi:10.3390/ma13030695 . . . . **209**

About the Editor

Antonio Boccaccio

Antonio Boccaccio holds an MS in Mechanical Engineering (graduated cum laude) (Politecnico di Bari, Italy, 2002) and a PhD in Bioengineering (Politecnico di Milano, Italy, 2006). Since December 2019, he has been the Associate Professor (ING-IND/15 Design Methods for Industrial Engineering) at the Politecnico di Bari. In 2005, he was a Visiting Research Scholar at the Centre for Bioengineering, Trinity College, Dublin. Professor Boccaccio's research interests are morphological optimization of biomaterials, modeling and simulation of biomedical devices and mechanobiological processes, and optical techniques for reverse engineering. He has authored more than 120 publications (56 peer-reviewed ISI journal papers, 3 editorials, 1 letter, 2 national journal papers, 10 book chapters, 3 invited lectures, 49 conference papers) and 1 European patent. He is a member of the Editorial Board of three internationally reputed journals and serves as a reviewer for more than 40 ISI journals. In 2012, he has been awarded the Fylde Electronics Prize for the best paper published in 2010 in the journal *Strain* by the British Society for Strain Measurement. He also ranked first (among 40 applicants) in a university competition for post-doctoral fellowships held in the Politecnico di Bari. In 2021, he received the Progetto Ingegneria award as the "Professor Most Voted" by the students of the degree course in Industrial Design, Politecnico di Bari.

Preface to "Design of Materials for Bone Tissue Scaffolds"

The strong growth recently experienced by the manufacturing technologies, along with the development of innovative biocompatible materials, has allowed the fabrication of high-performing scaffolds for bone tissue engineering. The design process of materials for bone tissue scaffolds presently represents an issue of crucial importance and is being studied by many researchers throughout the world. A number of studies have been conducted, aimed at identifying the optimal material, geometry, and surface that the scaffold must possess to stimulate the formation of the largest amounts of bone in the shortest time possible. This book presents a collection of 10 research articles and 2 review papers describing numerical and experimental design techniques definitively aimed at improving the scaffold performance, shortening the healing time, and increasing the success rate of the scaffold implantation process.

Antonio Boccaccio
Editor

Editorial

Design of Materials for Bone Tissue Scaffolds

Antonio Boccaccio

Dipartimento di Meccanica, Matematica e Management, Politecnico di Bari, 70125 Bari, Italy; antonio.boccaccio@poliba.it; Tel.: +39-080-5963393

Abstract: The strong impulse recently experienced by the manufacturing technologies as well as the development of innovative biocompatible materials has allowed the fabrication of high-performing scaffolds for bone tissue engineering. The design process of materials for bone tissue scaffolds represents, nowadays, an issue of crucial importance and the object of study of many researchers throughout the world. A number of studies have been conducted, aimed at identifying the optimal material, geometry, and surface that the scaffold must possess to stimulate the formation of the largest amounts of bone in the shortest time possible. This book presents a collection of 10 research articles and 2 review papers describing numerical and experimental design techniques definitively aimed at improving the scaffold performance, shortening the healing time, and increasing the success rate of the scaffold implantation process.

Keywords: bone tissue engineering; porous materials; bone regeneration

Scaffolds for bone tissue engineering are porous materials that are used to reconstruct large dimensions bone defects. The ideal scaffold should satisfy to the following three principal requirements: (1) it should exhibit a structural response that is adequate and as close as possible to that of the tissues adjacent to the fracture site; (2) it should be biocompatible and biodegradable; (3) it should possess adequate surfaces capable of promoting the adhesion of mesenchymal stem cells, their proliferation and their subsequent osteogenic differentiation [1]. It is commonly known that the rate of bone tissue regeneration and the cellular response is significantly influenced by: (a) the scaffold mechanical behavior, which is, in turn, a function of the scaffold micro-architecture and of the mechanical properties of the material it is made from [2,3]; (b) the surface roughness status and the biological/chemical response of the scaffold/tissue interface surfaces to external factors [4]. The adhesion of stem cells to the scaffold surface as well as the tissue differentiation process occurring in the scaffold pores are regulated by very complex mechanobiological mechanisms taking place at both the micro- (i.e., some micrometers, approximately the dimension of a stem cell) and macro- (i.e., some hundreds of micrometers, corresponding to the typical dimensions of scaffold pores) levels, respectively [5–9]. The scaffold surface must be adequately structured to favor the adhesion of stem cells and their consequent differentiation. Similarly, the scaffold architecture must be properly shaped, and the scaffold material must be adequately designed to trigger favorable biophysical stimuli, leading to the formation of the bony tissue.

Many studies have recently been conducted to investigate the optimal manufacturing technologies that can be used to fabricate "smart and custom" scaffolds capable not only of guaranteeing the above-mentioned requirements, but also of satisfying the specific requests of the specific patient in whom it will be implanted [5]. One of the most recent research lines, in fact, has been focused on the design of "personalized" scaffolds that better suit the anthropometric features of the patient, thus allowing to achieve a successful follow-up in the shortest possible time [10]. Different studies have recently been published with the aim of better understanding the relationship between the scaffold geometry/material properties and the consequent mechanobiological phenomena taking place inside the scaffold

Citation: Boccaccio, A. Design of Materials for Bone Tissue Scaffolds. *Materials* **2021**, *14*, 5985. https://doi.org/10.3390/ma14205985

Received: 17 September 2021
Accepted: 6 October 2021
Published: 12 October 2021

Publisher's Note: MDPI stays neutral with regard to jurisdictional claims in published maps and institutional affiliations.

Copyright: © 2021 by the author. Licensee MDPI, Basel, Switzerland. This article is an open access article distributed under the terms and conditions of the Creative Commons Attribution (CC BY) license (https://creativecommons.org/licenses/by/4.0/).

during the regeneration process. However, no clear explanations are yet available on the relationship existing between the mechanical/chemical environment and the consequent biological response of tissues occupying the scaffold pores. This Special Issue attempts to bridge the gap and to give a possible response to the open questions.

Most of the studies of the Special Issue developed innovative materials favoring the formation of new bone in the fracture site where the scaffold is implanted [11–16]. Three papers investigate the issues related to the geometry/dimensions that the scaffold pores must possess to guarantee an adequate mechanobiological response [10,17,18]. Finally, three articles deal with more clinical/applicative aspects [19–21].

The studies investigating innovative materials concern not only the material the scaffold is made from, but also all the materials in presence of which mesenchymal stem cells can be put to favor their adhesion, proliferation, and differentiation. In detail, Nicoara et al. [12], synthesized and characterized two types of materials—with antibacterial properties provided by silver nanoparticles (AgNPs)—based on hydroxyapatite and bacterial cellulose, that are known to possess excellent biocompatibility and bioactivity properties and are, hence, particularly suited to be used in the field of bone tissue engineering. The obtained composite materials were found to have a homogenous porous structure, a high water absorption capacity, and a considerable antimicrobial effect due to silver nanoparticles embedded in the polymer matrix. The fabrication of a composite bone cement made of graphene oxide and poly(methyl methacrylate) was described by Krukiewicz et al. [14], who investigated the potential of this cement to enhance the osteogenic differentiation of human primary mesenchymal stem and progenitor cells. Bastos et al. [15] developed an advanced three-dimensional (3D) biomaterial by integrating bioactive factors, such as lactoferrin and hydroxyapatite, within gellan gum spongy-like hydrogels. The authors demonstrated that that gellan gum spongy-like hydrogels gathered favorable 3D bone-like microenvironment with an increased human adipose-derived stem cells viability. Ishida et al. [16], evaluated starfish-derived β-tricalcium phosphate obtained by phosphatization of starfish-bone-derived porous calcium carbonate as a potential bone substitute material. They concluded that starfish-derived β-tricalcium phosphate may be effective for bone regeneration applications, such as in the treatment of fractures and bone loss. The osteoblastic features of adult mesenchymal stem cells integrated with 3D-printed polycarbonate scaffolds differentiated in the presence of oligostilbenes, such as resveratrol and polydatin, were investigated by Posa et al. [13]. They found that both resveratrol and polydatine stimulate the adhesion of the mesenchymal stem cells to the bone matrix protein osteopontin via $\alpha_V\beta_3$ integrin and, specifically, polydatine treatment prompted a greater reorganization of this integrin in focal adhesion sites. The effects of a titanium surface coated with polylysine homopolymers on the cell growth of dental pulp stem cells and keratinocytes was investigated by Contaldo et al. [11]. They found an increase in cell growth for both cellular types cultured with polylysine-coated titanium compared to cultures without titanium and those without coating.

Very interesting are also the studies investigating the geometry of the scaffold pores, as well as the issues related to the structural response to mechanical loads and the scaffold porosity. Percoco et al. [17] and Rodríguez-Montaño et al. [10], using the mechanoregulation model by Prendergast et al. [22], determined the optimal dimensions that the pores of scaffolds 3D printed with the FDM technique and including spherical pores, respectively, must possess. In this model, the fracture site is modelled as a biphasic poroelastic material, and the biophysical stimulus that triggers the osteogenic differentiation of the mesenchymal stem cells is hypothesized to be a function of the octahedral shear strain and of the interstitial fluid flow measured in the regenerating tissue. The authors, by using this model, defined, via an optimization algorithm, the optimal dimensions of pores for different load values acting on the scaffold [10,17]. Martinez-Marquez et al. [18], in their review paper, used the quality by design system to explore the quality target product profile and ideal quality attributes of additively manufactured titanium porous scaffolds for bone regeneration with a biomimetic approach. The systematic literature review presented an overview of the reported properties in research studies of fully porous titanium bone implants fabricated with additive manufacturing published in the last two decades. Unit cell geometry, porosity, elastic modulus, compressive yield strength, ultimate compressive yield strength, and compressive fatigue strength were systematically reviewed and benchmarked against the proposed ideal quality attributes.

The studies dealing with applicative/clinical aspects investigate very wide and interesting topics. The effects of chronic alcoholism on the repair of bone defects associated with xenograft was investigated by German et al. [21]. The interesting review paper by Stokovic et al. [19] summarizes the bone regeneration strategies and the animal models used for the initial, intermediate, and advanced evaluation of promising therapeutical solutions for new bone formation and repair. Dentistry issues were investigated by Grassi et al. [20], who evaluated the clinical success of horizontal ridge augmentation in severely atrophic maxilla using freeze-dried, custom-made bone harvested from the tibial hemiplateau of cadaver donors.

All the papers of the Special Issue were submitted to peer review, and thanks to the help of the reviewers, the quality of all the manuscripts was significantly improved. My special thanks go, therefore, to the authors for their excellent contribution, to the reviewers, for their invaluable help, as well as to the editorial staff of *Materials*, in particular to Ariel Zhou, Section Managing Editor for her kind assistance, competence and patience.

Funding: I thank the Italian Ministry of Education, University and Research under the Programme: (1) PON R&I 2014–2020 and FSC (Project 'CONTACT', ARS01_01205); (2) 'Department of Excellence' Legge 232/2016 (Grant No. CUP - D94I18000260001), for the funding received.

Conflicts of Interest: The author declares no conflict of interest.

References

1. Tariverdian, T.; Sefat, F.; Gelinsky, M.; Mozafari, M. *Scaffold for Bone Tissue Engineering*; Elsevier Ltd.: Amsterdam, The Netherlands, 2019; ISBN 9780081025635.
2. Byrne, D.P.; Lacroix, D.; Planell, J.A.; Kelly, D.J.; Prendergast, P.J. Simulation of tissue differentiation in a scaffold as a function of porosity, Young's modulus and dissolution rate: Application of mechanobiological models in tissue engineering. *Biomaterials* **2007**, *28*, 5544–5554. [CrossRef]
3. Rodríguez-Montaño, Ó.L.; Cortés-Rodríguez, C.J.; Uva, A.E.; Fiorentino, M.; Gattullo, M.; Monno, G.; Boccaccio, A. Comparison of the mechanobiological performance of bone tissue scaffolds based on different unit cell geometries. *J. Mech. Behav. Biomed. Mater.* **2018**, *83*, 28–45. [CrossRef]
4. Zadpoor, A.A. Bone tissue regeneration: The role of scaffold geometry. *Biomater. Sci.* **2015**, *3*, 231–245. [CrossRef]
5. Bose, S.; Roy, M.; Bandyopadhyay, A. Recent advances in bone tissue engineering scaffolds. *Trends Biotechnol.* **2012**, *30*, 546–554. [CrossRef]
6. Adachi, T.; Osako, Y.; Tanaka, M.; Hojo, M.; Hollister, S.J. Framework for optimal design of porous scaffold microstructure by computational simulation of bone regeneration. *Biomaterials* **2006**, *27*, 3964–3972. [CrossRef]
7. Rodríguez-Montaño, Ó.L.; Cortés-Rodríguez, C.J.; Naddeo, F.; Uva, A.E.; Fiorentino, M.; Naddeo, A.; Cappetti, N.; Gattullo, M.; Monno, G.; Boccaccio, A. Irregular Load Adapted Scaffold Optimization: A Computational Framework Based on Mechanobiological Criteria. *ACS Biomater. Sci. Eng.* **2019**, *5*, 5392–5411. [CrossRef]
8. Sandino, C.; Checa, S.; Prendergast, P.J.; Lacroix, D. Simulation of angiogenesis and cell differentiation in a CaP scaffold subjected to compressive strains using a lattice modeling approach. *Biomaterials* **2010**, *31*, 2446–2452. [CrossRef] [PubMed]
9. Sandino, C.; Planell, J.A.; Lacroix, D. A finite element study of mechanical stimuli in scaffolds for bone tissue engineering. *J. Biomech.* **2008**, *41*, 1005–1014. [CrossRef] [PubMed]
10. Rodríguez-Montaño, Ó.L.; Cortés-Rodríguez, C.J.; Uva, A.E.; Fiorentino, M.; Gattullo, M.; Manghisi, V.M.; Boccaccio, A. An algorithm to optimize the micro-geometrical dimensions of scaffolds with spherical pores. *Materials* **2020**, *13*, 4062. [CrossRef] [PubMed]
11. Contaldo, M.; De Rosa, A.; Nucci, L.; Ballini, A.; Malacrinò, D.; La Noce, M.; Inchingolo, F.; Xhajanka, E.; Ferati, K.; Bexheti-Ferati, A.; et al. Titanium functionalized with polylysine homopolymers: In vitro enhancement of cells growth. *Materials* **2021**, *14*, 3735. [CrossRef]
12. Nicoara, A.I.; Stoica, A.E.; Ene, D.I.; Vasile, B.S.; Holban, A.M.; Neacsu, I.A. In situ and ex situ designed hydroxyapatite: Bacterial cellulose materials with biomedical applications. *Materials* **2020**, *13*, 4793. [CrossRef]
13. Posa, F.; Di Benedetto, A.; Ravagnan, G.; Cavalcanti-Adam, E.A.; Muzio, L.L.; Percoco, G.; Mori, G. Bioengineering bone tissue with 3d printed scaffolds in the presence of oligostilbenes. *Materials* **2020**, *13*, 4471. [CrossRef]

14. Krukiewicz, K.; Putzer, D.; Stuendl, N.; Lohberger, B.; Awaja, F. Enhanced Osteogenic Differentiation of Human Composite Scaffolds. *Materials* **2020**, *13*, 1–12.
15. Bastos, A.R.; Maia, F.R.; Pina, S.; Rodrigues, T.; Sousa, F.; Oliveira, J.M.; Cornish, J. Hydrogels for Bone Tissue Engineering. *Materials* **2019**, *12*, 2074. [CrossRef] [PubMed]
16. Ishida, H.; Haniu, H.; Takeuchi, A.; Ueda, K.; Sano, M.; Tanaka, M.; Takizawa, T.; Sobajima, A.; Kamanaka, T.; Saito, N. In Vitro and In Vivo Evaluation of Starfish Bone-Derived β-Tricalcium Phosphate as a Bone Substitute Material. *Materials* **2019**, *12*, 1881. [CrossRef]
17. Percoco, G.; Uva, A.E.; Fiorentino, M.; Gattullo, M.; Manghisi, V.M.; Boccaccio, A. Mechanobiological approach to design and optimize bone tissue scaffolds 3D printed with fused deposition modeling: A feasibility study. *Materials* **2020**, *13*, 648. [CrossRef] [PubMed]
18. Martinez-Marquez, D.; Delmar, Y.; Sun, S.; Stewart, R.A. Exploring macroporosity of additively manufactured titanium metamaterials for bone regeneration with quality by design: A systematic literature review. *Materials* **2020**, *13*, 4794. [CrossRef]
19. Stokovic, N.; Ivanjko, N.; Maticic, D.; Luyten, F.P.; Vukicevic, S. Bone morphogenetic proteins, carriers, and animal models in the development of novel bone regenerative therapies. *Materials* **2021**, *14*, 3513. [CrossRef]
20. Grassi, F.R.; Grassi, R.; Vivarelli, L.; Dallari, D.; Govoni, M.; Nardi, G.M.; Kalemaj, Z.; Ballini, A. Design techniques to optimize the scaffold performance: Freeze-dried bone custom-made allografts for maxillary alveolar horizontal ridge augmentation. *Materials* **2020**, *13*, 1393. [CrossRef]
21. German, I.J.S.; Pomini, K.T.; Bighetti, A.C.C.; Andreo, J.C.; Reis, C.H.B.; Shinohara, A.L.; Rosa, G.M.; de Bortoli Teixeira, D.; de Oliveira Rosso, M.P.; Buchaim, D.V.; et al. Evaluation of the use of an inorganic bone matrix in the repair of bone defects in rats submitted to experimental alcoholism. *Materials* **2020**, *13*, 695. [CrossRef]
22. Prendergast, P.J.; Huiskes, R.; Søballe, K. Biophysical stimuli on cells during tissue differentiation at implant interfaces. *J. Biomech.* **1997**, *30*, 539–548. [CrossRef]

Communication

An Algorithm to Optimize the Micro-Geometrical Dimensions of Scaffolds with Spherical Pores

Óscar Libardo Rodríguez-Montaño [1,2], Carlos Julio Cortés-Rodríguez [1], Antonio Emmanuele Uva [2], Michele Fiorentino [2], Michele Gattullo [2], Vito Modesto Manghisi [2] and Antonio Boccaccio [2,*]

1. Departamento de Ingeniería Mecánica y Mecatrónica, Universidad Nacional de Colombia, 111321 Bogotá, Colombia; olrodriguezm@unal.edu.co (Ó.L.R.-M.); cjcortesr@unal.edu.co (C.J.C.-R.)
2. Dipartimento di Meccanica, Matematica e Management, Politecnico di Bari, 70125 Bari, Italy; antonio.uva@poliba.it (A.E.U.); michele.fiorentino@poliba.it (M.F.); michele.gattullo@poliba.it (M.G.); vitomodesto.manghisi@poliba.it (V.M.M.)
* Correspondence: antonio.boccaccio@poliba.it; Tel.: +39-080-5963393

Received: 14 August 2020; Accepted: 11 September 2020; Published: 13 September 2020

Abstract: Despite the wide use of scaffolds with spherical pores in the clinical context, no studies are reported in the literature that optimize the micro-architecture dimensions of such scaffolds to maximize the amounts of neo-formed bone. In this study, a mechanobiology-based optimization algorithm was implemented to determine the optimal geometry of scaffolds with spherical pores subjected to both compression and shear loading. We found that these scaffolds are particularly suited to bear shear loads; the amounts of bone predicted to form for this load type are, in fact, larger than those predicted in other scaffold geometries. Knowing the anthropometric characteristics of the patient, one can hypothesize the possible value of load acting on the scaffold that will be implanted and, through the proposed algorithm, determine the optimal dimensions of the scaffold that favor the formation of the largest amounts of bone. The proposed algorithm can guide and support the surgeon in the choice of a "personalized" scaffold that better suits the anthropometric characteristics of the patient, thus allowing to achieve a successful follow-up in the shortest possible time.

Keywords: geometry optimization; computational mechanobiology; bone tissue engineering; python code; parametric CAD (Computer Aided Design) model

1. Introduction

One of the main issues recently investigated in the field of bone tissue engineering and that has received substantial attention is the identification of the optimal geometry of bony tissue scaffolds to support the numerous cellular activities involved in bone formation and regeneration [1]. Scaffolds are porous structures that mainly perform a dual function: transporting nutrients, waste, and oxygen, and a structural function consisting of transferring the load to the cells and regenerated tissues occupying their pores and to the adjacent tissues where they are implanted [2,3]. A large number of porous topologies have been studied from both the theoretical and the experimental point of view, but there is not yet a consensus between researchers regarding the geometry that the "optimal" scaffold should possess to maximize the amounts of regenerated bone [4]. However, some "general" guidelines are commonly accepted in the literature such as the range of the dimensions that pores have to possess to favor the regeneration process [5].

In general, bone tissue scaffolds can be classified into two principal categories: irregular and regular. Regular scaffolds are fabricated using advanced manufacturing processes such as additive layer manufacturing (ALM) that allow controlling with high precision the specific dimension of the single unit cell the scaffold is made from. The irregular scaffolds are fabricated with conventional physical-chemical processes that allow controlling the average dimensions of the scaffold microarchitecture only on a statistical base [6]. A typical advantage of regular structures is the regularity of the scaffold domain that implies the regularity of the physical environment and hence the regularity of the mechanical stimulus acting on the regenerating tissue.

A very interesting scaffold topology is that including spherical pores. It is commonly known that the adhesion and differentiation of stem cells take place more easily on curved surfaces, especially on concave surfaces [5,7]. Scaffold topologies including spherical pores were recently produced with ALM techniques [8]. Spherical pores are also included in previously explored scaffold geometries such as FCC (face-centered cubic), BCC (body-centered cubic) [9,10], and Schwartz-P primitives [11–13]. However, no studies are reported in the literature optimizing the geometry of scaffolds with spherical pores, with the scope of maximizing the amounts of neo-formed bone. Here we aim to bridge this gap. We modeled the scaffold and the tissues occupying it as biphasic poroelastic materials and computed the biophysical stimulus acting on the tissue inside the scaffold pores according to the model of Prendergast et al. [14], as a function of the octahedral shear strain and the interstitial fluid flow. The objective of this study was to identify the optimal geometrical parameters of a regular scaffold with spherical pores and cylindrical interconnections that maximize the amounts of neo-formed bone. We found that this scaffold topology is particularly suited to bear shear loads. The proposed model fits well the requirements of so-called Precision Medicine (i.e., the branch of Medicine that studies personalized medical solutions for the specific requirements of the patient) and tries to answer the question about the optimal scaffold micro-geometry to achieve a successful follow-up in the shortest possible time.

2. Materials and Methods

2.1. Unit Cell Geometry

The parametric model of a scaffold occupying a cubic volume of side $L = 2.548$ mm and including $4 \times 4 \times 4 = 64$ unit cells was developed. The same scaffold dimensions were utilized in previous studies [15,16]. The general purpose software Abaqus (version 6.12, Dassault Systèmes, Vélizy-Villacoublay, France) was utilized for both the parametric geometry modeling and the finite element analysis. Each unit cell is a hexahedron with a spherical cavity and cylindrical interconnections oriented along the orthogonal directions of the coordinate axes. It can be obtained as a Boolean subtraction of the volume of a sphere with cylinders from a cubic volume with the side $L_{uc} = L/4$ (Figure 1). Depending on the diameter of the spherical surface D_s, two different unit cell topologies can be designed: a "small" (S) topology where $0 < D_s \leq L_{uc}$ and a "large" (L) topology where $L_{uc} < D_s < L_{uc} \times \sqrt{2}$ (Figure 2). Obviously, spherical diameters $D_s > L_{uc} \times \sqrt{2}$ are not allowed, as the geometry deriving from such an assumption would lead to a scaffold unit cell completely different with respect to that hypothesized. Regarding the diameters of cylinders D_c, other constraints must be respected depending on the specific topology. In the case of Topology (S), the diameter of cylinders must satisfy the following inequality:

$$0 < D_c \leq D_s / \sqrt{2}, \tag{1}$$

In the section views obtained with a plane cutting the unit cell in half (Figure 3a), the figure of a square (represented with a dashed line, Figure 3) can be traced as the intersection of the edges of the cylinders. If this square is included within the edge of the spherical surface (highlighted in blue, Figure 3), the inequality (1) is verified. Inside the unit cell, a unique spherical surface can be identified that is interrupted by the cylindrical surfaces (Figure 3b). When the vertices of the square touch the spherical edge, the condition

$$D_c = D_s / \sqrt{2}, \tag{2}$$

is reached. Finally, when the vertices of the square go beyond the spherical edges, only isolated (i.e., $D_c > D_s/\sqrt{2}$) or no (i.e., $D_c \gg D_s/\sqrt{2}$) portions of spherical surface can be identified, and the geometry of the unit cell changes completely with respect to that hypothesized, which leads to the change in the scaffold connectivity.

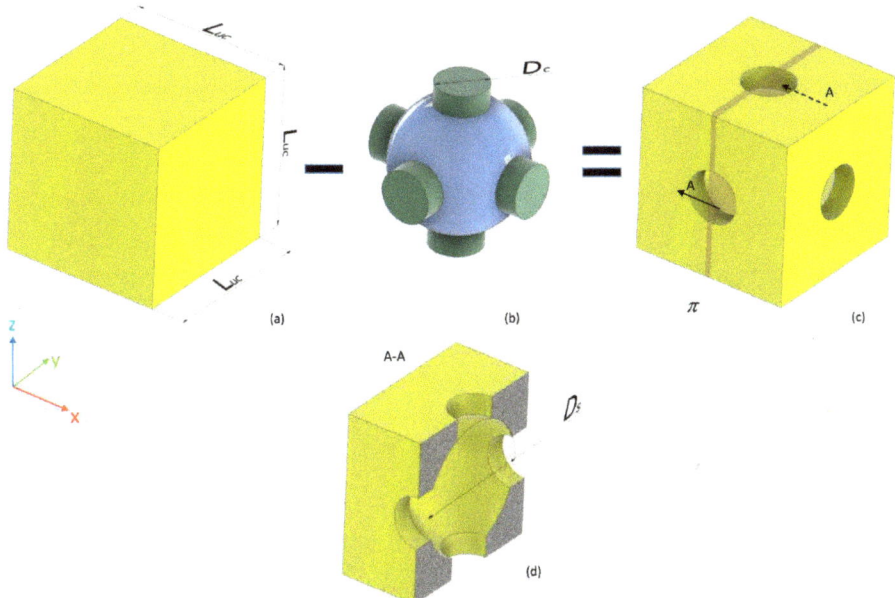

Figure 1. To build the scaffold unit cell (**c**), a boolean subtraction was carried out between a cubic volume (side $L_{uc} = L/4$) (**a**) and the volume of a sphere (highlighted in blue) with cylinders (highlighted in green) oriented orthogonally according to the coordinate axes (**b**). The section A-A view (**d**) with the plane π (**c**), shows how the unit cell is interiorly made.

In the case of Topology (L), the diameter of cylindrical surfaces D_c must satisfy the following inequality

$$\sqrt{(D_s^2 - L_{uc}^2)} < D_c \leq D_s/\sqrt{2}, \tag{3}$$

In fact, to guarantee the "coherence" of the hypothesized scaffold geometry, the cylindrical diameter must be greater than the length of the chord C obtained by the intersection of the spherical edge with the edge of the cylindrical surface (Figure 4). The length of the chord is given by

$$C = \sqrt{(D_s^2 - L_{uc}^2)}, \tag{4}$$

The considerations regarding the figure of the square that can be traced in the section view as the intersection of the cylindrical edges continue to remain valid also in the case of the Topology (L) and, consequently, lead to define the upper limit for D_c that must be $D_c \leq D_s/\sqrt{2}$. Table 1 summarizes the constraint equations that D_s and D_c must satisfy to guarantee that the unit cell geometry remains the same, thus conserving its "intrinsic" coherence, for the variable values that D_s and D_c can assume.

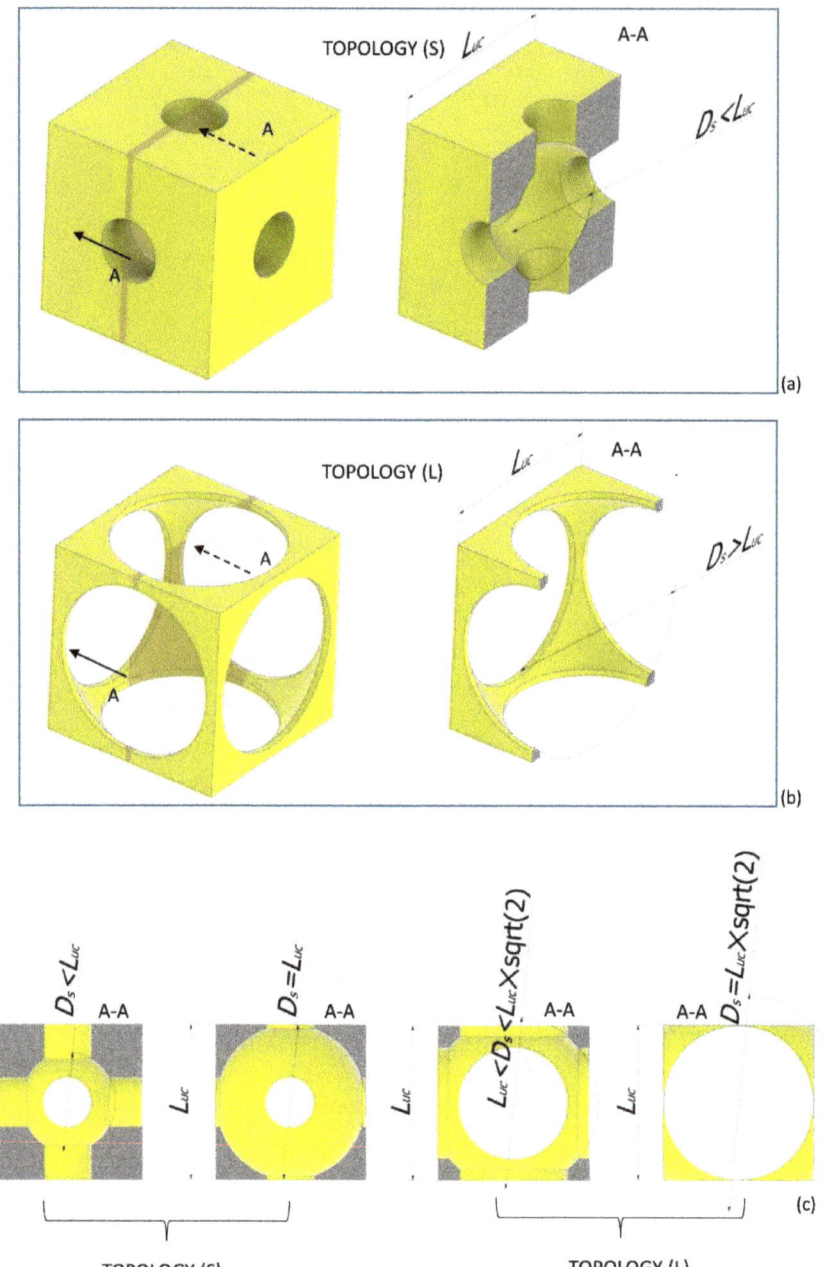

Figure 2. Two different topologies can be built for the scaffold unit cell: "small" (S) (**a**) and "large" (L) (**b**). Topology (S) includes a spherical surface with $0 < D_s \leq L_{uc}$ (**c**); Topology (L) includes a spherical surface with $L_{uc} < D_s \leq \sqrt{2} \times L_{uc}$ (**c**).

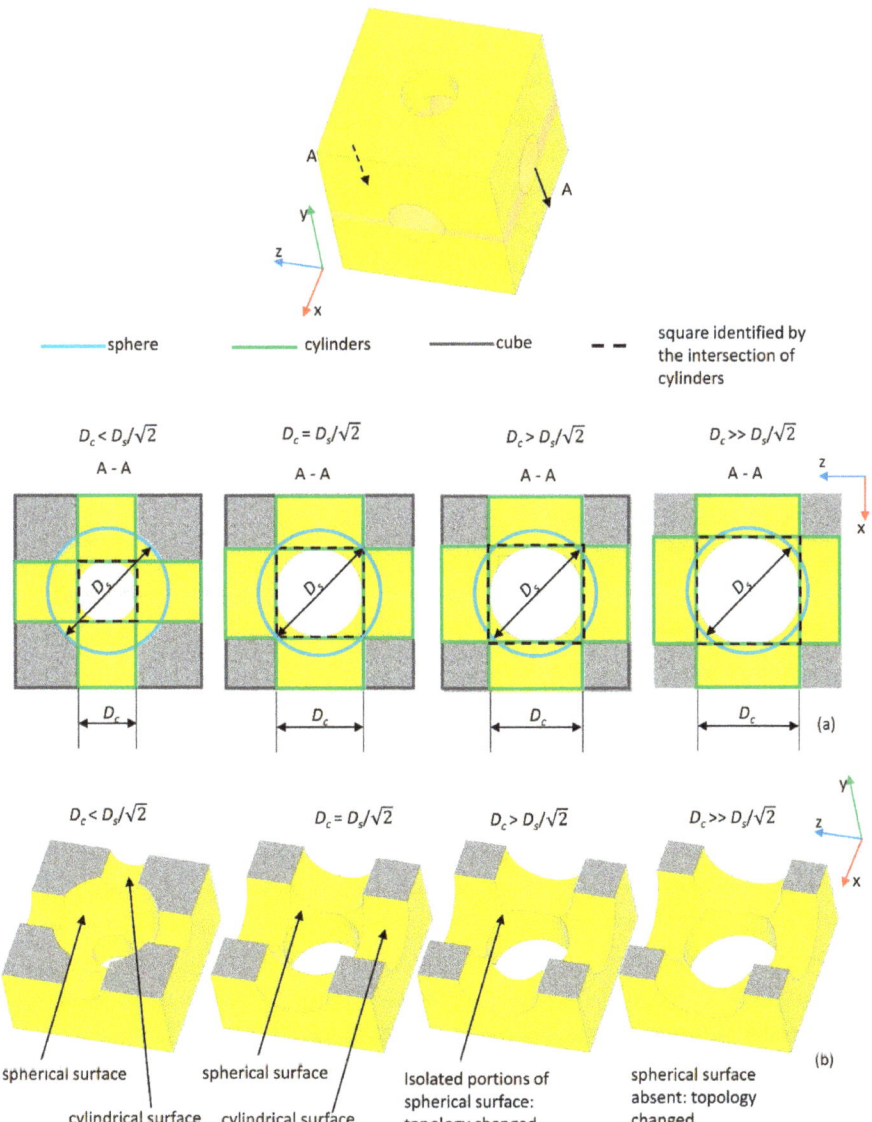

Figure 3. (**a**) Section views—in the plane x–z—of the scaffold unit cell (topology (S) with indicated edges of the primitives (cube, cylinders, and sphere) utilized. When the square obtained by the intersection of the cylinders touches with its vertices, the spherical edge (in blue), the limit condition $D_c = D_s/\sqrt{2}$ is reached. For $D_c > D_s/\sqrt{2}$, the topology of the unit cell changes. (**b**) Section views—in the three-dimensional space—of the unit cell obtained for different values of D_s and D_c.

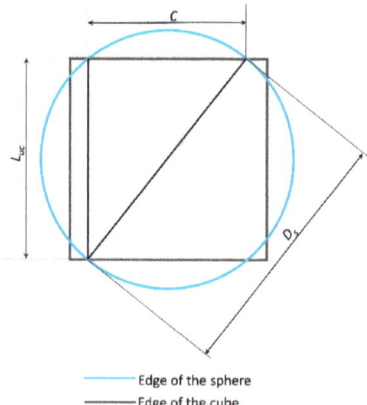

Figure 4. Schematic utilized to determine the equation constraint that the diameter of cylinders D_c must satisfy in the case of Topology (L).

Table 1. Constraint equations that the diameter of the sphere D_s and the cylinders D_c must satisfy to guarantee the coherence of the scaffold geometry.

Constraint Equation for D_s	Topology	Constraint Equation for D_c
if $0 < D_s \leq L_{uc}$ →	Small topology (S) →	$0 < D_c \leq D_s/\sqrt{2}$
if $L_{uc} < D_s \leq L_{uc} \times \sqrt{2}$ →	Large topology (L) →	$\sqrt{(D_s^2 - L_{uc}^2)} < D_c \leq D_s/\sqrt{2}$

2.2. Scaffold Model and Applied Boundary and Loading Conditions

The unit cell described above was mirrored with respect to different planes and replicated 64 times to generate the geometry of the entire scaffold (Figure 5). The model includes also the granulation tissue, highlighted in red (Figure 5), occupying the scaffold pores. Both the scaffold and the granulation tissue were modeled as biphasic poroelastic materials with the same material properties (Table 1) as those utilized in previous studies [15,17,18].

A rigid plate (highlighted in blue, Figure 5d,e) was fixed at the upper face of the scaffold-granulation tissue system using a tie constraint to uniformly transfer the load. A tie constraint between the scaffold and granulation tissue was also established to prevent any relative displacement between these two materials. On the bottom surface of the model, an encastre boundary condition was fixed, while for the outer surfaces of the granulation tissue, a pore pressure equal to zero was set to allow, according to Byrne et al. [19], the free exudation of fluid. Two different loading conditions were hypothesized: a compression (Figure 5d) and a shear (Figure 5e) load. The values of load per unit area F_{UA} hypothesized in this study were the same as those utilized in a previous article [16]: in the case of compression load, 0.05, 0.1, 0.5, 1.0, and 1.5 MPa, and in the case of shear load, 0.01, 0.05, 0.1, 0.2, and 0.5 MPa. C3D4P tetrahedral elements available in Abaqus® were used to discretize the model. The average element size and the maximum deviation factor were set at 50 μm and 0.01, respectively.

A python script was generated that allows automatically (i) building the scaffold and the granulation tissue geometry; (ii) applying the boundary and the loading conditions; (iii) discretizing the model into finite elements; and (iv) running the finite element analyses. This script was then incorporated within an optimization code written in Matlab (Version R2016b, MathWorks, Natick, MA, USA) that, based on mechanobiological criteria deriving from the model of Prendergast et al. [14], allows the optimal scaffold geometry to be predicted.

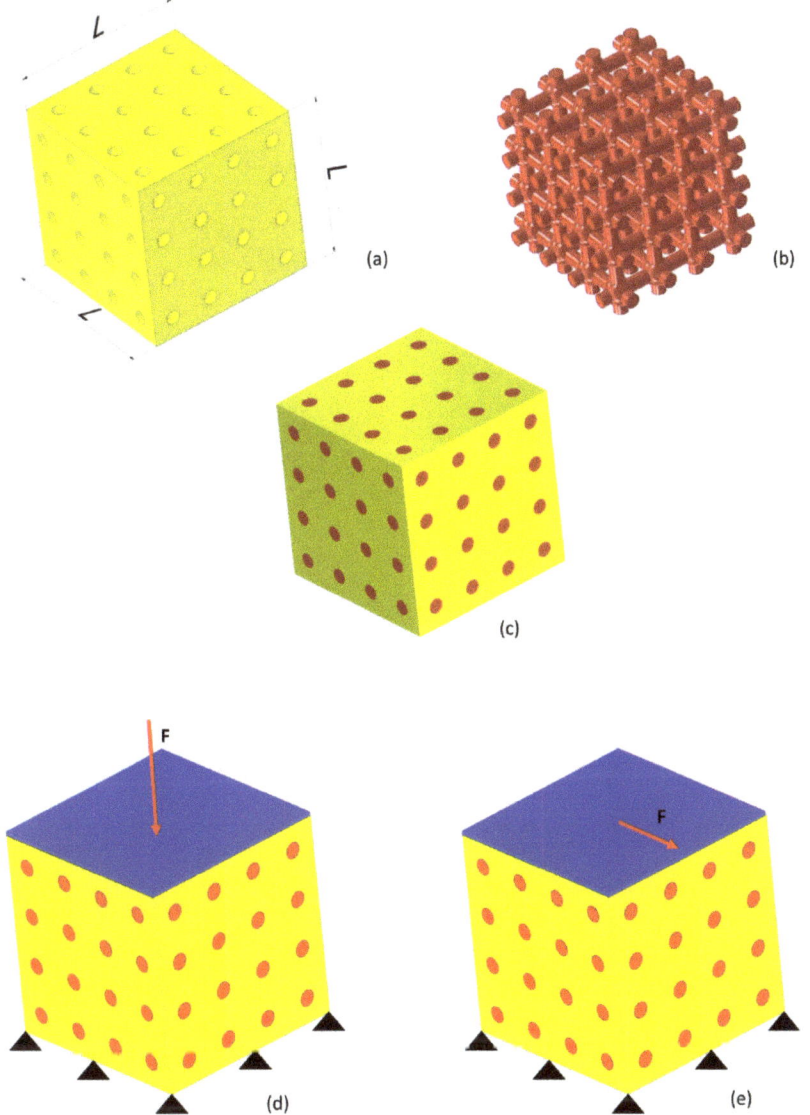

Figure 5. The CAD models of scaffold (**a**) and granulation tissue (**b**) were assembled to generate the model (**c**) utilized in the study. Two different boundary and loading conditions were hypothesized to act on the model: a compression load **F** ($|\mathbf{F}| = F_{UA} \times L \times L$) on the upper surface and an encastre on the lower one (**d**); a shear load **F** ($|\mathbf{F}| = F_{UA} \times L \times L$) on the upper surface and an encastre on the lower surface (**e**).

2.3. A Brief Outline of the Mechano-Regulation Model Implemented to Determine the Scaffold Optimal Geometry

Once the scaffold is implanted in the region with bone deficiency, mesenchymal stem cells (MSCs) migrate from the adjacent tissues, thus invading the scaffold. Therefore, MSCs start their differentiation process. The model of Prendergast et al. [14] assumes that the biophysical stimulus S that triggers the differentiation process in the fracture domain is a function of the octahedral shear strain and of the interstitial fluid flow acting on the mesenchymal tissue. Depending on the values that S assumes, differentiation into different phenotypes, such as fibroblasts, chondrocytes, or osteoblasts, will be stimulated. The ranges of the biophysical stimulus S that determine the fate of the MSCs are described in the following inequalities:

$$\begin{aligned} S > 3 &\rightarrow \text{Fibroblasts (Fibrous tissue)} \\ 1 < S < 3 &\rightarrow \text{Chondrocytes (Cartilage)} \\ 0.53 < S < 1 &\rightarrow \text{Osteoblasts (Immature bone)} \\ 0.01 < S < 0.53 &\rightarrow \text{Osteoblasts (Mature bone)} \\ 0 < S < 0.01 &\rightarrow \text{Bone resorption} \end{aligned} \tag{5}$$

Further details on the mechano-regulation algorithm can be found in previous studies [20,21].

2.4. Optimization Algorithm

The optimization algorithm aims to identify the scaffold geometry that allows maximizing the amounts of neo-formed bone for each value of force per unit area F_{UA} hypothesized in the study (Figure 6).

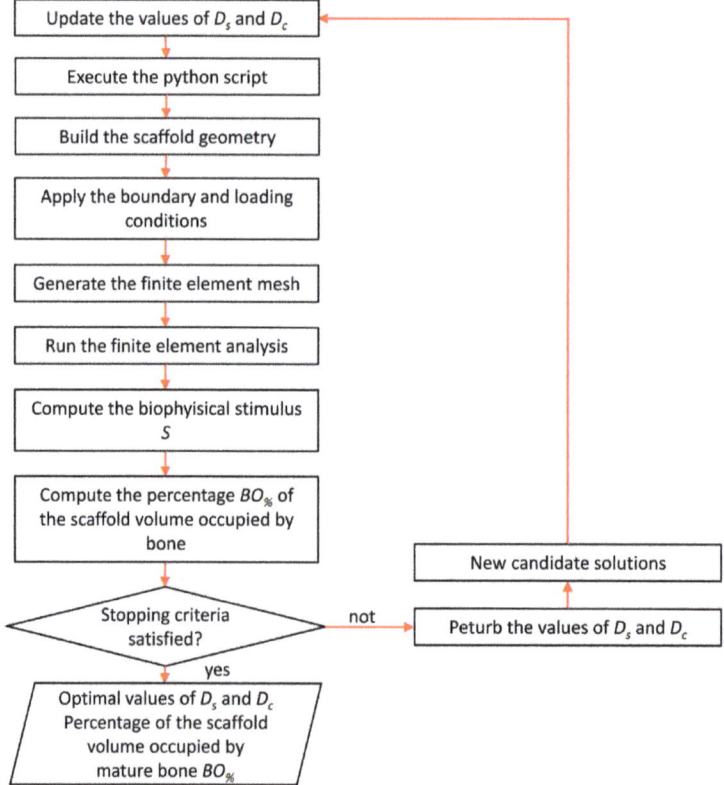

Figure 6. Schematic of the optimization algorithm implemented to determine the optimal scaffold geometry.

In detail, the algorithm, written in Matlab, employs the *fmincon* function from the Matlab optimization toolbox to determine the optimal values of the design variables D_s and D_c that maximize $BO_\%$, the percentage of the scaffold volume occupied by mature bone. In each optimization cycle, the values of D_s and D_c are perturbed and entered into a python script. This script is given in input to Abaqus, which builds the model, applies the boundary and loading conditions, generates the mesh, and runs the finite element analysis. Then, the algorithm reads the results of the FEM analysis, computes the biophysical stimulus S, and compares it with the boundary values reported in the inequalities (5). At this point, it computes $BO_\%$, the percentage of the scaffold volume occupied by mature bone, as the ratio between the volume of the elements with S that satisfy the inequality $0.01 < S < 0.53$, and the total volume of the scaffold $L \times L \times L$. The algorithm perturbs so many times the values of D_s and D_c until the maximum value of $BO_\%$ is determined. Once this occurs, the optimization algorithm stops and outputs the predicted optimal values of the design variables D_s and D_c as well as the value of the percentage $BO_\%$, which represents the maximum percentage of the scaffold volume that can be occupied by bone for a given load value. During the optimization process, D_s and D_c can assume variable values concerning both (L) and (S) Topology but must always satisfy the constraint equations summarized in Table 1.

All the optimization analyses were conducted on an HP XW6600-Intel®Xeon®DualProcessor E5-5450 3 GHz–32 Gb RAM workstation (Intel Corporation, Mountain View, CA, USA) and required approximately 1500 h of computation.

3. Results and Discussion

The optimized scaffold geometries predicted by the proposed algorithm in the case of compression load present spherical pores and cylindrical interconnections that become smaller for increasing values of the load (Figure 7). This can be explained with the argument that as the load increases, the biophysical stimulus acting on the mesenchymal tissue increases too, thus favoring the formation of soft tissues like cartilage and fibrous tissue. Hence, the algorithm to counterbalance this tends to increase the scaffold stiffness by decreasing the dimensions of the spherical pores and the cylindrical connections (Figure 7a,b). Comparing the percentages $BO_\%$ with those predicted in a previous study [20] for regular scaffolds based on a hexahedron unit cell with elliptic and rectangular extrusions, we found that scaffolds with rectangular extrusions perform always better than those with spherical pores. Conversely, those with elliptic extrusions work better than the scaffolds with spherical pores only for high load values (Figure 7c). When the load is high, in fact, elliptic and rectangular extrusions tend to orientate according to the load direction, which makes the scaffold more "suited" to bear and transfer the compression load acting on it.

The optimal geometries predicted in the case of shear load present pores with dimensions that get increasingly smaller as we move towards higher load values (Figure 8a,b). Interestingly, in this case, the scaffold with spherical pores performs, for all the hypothesized values of shear load, better than those with elliptic and rectangular extrusions (Figure 8c).

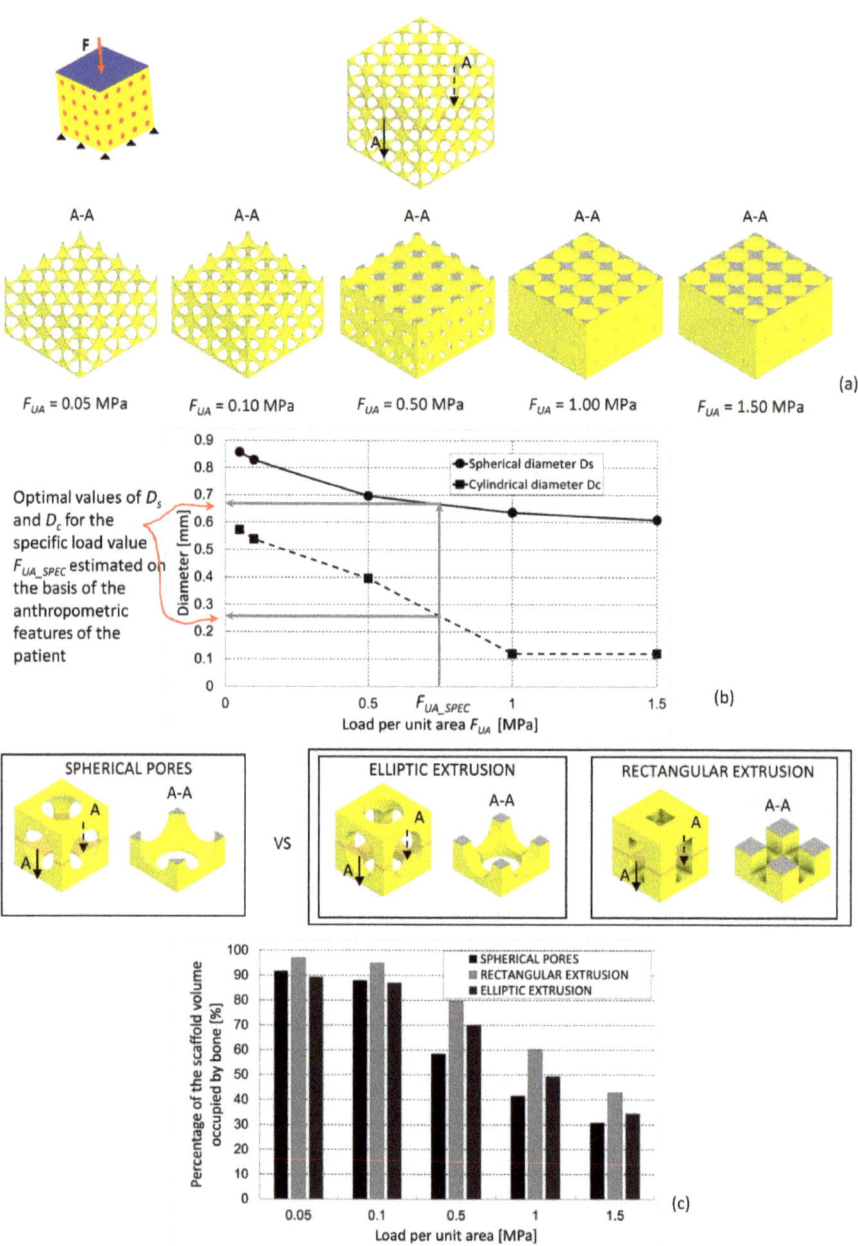

Figure 7. (**a**) Optimized scaffold geometries (section views A-A), (**b**) optimal values of D_s and D_c, and (**c**) percentage of the scaffold volume occupied by mature bone, predicted by the optimization algorithm for different values of the compression load. The percentages of bone are compared with those predicted for scaffolds with hexahedron unit cells including elliptic and rectangular extrusions [20].

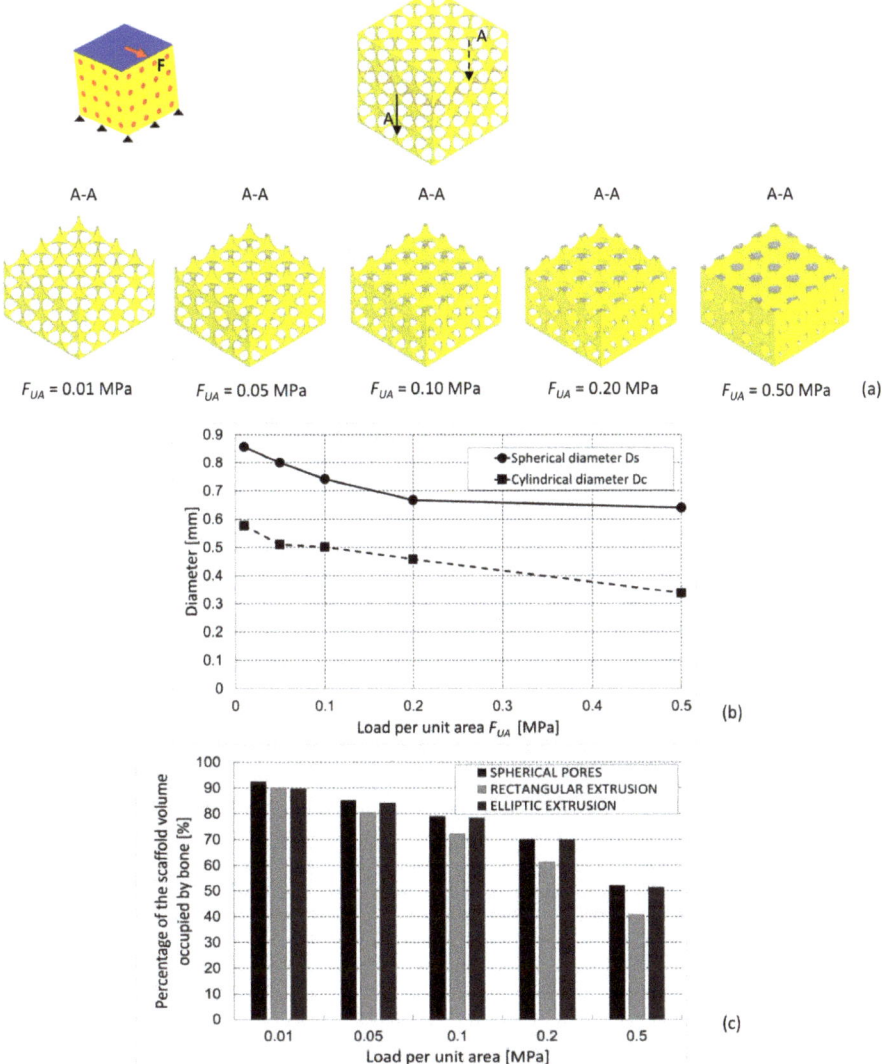

Figure 8. (**a**) Optimized scaffold geometries (section views A-A), (**b**) optimal values of D_s and D_c, and (**c**) percentage of the scaffold volume occupied by mature bone, predicted by the optimization algorithm for different values of the shear load. The percentages of bone are compared with those predicted for scaffolds with hexahedron unit cells including elliptic and rectangular extrusions [20].

In general, the biophysical stimulus S acting on the mesenchymal tissue assumes higher values in the proximity of the spherical pores, while smaller values are observed in the proximity of the cylindrical interconnections (Figure 9). The regularity of the scaffold geometry leads to a regular distribution of the biophysical stimulus that is repeated with approximately the same characteristics as many times as the cells of the scaffold. Such a spatial distribution demonstrates that the biophysical stimulus depends on the scaffold geometry and on how this transfers the load to the mesenchymal tissue.

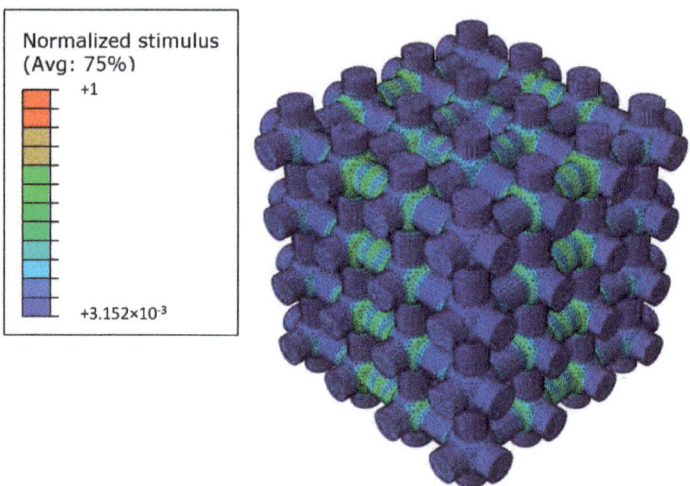

Figure 9. Spatial distribution of the normalized biophysical stimulus S/S_{max} computed for a scaffold ($D_S = 0.425$ mm $D_c = 0.275$ mm) subjected to the compression load of $F_{UA} = 0.5$ MPa.

The proposed study has some limitations. First, the model includes a spherical pore the diameter of which was optimized based on the mechanobiological model of Prendergast et al. [14]. As demonstrated in previous studies [16,20], scaffolds oriented according to the load direction perform better than those without a specific orientation [18]. To make the proposed geometry "oriented" according to the load direction, the spherical surface should be changed with that of prolate or oblate spheroids. In this case, the number of variables to optimize are two: the minor and the major axis of the spheroid. With this strategy, the spheroidal surface would properly orient, thus making the scaffold more "suited" to bear and transfer the load acting on it [22–24]. This topic will be the objective of future studies. Second, a clear and direct experimental study that demonstrates the correctness of the predictions of the proposed model is, at the moment, lacking. In general, it is difficult to systemically study the effects of scaffold geometry on the process of bone tissue regeneration. The identification of the geometrical features that principally affect the tissue differentiation process occurring in a scaffold requires the systematic study of different scaffold geometries. However, at the moment, no such studies are available in the literature [7]. Third, a simplified hypothesis was followed regarding the diffusion of mesenchymal stem cells once the scaffold is implanted. The event in which the MSCs migrate from the adjacent tissues and invade the scaffold could not take place *sic et simpliciter*. In fact, once a scaffold is implanted, it will be most likely infiltrated with blood, which clots within a few minutes, thus clogging the pores of the scaffolds. Moreover, other cells such as connective tissue fibroblasts could compete with MSCs to colonize the scaffolds. However, in the case where MSCs are the only cells entering the scaffold, having a highly osteogenic microarchitecture, once the new bone is deposited, it will prevent further MSCs inwards migration and bone ingrowth. Studies on the transient phase of the MSCs migration and diffusion through the scaffold should be carried out in the future. Fourth, the proposed algorithm allows to determine the optimal dimensions of the spherical pores and the cylindrical interconnections. However, this poses relevant technological issues in the sense that the proposed approach requires the implementation of additive manufacturing techniques that must guarantee adequate precision for the produced scaffolds. Stereolithography is one of the most powerful and versatile additive manufacturing techniques [25]. It has the highest fabrication accuracy, which ranges from 1.2 to 200 μm [26]. Fused deposition modelling (FDM) was demonstrated to have the lowest precision [27]. The experimental tests previously conducted with FDM demonstrated that this technique is suitable to build accurate scaffold samples only in the cases where the strand diameter is close to the nozzle diameter. Conversely, when a large difference exists, large fabrication errors can be committed on the diameter of the filaments [17]. Scaffolds fabricated with selective laser sintering (SLS) show dimensional deviations—with respect to the nominal dimensions—up to 7.5% [28]. Fifth, the scaffold model investigated has rather small dimensions with respect to those of the scaffolds commonly used in the clinical context. In principle, using a larger scaffold model is possible but poses serious issues of computational power. Sixth, the time variable was not included in the proposed algorithm, i.e., we do not simulate how the bone regeneration process takes place in the scaffold and optimize the scaffold geometry based on the "picture" taken at the instant of time zero, after its implantation. In reality, the inclusion of the time variable requires very high computational power and a computational time tremendously longer than the time required to perform the optimization analyses carried out in this study. In fact, for each candidate geometrical solution, the algorithm should ideally predict how the bony tissue growths and how the scaffold dissolves. This series of analysis cycles should be repeated as many times as the cycles required by the optimization algorithm, which leads to computational times at least two orders of magnitude larger than those required in this study. Increases in computational power will ultimately allow simulating the bone regeneration and the scaffold dissolution processes to optimize the scaffold geometry on a temporal perspective as well as modelling scaffolds with dimensions closer to those actually employed in clinical practice.

Despite these limitations, the proposed model shows a mechanical behavior consistent with that of spongy bone. In fact, if we compute the ratio E_{app}/E, where E_{app} is the "apparent" Young's modulus of the scaffold considered in its entirety and $E = 1000$ MPa is Young's modulus of the material the scaffold is made from (Table 2), we find values falling within the variability range of this ratio experimentally measured for cancellous bone (Figure 10).

Table 2. Material properties utilized in the model of scaffold and granulation tissue [15,17,18].

Material Property	Granulation Tissue	Scaffold
Young's modulus [MPa]	0.2	1000
Poisson's ratio	0.167	0.3
Permeability [m4/(Ns)]	1×10^{-14}	1×10^{-14}
Porosity	0.8	0.5
Bulk modulus grain [MPa]	2300	13,920
Bulk modulus fluid [MPa]	2300	2300

To compute the ratio E_{app}/E, three different finite element models of the sole scaffold (i.e., the granulation tissue was removed) were built, with the following pairs of D_s and D_c values expressed in millimeters [mm]: ($D_s = 0.85$; $D_c = 0.55$), ($D_s = 0.75$; $D_c = 0.5$), ($D_s = 0.65$; $D_c = 0.45$), which are close to the typical dimensions of pores commonly adopted in scaffolds for bony tissue [29,30]. These models were clamped on the lower base and subjected to a compression load of $F_{UA} = 0.1$ MPa. The displacement u_2 (Figure 10a) produced by the load was computed with Abaqus and used to determine the apparent Young's modulus as:

$$E_{app} = F_{UA} \times L/u_2, \qquad (6)$$

Interestingly, the values of the ratio predicted numerically are consistent with those measured experimentally [31,32] on samples of human spongy bone (Figure 10b). Furthermore, if we compute for the three models described above the scaffold volume fraction V_f, i.e., the ratio between the volume of the scaffold V_s and the total volume of the model $V_{tot} = L \times L \times L$, we find values that are consistent with those experimental reported by Snyder and Hayes [33] and measured for human spongy bone (Figure 10c).

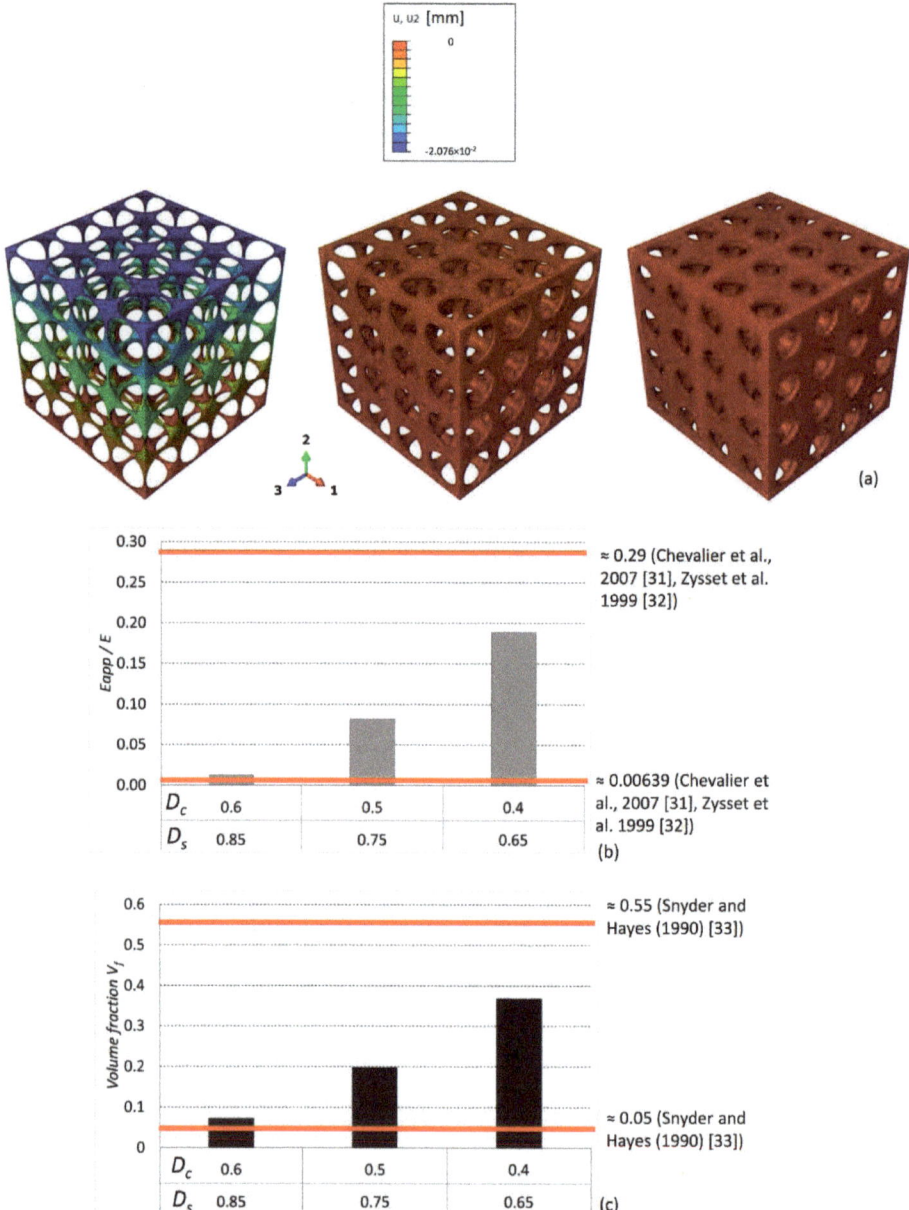

Figure 10. (a) u_2 displacement field of the scaffold models subjected to a compression load of 0.1 MPa. (b) Values of the ratio E_{app}/E computed for the three models and compared with those experimentally measured (represented with the red lines) for cancellous bone. (c) Scaffold volume fraction values compared with the volume fraction of human spongy bone.

The proposed model fits well the requirements of so-called Precision Medicine. The optimization algorithm presented in this article represents a possible approach to try to identify, given the specific patient with her/his specific anthropometric characteristics (i.e., macroscopic characteristics of the patient, such as weight, height, and geometric parameters of posture, that is, all the characteristics that allow identifying the boundary and loading conditions that act on a given anatomical region when a specific activity is performed), which are the optimal dimensions of the scaffold micro-geometry to achieve a successful follow-up with the formation of the largest amounts of bone in the shortest possible time? In fact, if one knows the anthropometric characteristics of the patient, they can hypothesize the possible value of load acting on the scaffold that will be implanted, and through diagrams such as those shown in Figures 7b and 8b, they can determine the optimal dimensions of the scaffold that favor the formation of the largest amounts of bone (Figure 7b). Furthermore, the proposed approach can support the surgeon in the choice of the best scaffold to implant in the specific fracture site of the patient. In fact, the surgeon has nowadays a very large range of scaffold geometries available on the market and hence has to choose the most suitable one for the specific requirements of the patient. For example, if, based on the anthropometric characteristics and the anatomical region of the fracture site, it is found that the scaffold will be subjected mainly to compression loading, the surgeon will choose the scaffold with rectangular extrusions (Figure 7c). If, on the other hand, it is found that the acting load will be mainly shear, then the surgeon will choose the scaffold with spherical pores (Figure 8c).

4. Conclusions

In this study, using a mechanobiology-based optimization algorithm, we computed the optimal dimensions of the micro-architecture of scaffolds including spherical pores and cylindrical interconnections. The optimization algorithm perturbs the scaffold geometry until the specific dimensions that favor the formation of the largest amounts of bone are identified. The proposed algorithm can guide and support the surgeon in the choice of a "personalized" scaffold that better suits the anthropometric characteristics of the patient, thus allowing to achieve a successful follow-up in the shortest possible time.

Author Contributions: Conceptualization, Ó.L.R.-M. and A.B.; methodology, Ó.L.R.-M. and A.B.; software, Ó.L.R.-M.; validation, A.B.; formal analysis, Ó.L.R.-M. and A.B.; writing—original draft preparation, Ó.L.R.-M. and A.B.; writing—review and editing, Ó.L.R.-M., A.B., C.J.C.-R., A.E.U., M.F., M.G., and V.M.M.; visualization, C.J.C.-R., A.E.U., M.F., M.G., and V.M.M.; project administration, C.J.C.-R. and Ó.L.R.-M.; funding acquisition, C.J.C.-R. and Ó.L.R.-M. All authors have read and agreed to the published version of the manuscript.

Funding: This research has been made possible by the collaboration between the Universidad Nacional de Colombia and Polytechnic University of Bari and was supported by the grant 647-2015 from the Colombian Ministry of Science, Technology and Innovation (MINCIENCIAS).

Conflicts of Interest: The authors declare no conflict of interest.

References

1. Ghassemi, T.; Shahroodi, A.; Ebrahimzadeh, M.H.; Mousavian, A.; Movaffagh, J.; Moradi, A. Current concepts in scaffolding for bone tissue engineering. *Arch. Bone Jt. Surg.* **2018**, *6*, 90–99. [PubMed]
2. Castro, A.P.G.; Lacroix, D. Micromechanical study of the load transfer in a polycaprolactone–collagen hybrid scaffold when subjected to unconfined and confined compression. *Biomech. Model. Mechanobiol.* **2018**, *17*, 531–541. [CrossRef] [PubMed]
3. Castro, A.P.G.; Pires, T.; Santos, J.E.; Gouveia, B.P.; Fernandes, P.R. Permeability versus design in TPMS scaffolds. *Materials* **2019**, *12*, 1313. [CrossRef] [PubMed]
4. Gleadall, A.; Visscher, D.; Yang, J.; Thomas, D.; Segal, J. Review of additive manufactured tissue engineering scaffolds: Relationship between geometry and performance. *Burn. Trauma* **2018**, *6*. [CrossRef] [PubMed]
5. Abbasi, N.; Ivanovski, S.; Gulati, K.; Love, R.M.; Hamlet, S. Role of offset and gradient architectures of 3-D melt electrowritten scaffold on differentiation and mineralization of osteoblasts. *Biomater. Res.* **2020**, *24*, 2. [CrossRef] [PubMed]
6. Kundu, J.; Pati, F.; Shim, J.H.; Cho, D.W. *Rapid Prototyping Technology for Bone Regeneration*; Woodhead Publishing Limited: Cambridge, UK, 2014; ISBN 9780857095992.

7. Zadpoor, A.A. Bone tissue regeneration: The role of scaffold geometry. *Biomater. Sci.* **2015**, *3*, 231–245. [CrossRef]
8. Kelly, C.N.; Miller, A.T.; Hollister, S.J.; Guldberg, R.E.; Gall, K. Design and Structure–Function Characterization of 3D Printed Synthetic Porous Biomaterials for Tissue Engineering. *Adv. Healthc. Mater.* **2018**, *7*, 1–16. [CrossRef]
9. Sanz-Herrera, J.A.; Doblaré, M.; García-Aznar, J.M. Scaffold microarchitecture determines internal bone directional growth structure: A numerical study. *J. Biomech.* **2010**, *43*, 2480–2486. [CrossRef]
10. Ly, H.B.; Le Droumaguet, B.; Monchiet, V.; Grande, D. Designing and modeling doubly porous polymeric materials. *Eur. Phys. J. Spec. Top.* **2015**, *224*, 1689–1706. [CrossRef]
11. Soro, N.; Attar, H.; Wu, X.; Dargusch, M.S. Investigation of the structure and mechanical properties of additively manufactured Ti-6Al-4V biomedical scaffolds designed with a Schwartz primitive unit-cell. *Mater. Sci. Eng. A* **2019**, *745*, 195–202. [CrossRef]
12. Ambu, R.; Morabito, A.E. Porous scaffold design based on minimal surfaces: Development and assessment of variable architectures. *Symmetry* **2018**, *10*, 361. [CrossRef]
13. Ambu, R.; Morabito, A.E. Modeling, assessment, and design of porous cells based on schwartz primitive surface for bone scaffolds. *Sci. World J.* **2019**, *2019*, 7060847. [CrossRef] [PubMed]
14. Prendergast, P.J.; Huiskes, R.; Søballe, K. Biophysical stimuli on cells during tissue differentiation at implant interfaces. *J. Biomech.* **1997**, *30*, 539–548. [CrossRef]
15. Boccaccio, A.; Fiorentino, M.; Uva, A.E.; Laghetti, L.N.; Monno, G. Rhombicuboctahedron unit cell based scaffolds for bone regeneration: Geometry optimization with a mechanobiology–driven algorithm. *Mater. Sci. Eng. C* **2018**, *83*, 51–66. [CrossRef] [PubMed]
16. Rodríguez-Montaño, Ó.L.; Cortés-Rodríguez, C.J.; Naddeo, F.; Uva, A.E.; Fiorentino, M.; Naddeo, A.; Cappetti, N.; Gattullo, M.; Monno, G.; Boccaccio, A. Irregular Load Adapted Scaffold Optimization: A Computational Framework Based on Mechanobiological Criteria. *ACS Biomater. Sci. Eng.* **2019**, *5*, 5392–5411. [CrossRef]
17. Percoco, G.; Uva, A.E.; Fiorentino, M.; Gattullo, M.; Manghisi, V.M.; Boccaccio, A. Mechanobiological approach to design and optimize bone tissue scaffolds 3D printed with fused deposition modeling: A feasibility study. *Materials* **2020**, *13*, 648. [CrossRef]
18. Rodríguez-Montaño, Ó.L.; Cortés-Rodríguez, C.J.; Uva, A.E.; Fiorentino, M.; Gattullo, M.; Monno, G.; Boccaccio, A. Comparison of the mechanobiological performance of bone tissue scaffolds based on different unit cell geometries. *J. Mech. Behav. Biomed. Mater.* **2018**, *83*, 28–45. [CrossRef]
19. Byrne, D.P.; Lacroix, D.; Planell, J.A.; Kelly, D.J.; Prendergast, P.J. Simulation of tissue differentiation in a scaffold as a function of porosity, Young's modulus and dissolution rate: Application of mechanobiological models in tissue engineering. *Biomaterials* **2007**, *28*, 5544–5554. [CrossRef]
20. Boccaccio, A.; Uva, A.E.; Fiorentino, M.; Lamberti, L.; Monno, G. A mechanobiology-based algorithm to optimize the microstructure geometry of bone tissue scaffolds. *Int. J. Biol. Sci.* **2016**, *12*, 1–17. [CrossRef]
21. Boccaccio, A.; Uva, A.E.; Fiorentino, M.; Mori, G.; Monno, G. Geometry design optimization of functionally graded scaffolds for bone tissue engineering: A mechanobiological approach. *PLoS ONE* **2016**, *11*, e0146935. [CrossRef]
22. Tanaka, M.; Matsugaki, A.; Ishimoto, T.; Nakano, T. Evaluation of crystallographic orientation of biological apatite in vertebral cortical bone in ovariectomized cynomolgus monkeys treated with minodronic acid and alendronate. *J. Bone Miner. Metab.* **2016**, *34*, 234–241. [CrossRef] [PubMed]
23. Noyama, Y.; Nakano, T.; Ishimoto, T.; Sakai, T.; Yoshikawa, H. Design and optimization of the oriented groove on the hip implant surface to promote bone microstructure integrity. *Bone* **2013**, *52*, 659–667. [CrossRef] [PubMed]
24. Liebi, M.; Georgiadis, M.; Menzel, A.; Schneider, P.; Kohlbrecher, J.; Bunk, O.; Guizar-Sicairos, M. Nanostructure surveys of macroscopic specimens by small-angle scattering tensor tomography. *Nature* **2015**, *527*, 349–352. [CrossRef] [PubMed]
25. Melchels, F.P.W.; Feijen, J.; Grijpma, D.W. A review on stereolithography and its applications in biomedical engineering. *Biomaterials* **2010**, *31*, 6121–6130. [CrossRef]
26. Li, J.; Chen, M.; Fan, X.; Zhou, H. Recent advances in bioprinting techniques: Approaches, applications and future prospects. *J. Transl. Med.* **2016**, *14*, 271. [CrossRef]

27. Liao, W.; Xu, L.; Wangrao, K.; Du, Y.; Xiong, Q.; Yao, Y. Three-dimensional printing with biomaterials in craniofacial and dental tissue engineering. *PeerJ* **2019**, *2019*, e727. [CrossRef]
28. Singh, J.P.; Pandey, P.M. Fitment study of porous polyamide scaffolds fabricated from selective laser sintering. *Procedia Eng.* **2013**, *59*, 59–71. [CrossRef]
29. Pircher, N.; Fischhuber, D.; Carbajal, L.; Strauß, C.; Nedelec, J.M.; Kasper, C.; Rosenau, T.; Liebner, F. Preparation and Reinforcement of Dual-Porous Biocompatible Cellulose Scaffolds for Tissue Engineering. *Macromol. Mater. Eng.* **2015**, *300*, 911–924. [CrossRef]
30. Diego, R.B.; Olmedilla, M.P.; Aroca, A.S.; Ribelles, J.L.G.; Pradas, M.M.; Ferrer, G.G.; Sanchez, M.S. Acrylic scaffolds with interconnected spherical pores and controlled hydrophilicity for tissue engineering. *J. Mater. Sci.* **2005**, *40*, 4881–4887. [CrossRef]
31. Chevalier, Y.; Pahr, D.; Allmer, H.; Charlebois, M.; Zysset, P. Validation of a voxel-based FE method for prediction of the uniaxial apparent modulus of human trabecular bone using macroscopic mechanical tests and nanoindentation. *J. Biomech.* **2007**, *40*, 3333–3340. [CrossRef]
32. Zysset, P.K.; Edward Guo, X.; Edward Hoffler, C.; Moore, K.E.; Goldstein, S.A. Elastic modulus and hardness of cortical and trabecular bone lamellae measured by nanoindentation in the human femur. *J. Biomech.* **1999**, *32*, 1005–1012. [CrossRef]
33. Snyder, B.; Hayes, W. Multiaxial Structure-Property Relations in Trabecular Bone. In *Biomechanics of Diarthrodial Joints*; Springer: New York, NY, USA, 1990; Volume 2, pp. 31–59. ISBN 978-1-4612-8016-3.

© 2020 by the authors. Licensee MDPI, Basel, Switzerland. This article is an open access article distributed under the terms and conditions of the Creative Commons Attribution (CC BY) license (http://creativecommons.org/licenses/by/4.0/).

Article

Titanium Functionalized with Polylysine Homopolymers: In Vitro Enhancement of Cells Growth

Maria Contaldo [1,*,†], Alfredo De Rosa [1,†], Ludovica Nucci [1], Andrea Ballini [2,3], Davide Malacrinò [4], Marcella La Noce [5], Francesco Inchingolo [6], Edit Xhajanka [7], Kenan Ferati [8], Arberesha Bexheti-Ferati [8], Antonia Feola [9,‡] and Marina Di Domenico [3,10,*,‡]

[1] Multidisciplinary Department of Medical-Surgical and Dental Specialties, University of Campania Luigi Vanvitelli, Via Luigi de Crecchio, 6, 80138 Naples, Italy; alfredo.derosa@unicampania.it (A.D.R.); ludovica.nucci@unicampania.it (L.N.)
[2] Department of Biosciences, Biotechnologies and Biopharmaceutics, Campus Universitario Ernesto Quagliariello, University of Bari "Aldo Moro", 70125 Bari, Italy; andrea.ballini@uniba.it
[3] Department of Precision Medicine, University of Campania Luigi Vanvitelli, 80138 Naples, Italy
[4] Department of Research, Development and Quality Assessment, AISER SA, Rue du Rhone, 14 VH-1204 Genève, Switzerland; davide.malacrino@aiser.org
[5] Department of Experimental Medicine, Università Degli Studi della Campania Luigi Vanvitelli, Campania, 80138 Naples, Italy; marcella.lanoce@unicampania.it
[6] Department of Interdisciplinary Medicine, University of Medicine Aldo Moro, 70124 Bari, Italy; francesco.inchingolo@uniba.it
[7] Department of Dental Prosthesis, Medical University of Tirana, Rruga e Dibrës, U.M.T., 1001 Tirana, Albania; editxhajanka@yahoo.com
[8] Faculty of Medicine, University of Tetovo, 1220 Tetovo, North Macedonia; kenan.ferati@unite.edu.mk (K.F.); arberesha.ferati@unite.edu.mk (A.B.-F.)
[9] Department of Biology, University of Naples "Federico II", 80138 Naples, Italy; antonia.feola@unina.it
[10] Department of Biology, College of Science and Technology, Temple University, Philadelphia, PA 19122, USA
* Correspondence: maria.contaldo@unicampania.it (M.C.); marina.didomenico@unicampania.it (M.D.D.); Tel.: +39-32-0487-6058 (M.C.)
† These authors contributed equally to this work as co-first authors.
‡ These authors contributed equally to this work as co-last authors.

Abstract: In oral implantology, the success and persistence of dental implants over time are guaranteed by the bone formation around the implant fixture and by the integrity of the peri-implant mucosa seal, which adheres to the abutment and becomes a barrier that hinders bacterial penetration and colonization close to the outer parts of the implant. Research is constantly engaged in looking for substances to coat the titanium surface that guarantees the formation and persistence of the peri-implant bone, as well as the integrity of the mucous perimeter surrounding the implant crown. The present study aimed to evaluate in vitro the effects of a titanium surface coated with polylysine homopolymers on the cell growth of dental pulp stem cells and keratinocytes to establish the potential clinical application. The results reported an increase in cell growth for both cellular types cultured with polylysine-coated titanium compared to cultures without titanium and those without coating. These preliminary data suggest the usefulness of polylysine coating not only for enhancing osteoinduction but also to speed the post-surgery mucosal healings, guarantee appropriate peri-implant epithelial seals, and protect the fixture against bacterial penetration, which is responsible for compromising the implant survival.

Keywords: cell growth; titanium; polylysine; dental implants; implantology; biomaterials; epithelial growth

1. Introduction

Dental implants are multi-material prostheses that replace tooth roots with screw-like metal fixtures surgically inserted into the edentulous bone that are connected by the abutment with an artificial crown that replaces the missing tooth, looking and acting identical to the real one (Figure 1).

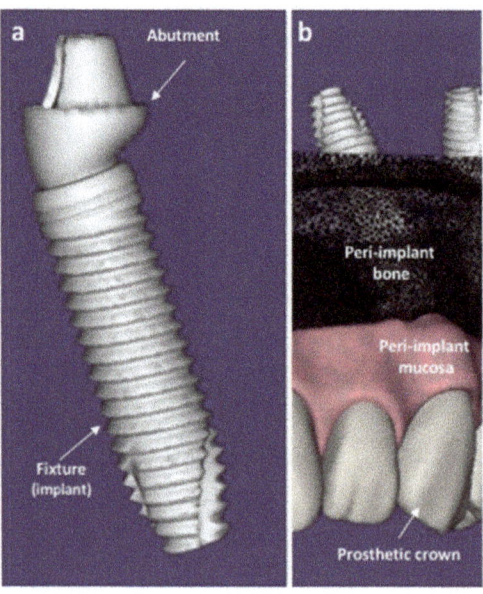

Figure 1. Schematic representation of a dental implant (**a**) and the oral structures (peri-implant mucosa and peri-implant bone) surrounding the fixture (**b**), surgically inserted in the bone. Original figures made by D.M. with SOLIDWORKS ® (CSWP-MBD Version, 2021, SolidWorks, Dassault systems, Waltham, MA, USA).

Dental implants fixtures are generally composed of biomedical titanium and its alloys [1], as they are biocompatible as well as resistant to corrosion and strength [2].

Numerous surgical protocols and variables may affect dental implant placement, and, over the years, novel implantology procedures have been constantly proposed [3–6].

The main sign of the success of a dental implant is its capability to integrate its shape with the bone and to induct the formation of novel bone around it; these properties are defined as "osseointegration"—"the close contact between the bone and an implant material in histological sections" [7,8]—and "osteoinduction"—the ability to induce the osteogenesis of new mineralized bone around the implant surfaces, thus firmly blocking the fixture within the bones of the jaws [9].

In addition to different surgical protocols [10], geometry modifications [5,11] and various surface treatments for increasing surface roughness [1,12], such as acid-etching, grit-blasting, titanium plasma-spraying, or anodization [13], as well as the use of various coatings to make the titanium surface bioactive [14–17] are responsible for empowering the wettability, bone anchoring, and biomechanical stability between the implant–bone interfaces [3,6,10,12,18], thus increasing osteoinduction and osseointegration.

Among the coating substances, the polyaminoacid poly-L-lysine has been reported to be able to bridge the cell-adhesion trough covalent attachments to cysteine in the bone [19–21].

A model of study on the osteogenic effects of substances is the use of human dental pulp stem cells [22–24], which previously has been proven to be involved in bone–implant osseointegration [25–28]. In details, the role of induced pluripotent stem cells in dentistry has been recently discussed and the use of autologous dental-derived stem cells has been proposed for bone tissue regeneration, as less invasive and more predictable alternative to conventional tissue regenerative procedures [29].

Furthermore, the mechanisms underlying the potential effects of poly-L-lysine on these kinds of cells have been reported both in vitro [20] and in vivo on sheep animal models [21].

Bacterial-induced inflammation of the soft tissues surrounding the abutment is the main cause of failure of the osteointegration immediately after the fixture placement and during the years. To avoid bacterial penetration and contamination of the peri-implant bone, which is responsible for inflammation and bone loss, the integrity of the peri-implant seal is crucial [30]. Otherwise, peri-implant inflammation occurs, and the implant survival is compromised. [31,32].

Therefore, a good epithelial attachment between the implant and the peri-implant mucosa is fundamental to achieve and maintain the osteointegration [30,33–35], and it is essential to maintain an intact oral epithelial barrier, with no local and systemic risk factors, as bacterial plaque, to offer good resistance to mechanical stress that is both physiological and pathological.

The present work aimed to confirm the in vitro effects of titanium functionalized with a poly-L-lysine coating on human dental pulp stem cells (hDPSCs), which are responsible for osteogenesis, and evaluate analogues effects on keratinocyte cell lines (HaCaT), which are responsible for epithelial attachment of the mucosa surrounding the abutment, to hypothesize a potential improvement of implant osteointegration and the potential use of poly-L-lysine for rapid mucosal healing after the implant placement and during years to preserve the health of peri-implant mucosa.

2. Materials and Methods

Machined clean square plates (1 cm × 1 cm in size; 0.2 mm thick) made up of 5-Ti-6Al-4V ELI alloy (Klein s.r.l., Milan, Italy) (Figure 2) were sterilized with ethanol 70%, dried under a fume hood, and used in six types of experiments: hDPSCs cultures alone (standard condition), hDPSCs cultures with titanium, and hDPSCs cultures poly-L-lysine-coated titanium (Figure 3a); HaCaT immortalized human keratinocyte line cultures alone, HaCaT cultures with titanium, and HaCaT cultured with poly-L-lysine-coated titanium (Figure 3b). In each experiment, cell viability and proliferation were assessed, as reported below. Sterilized titanium plates were coated with poly-L-lysine incubating at 37 °C for 30 min with a solution containing 0.01% poly-L-lysine and then dried and washed twice with sterile water. After this, cells were cultured on the disks.

2.1. hDPSCs Culture and Growth Curve

Experimental procedures were conducted following our previous experience in the field and according to the manufacturer's specifications [22,23,25,36–40].

Each patient or guardian gave informed consent to tooth extraction obtained with piezo-surgery technology, which was in accordance with the Declaration of Helsinki, for re-use of biospecimens in research applications. Moreover, the study was approved by the Independent Ethical Committee of University Hospital of Bari, Italy (protocol number 155/2021, 27 January 2021). With the purpose to preserve dental tissues for consequent cell isolation and expansion, piezo-surgery technology enables selective tissue cutting, and consequently, tooth buds or embedded third molars can effortlessly be removed from bones with slight wound to periodontal fibers or bud follicles.

In addition, tooth extraction, especially by piezosurgery technique, can be considered less invasive in comparison to bone marrow or other tissues biopsy [22].

Briefly, the pulp was removed and immersed for 1 h at 37 °C in a digestive solution of 3 mg/mL of type I collagenase and 4 mg/mL of dispase in PBS (phosphate buffered saline) containing 40 mg/mL of gentamicin. Once digested, the solution was filtered through 70 μm Falcon strainers (Becton & Dickinson, Franklin Lakes, NJ, USA). Cells were cultured in standard medium consisting of Dulbecco's modified Eagle's medium (DMEM) with 100 units/mL of penicillin, 100 mg/mL of streptomycin, and 200 mM l-glutamine (all purchased from Gibco), supplemented with 10% fetal bovine serum (FBS) (Invitrogen,

Waltham, MA, USA). Cells were maintained in a humidified atmosphere under 5% CO_2 at 37 °C, and the media were changed twice a week.

At first passage of culture, cells were seeded at a density of 150.00 cells/titanium implant—with and without poly-L-lysine homopolimers coating—and in standard condition. After 1 h of incubation in 100 µL of culture medium to allow cell attachment, the cell implants and cells cultured without implants were incubated in DMEM at 10% of FBS (fetal bovine serum) into an incubator at 37 °C in a humidified atmosphere consisting of 5% CO_2 and 95% O_2 for 24, 48, and 72 h. For each time, an aliquot of cell suspension was diluted with 0.4% trypan blue (Sigma Aldrich, St. Louis, MO, USA), pipetted onto a haemocytometer, and counted under a microscope at 200× magnification. Live cells excluded the dye, whereas dead cells admitted the dye and were consequently stained intensely with trypan blue. The number of viable cells for each experimental condition was counted and represented on a linear graph.

MTT Analyses

In order to evaluate the cytotoxicity of titanium implants on cells, MTT assay (3-(4,5-dimethylthiazol-2-yl)-2,5-diphenyltetrazolium bromide) was used [41,42]. Cells, at a density of 300.00 cells/implant with and without poly-L-lysine coating and cells cultured in standard condition (hDPSCs cultured in tissue culture polystyrene (TCP) without titanium and polylysine) were plated in DMEM at 10% FBS for 24, 48, and 72 h. After each time point, the medium was removed, and 200 µL of MTT (Sigma, Milan, Italy) solution (5 mg/mL in DMEM without phenol red) and 1.8 mL of DMEM were added. Four hours later, the formazan precipitate was dissolved in 100 µL dimethyl sulfoxide, and then, the absorbance was measured in an ELISA reader (Thermo Molecular Devices Co., Union City, NJ, USA) at 550 nm. The mean and the standard deviations were obtained from three different experiments of the same specimen.

Element Number	Element Symbol	Element Name	Atomic Conc.	Weight Conc.
22	Ti	Titanium	52.12	74.52
6	C	Carbon	21.65	7.77
8	O	Oxygen	15.86	7.58
13	Al	Aluminium	7.87	6.34
23	V	Vanadium	2.49	3.78

Figure 2. On top, the machined clean titanium plate at SEM. FOV: 134 µm, Mode: 15 kV—Point, Detector: BSD Full. On bottom, the chemical composition analysis of the titanium surface, in spot 1, pointed by a cross in the figure.

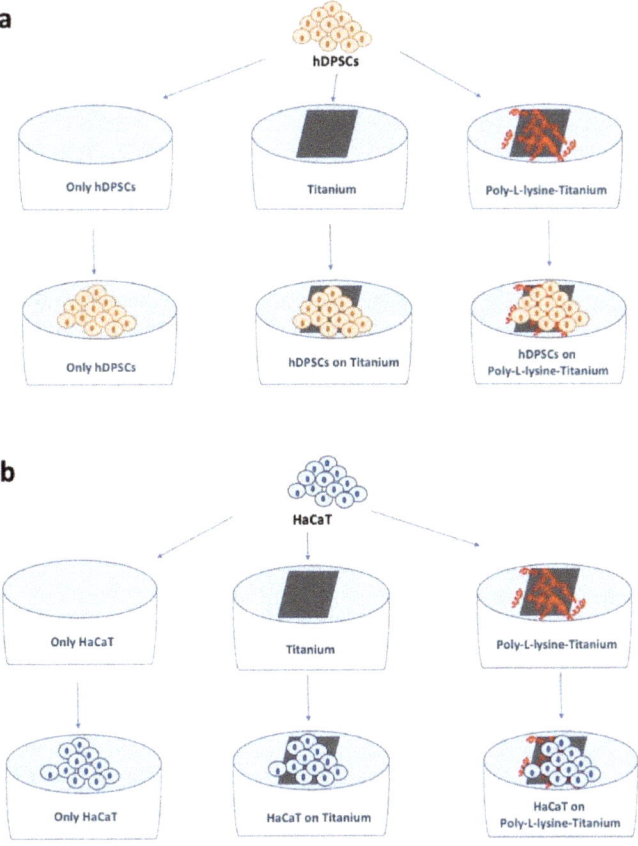

Figure 3. Schematic representation of the experiments. (**a**) hDPSCs cultured alone, on titanium plates, and on titanium plates coated with poly-L-lyisine. (**b**) HaCaT cells cultured alone, on titanium plates, and on titanium plates coated with poly-L-lyisine.

2.2. HaCaT Cells Culture and Growth Curve

HaCaT were cultured in complete culture medium consisting of DMEM (Sigma D5796, Sigma Aldrich, St. Louis, MO, USA) with 1% penicillin/streptomycin (Sigma P0781, Sigma Aldrich, St. Louis, MO, USA), 2 mM glutamine (Sigma G7513, Sigma Aldrich, St. Louis, MO, USA), and supplemented with 10% fetal bovine serum (Sigma F7524, Sigma Aldrich, St. Louis, MO, USA) [43]. All procedures were performed under sterile conditions under a NuAire laminar flow biological hood.

The cultures were expanded in plates every three days in an incubator under 5% CO_2 at 37 °C (RH = 95%), until the required number of cells was reached. Then, 5×10^5 cells were subsequently transferred to plates containing titanium alone and titanium with poly-L-lysine coating to promote engraftment.

In order to highlight cell clones adhering to titanium, after 48 h, the titanium plates with and without poly-L-lysine coating were removed from the culture and, after suitable washing with PBS twice, they were placed in plates containing only fresh culture medium to observe cell viability.

The cytotoxicity check was performed by culturing the cells in the absence and presence of the titanium plate (with and without poly-L-lysine coating) and by evaluating their viability after replacement in a new fresh medium.

After 72 h, all cells were trypsinized, collected, and evaluated for viability according to tryp blue method using the Burker chamber count (Invitrogen, Milan, Italy).

2.3. Statistical Analyses

Student's test was used for statistical evaluation. A p-value < 0.05 was considered significant.

3. Results

3.1. hDPSCs Growth Curves Analyses

Cell growth analysis and viability staining with trypan blue showed that hDPSCs cultured in standard condition and on titanium with and without poly-L-lysine showed the same trend in growth; however, while the titanium alone slightly negatively affects the viability for cells ($p < 0.01$), the cell growth on the poly-L-lysine coated titanium was noticeably increased ($p < 0.001$) (Figure 4).

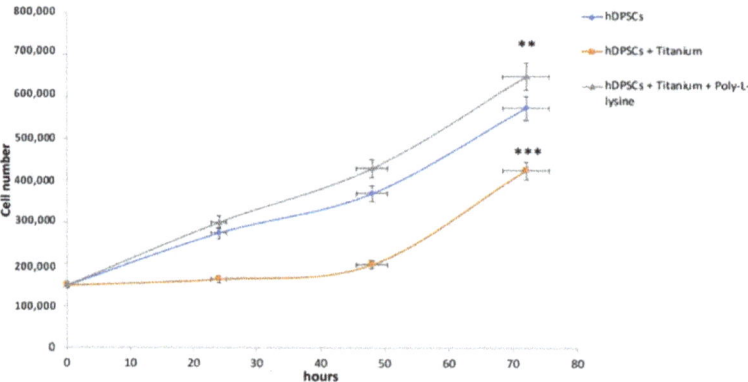

Figure 4. Cell growth analyses. Although hDPSCs cultured in standard condition and on titanium with and without poly-L-lysine showed the same trend in growth, in the culture with titanium coated with poly-L-lysine, the cell growth was higher than the hDPSCs alone and hDPSCs with only titanium. ** $p < 0.01$, *** $p < 0.001$ compared to the hDPSCs.

3.2. MTT Evaluation in hDPSCs

To evaluate how the titanium affected the viability and proliferation of hDPSCs, MTT analyses were performed. hDPSCs were cultured on titanium with and without poly-L-lysine coating for 24, 48, and 72 h. Results showed that titanium was not cytotoxic. In addition, there were no changes in terms of proliferation between cells cultured in standard condition and cells seeded on titanium, while the cells seeded on titanium coated with poly-L-lysine showed higher proliferation ($p < 0.001$) (Figure 5).

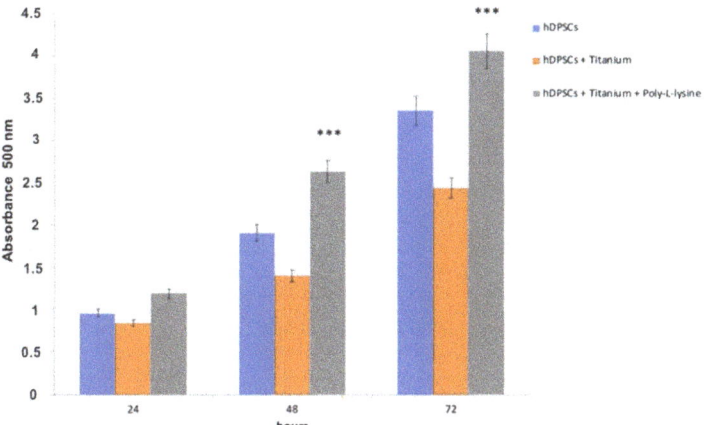

Figure 5. MTT evaluation. Titanium did not show cytotoxicity, and the proliferation of cells seeded on titanium was similar to those of cultured in standard condition, but in titanium coated with poly-L-liysine, cell proliferation was higher at 24, 48, and 72 h. *** $p < 0.001$ compared to the hDPSCs.

3.3. HaCaT Viability and Proliferation: Mucoproliferative Effects of Titanium

To evaluate the viability, HaCaT cells were cultured in the presence of titanium pre-coated with poly-L-lysine. As shown in Figure 6, the mere presence of titanium coated with poly-L-lysine did not affect HaCaT cell viability, as the cells at the bottom of the plate showed normal health morphology [44]. Moreover, cells cultured in the presence of titanium showed an increase in cell proliferation of 40% after 72 h compared to plates containing cells in the absence of titanium. To further demonstrate the presence of live cells adherent on titanium, after 48 h of culture, the titanium plate was placed in a new well with only fresh medium, and even here, adherent keratinocytes were still appreciated. This evidence demonstrates that titanium was found to be a suitable substrate for the viability and growth of keratinocytes in the presence of a culture medium (Figure 6).

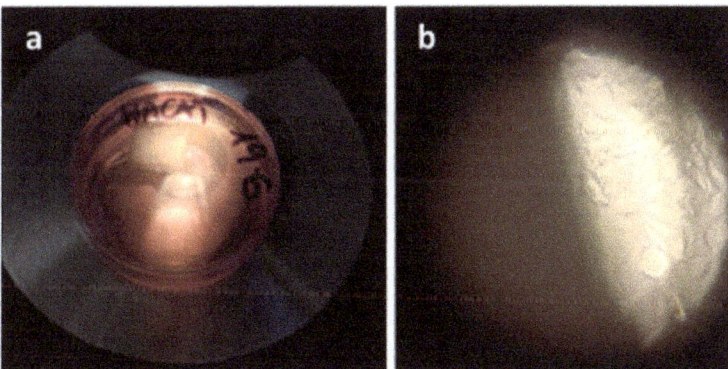

Figure 6. (**a**) HaCat cells culture on the titanium plate coated with poly-L-lysine. (**b**) After a further 48 h, the titanium plate with the adherent HaCaT cells was placed in a new well with a fresh medium, where vital and growing cells were appreciated adhering on the bottom of the well, as shown in figure (**b**) (optical microscopy, original magnification ×10).

4. Discussion

The titanium alloys used in dentistry are biocompatible and not cytotoxic, but their surface is also inert, thus not affecting positively the osteoinduction. To empower dental implant osteoinduction, the titanium surface can be functionalized by coating it with a series of bioactive compounds and substances. For this purpose, various coatings have been proposed: nanoparticles of silver, copper, and zinc, as sanitizing agents, and antibacterial and bioactive substances [14], such as quaternary ammonium ions and chlorhexidine, antibiotics, or antimicrobial peptides [15]; calcium-phosphate alone [16] or hydroxyapatite or octacalcium phosphate complexes [17]. These substances are used to make the titanium surface bioactive to improve osteoinduction, by adding, in some cases, antibacterial properties.

Poly-L-lysine is a polyaminoacid carrying positive charges, which increase cellular adhesion on different substrates, and it has been variously reported as an additional coating to titanium surfaces [19–21]. In 2005, Spoerke et al. [19] first reported that a nanotextured hybrid titanium coating made up of poly-L-lysine 14% by weight added to calcium phosphate was able to enhance the surface area of the implant and to potentiate the bioactivity of the calcium phosphate alone, by the presence of poly-L-lysine in bridging the cell-adhesion through covalent attachments to cysteine in the bone.

Different studies have reported the osteogenic effects of poly-L-lysine on dental-derived stem cells [22–24] and their involvement in bone–implant osteointegration [25–28]. In 2011, Galli et al. [20] described the potential mechanisms by which poly-L-lysine can enhance osteogenesis, thus reporting that hMSCs (human mesenchymal stem cells) and hDPSCs cultured on poly-L-lysine-treated titanium (Ti6Al4V) showed significantly higher expression of bone marker genes, produced a higher quantity of calcium deposits, and showed higher cell viability after 12 h of culture in comparison with the cells on the untreated titanium. These effects were allowed both by the poly-L-lysine positive charges and its interaction with β-integrin and other molecules from the extracellular matrix (as collagen I, fibronectin, and vitronectin) and their adhesion receptors on the studied cells, thus activating the intracellular signaling cascade responsible for the upregulation of osteogenic markers genes. Among the osteogenic markers activated, alkaline phosphatase is responsible for focal adhesion kinase (FAK) phosphorylation. While the unphosphorylated FAK is capable of blocking the mineral deposition, conversely, phosphorylated FAK (p-FAK) increased in the presence of titanium treated with poly-L-lysine and promoted calcium deposition, osteogenic differentiation, and bone maturation. In support of this mechanism, the same authors reported the presence of p-FAK only in cells treated with titanium-poly-L-lysine and from a twofold (at day 6) to eightfold (at day 25) increase of osteogenic differentiation markers in hMSCs and hDPSCs grown on titanium and poly-L-lysine compared to untreated hMSCs and hDPSCs [20]. In conclusion, poly-L-lysine seems to increase the p-FAK form, thus limiting its capability to block the mineral deposition and hence promoting the osteoblastic differentiation pathway and initiating mitogen-activated protein kinases, leading to osteogenic differentiation and bone maturation.

Four years later, in 2015, Varoni et al. confirmed the effect of poly-L-lysine coatings on titanium osseointegration by in vivo studies on sheep animal models [21]. Their results showed that cortical bone microhardness significantly improved in the presence of the poly-L-lysine coating by enhancing calcium deposition and implant early osseointegration in animals.

Little literature exists about the proliferative effects of poly-L-lysine on HaCaT cells; the work closest to highlighting this effect was the study by Renò et al. [45], but a complete and exhaustive explanation of the underlying mechanisms has not been reported yet. Renò et al. tested the efficacy of two different hydrogels synthesized by crosslinking gelatin with polylysine (positively charged) (HG1) and gelatin with polyglutamic acid (negatively charged) (HG2) as scaffolds for immortalized human keratinocytes (HaCaT) growth. They found that keratinocytes adhered both onto the HG1 and HG2 surface and were capable of proliferating, without toxicity, even if the cells displayed higher adhesion and proliferation onto HG2, forming a continuous and stratified epithelium after 7 days [45]. Further studies are necessary to elucidate the poly-L-lysine effects on epithelial cells and wound-healing processes in depth.

To prevent bacterial infections and facilitate the bone mineralization around the dental implants, recently, Guo et al. reported the synergistic effect of a composite coating made up of poly-L-lysine/sodium alginate and nanosilver [46], while Zhang et al. coated the titanium surfaces with a multilayer biofilm of ε-polylysine and arabic gum [47].

The present work tested the effects of poly-L-lysine-coated implant plates on the cell growth and cytotoxicity both on epithelial cells and dental-derived stem cells, in order (i) to confirm any proliferative effects on mesenchymal cells responsible for osteogenesis and (ii) to establish whether it exerts a potential similar muco-proliferative effect on cells of epithelial origin. For these purposes, a series of experiments were conducted on two different cell lines: epithelial (HaCaT) and mesenchymal (hDPSCs) cells.

Results unanimously have reported cell viability, lack of cytotoxicity, and a statistically significant improvement of the cell growth both for hDPSCs and HaCaT when cultured on poly-L-lysine-coated titanium plates, when compared with the cultures of the cells alone and those of the cells with uncoated titanium plates.

5. Conclusions

The oral cavity is always challenged by mechanical, chemical, and biological stimulations throughout life [48], and because the oral mucosa represents a protective barrier between the soft tissues and the external environment [49], it is essential to preserve its integrity and resistance to mechanical stress, both physiological and pathological, and to reduce irritating local factors such as bacterial plaque [30,50]. Oral dysbiosis and poor oral hygiene compromise the health of the peri-implant soft tissues. Furthermore, as in gingivitis and periodontitis, which are diseases responsible for gingival inflammation and bone loss strictly associated with bacterial plaque composition and bone diseases such as osteoporosis [51,52], peri-implant sites can be equally affected by their counterparts as well. These counterparts are called mucositis and peri-implantitis [31,32] which, respectively, lead to inflammation of the mucosa surrounding the abutment and the loss of bone around the fixture, thus compromising the stability of the implant in the bone, which is resorbed and decreased [53–56]. Furthermore, dysmetabolic diseases such as chronic hyperglycemia have been associated with periodontitis and peri-implantitis due to delayed and/or impaired wound healing for the activation of pathways linked to inflammation, oxidative stress, and cell apoptosis [57–59].

The present work is an exploratory study to confirm the bone proliferative effects of a poly-L-lysine coating on titanium and to establish analogue proliferative effects on keratinocytes and lack of cytotoxicity.

The results have confirmed the positive effects of poly-L-lysine on osteoinduction [20,21,28,42,53] and demonstrated a novel potential role also in promoting epithelial cell growth. It means that in clinical practice, a poly-L-lysine topic administration on the surgical mucosal site of a dental implant, could promote, accelerate, and ameliorate the formation of epithelial tissue around semi-submerged and on submerged implants, to favor more rapid healing of the surgical site after the fixture placement and to reinforce the epithelium surrounding the abutment during the remaining life of the implant, thus preventing mucositis and peri-implantitis arising from a loose gingival–implant thigh contact.

However, further in vivo studies are required to confirm the effects of titanium functionalized with a poly-L-lysine coating to improve implant osteointegration and to elucidate the mechanisms of action on keratinocytes and the in vivo efficacy of polylysine compounds in promoting epithelial cell growth and wound healing, as well as after the implant placement and during years to preserve the health of peri-implant mucosa, with particular attention to the aesthetic area [60].

Furthermore, the additional in vivo studies could be supported by non-invasive imaging techniques [61–65] as well as classical procedures, which could highlight and quantify the real histological and cytologic effects of poly-L-lysine on epithelial cell growth to enhance and/or support the wound healing not only at peri-implant sites but also for the treatment of oral lesions and injuries requiring the re-establishment of a healthy mucosal barrier [66–69] and the reduction of biofilm formation around the teeth and implants [70].

Author Contributions: Conceptualization, M.D.D., D.M. and A.D.R.; methodology, M.D.D., M.L.N. and A.F.; formal analysis, A.B.-F. and E.X.; data curation, M.D.D., M.C. and K.F.; writing—original draft preparation, M.C., L.N. and A.B.; writing—review and editing, M.C., M.L.N., D.M., A.F. and M.D.D.; supervision, F.I. and M.D.D.; funding acquisition, M.D.D. and A.D.R. All authors have read and agreed to the published version of the manuscript.

Funding: This research was partly funded by the SANIDENT S.R.L. Via Settembrini, Milan, Italy, in term partnership (from 29/01/2020 to 29/07/2020) with the Department of Precision Medicine, University of Campania "Luigi Vanvitelli", Naples, Italy.

Institutional Review Board Statement: The study was conducted according to the guidelines of the Declaration of Helsinki, and approved by the Independent Ethical Committee University Hospital of Bari, Italy, protocol number 155/2021.

Informed Consent Statement: Informed consent was obtained from all subjects involved in the study.

Data Availability Statement: Data sharing is not applicable to this article.

Conflicts of Interest: The authors declare no conflict of interest.

References

1. Majkowska-Marzec, B.; Tęczar, P.; Bartmański, M.; Bartosewicz, B.; Jankiewicz, B.J. Mechanical and Corrosion Properties of Laser Surface-Treated Ti13Nb13Zr Alloy with MWCNTs Coatings. *Materials* **2020**, *13*, 3991. [CrossRef] [PubMed]
2. Martinez-Marquez, D.; Delmar, Y.; Sun, S.; Stewart, R.A. Exploring Macroporosity of Additively Manufactured Titanium Metamaterials for Bone Regeneration with Quality by Design: A Systematic Literature Review. *Materials* **2020**, *13*, 4794. [CrossRef]
3. Inchingolo, A.D.; Inchingolo, A.M.; Bordea, I.R.; Xhajanka, E.; Romeo, D.M.; Romeo, M.; Zappone, C.M.F.; Malcangi, G.; Scarano, A.; Lorusso, F.; et al. The Effectiveness of Osseodensification Drilling Protocol for Implant Site Osteotomy: A Systematic Review of the Literature and Meta-Analysis. *Materials* **2021**, *14*, 1147. [CrossRef]
4. Fanali, S.; Tumedei, M.; Pignatelli, P.; Inchingolo, F.; Pennacchietti, P.; Pace, G.; Piattelli, A. Implant primary stability with an osteocondensation drilling protocol in different density polyurethane blocks. *Comput. Methods Biomech. Biomed. Engin.* **2021**, *24*, 14–20. [CrossRef]
5. Inchingolo, F.; Paracchini, L.; DE Angelis, F.; Cielo, A.; Orefici, A.; Spitaleri, D.; Santacroce, L.; Gheno, E.; Palermo, A. Biomechanical behaviour of a jawbone loaded with a prosthetic system supported by monophasic and biphasic implants. *Oral Implantol.* **2017**, *9*, 65–70. [CrossRef]
6. Bavetta, G.; Bavetta, G.; Randazzo, V.; Cavataio, A.; Paderni, C.; Grassia, V.; Dipalma, G.; Gargiulo Isacco, C.; Scarano, A.; De Vito, D.; et al. A Retrospective Study on Insertion Torque and Implant Stability Quotient (ISQ) as Stability Parameters for Immediate Loading of Implants in Fresh Extraction Sockets. *Biomed. Res. Int.* **2019**, *2019*, 9720419. [CrossRef]
7. Branemark, P.I.; Hansson, B.O.; Adell, R.; Breine, U.; Lindstrom, J.; Hallen, O.; Ohman, A. Osseointegrated implants in the treatment of the edentulous jaw. Experience from a 10-year period. *Scand. J. Plast. Reconstr. Surg. Suppl.* **1997**, *16*, 1–132.
8. Annunziata, M.; Guida, L. The Effect of Titanium Surface Modifications on Dental Implant Osseointegration. *Front. Oral Biol.* **2015**, *17*, 62–77.
9. Pawelec, K.; White, A.A.; Best, S.M. Properties and characterization of bone repair materials. In *Woodhead Publishing Series in Biomaterials, Bone Repair Biomaterials*, 2nd ed.; Pawelec, K.M., Planell, J.A., Eds.; Woodhead Publishing: Sawston, UK, 2019; pp. 65–102.
10. Antonelli, A.; Bennardo, F.; Brancaccio, Y.; Barone, S.; Femiano, F.; Nucci, L.; Minervini, G.; Fortunato, L.; Attanasio, F.; Giudice, A. Can Bone Compaction Improve Primary Implant Stability? An In Vitro Comparative Study with Osseodensification Technique. *Appl. Sci.* **2020**, *10*, 8623. [CrossRef]
11. Zemtsova, E.; Arbenin, A.; Valiev, R.; Smirnov, V. Modern techniques of surface geometry modification for the implants based on titanium and its alloys used for improvement of the biomedical characteristics. In *Titanium in Medical and Dental Applications*; Elsevier: Amsterdam, The Netherlands, 2018; pp. 115–145.
12. Jemat, A.; Ghazali, M.J.; Razali, M.; Otsuka, Y. Surface Modifications and Their Effects on Titanium Dental Implants. *Biomed. Res. Int.* **2015**, *2015*, 791725. [CrossRef]
13. Le Guéhennec, L.; Soueidan, A.; Layrolle, P.; Amouriq, Y. Surface treatments of titanium dental implants for rapid osseointegration. *Dent. Mater.* **2007**, *23*, 844–854. [CrossRef] [PubMed]
14. Shimabukuro, M. Antibacterial Property and Biocompatibility of Silver, Copper, and Zinc in Titanium Dioxide Layers Incorporated by One-Step Micro-Arc Oxidation: A Review. *Antibiotics* **2020**, *9*, 716. [CrossRef]

15. Garaicoa, J.L.; Bates, A.M.; Avila-Ortiz, G.; Brogden, K.A. Antimicrobial Prosthetic Surfaces in the Oral Cavity-A Perspective on Creative Approaches. *Microorganisms* **2020**, *8*, 1247. [CrossRef]
16. Kulkarni Aranya, A.; Pushalkar, S.; Zhao, M.; LeGeros, R.Z.; Zhang, Y.; Saxena, D. Antibacterial and bioactive coatings on titanium implant surfaces. *J. Biomed. Mater. Res. A* **2017**, *105*, 2218–2227. [CrossRef]
17. Barrere, F.; Layrolle, P.; van Blitterswijk, C.; de Groot, K. Biomimetic coatings on titanium: A crystal growth study of octacalcium phosphate. *J. Mater. Sci. Mater. Med.* **2001**, *12*, 529–534. [CrossRef]
18. Li, J.; Jansen, J.A.; Walboomers, X.F.; van den Beucken, J.J. Mechanical aspects of dental implants and osseointegration: A narrative review. *J. Mech. Behav. Biomed. Mater.* **2020**, *103*, 103574. [CrossRef] [PubMed]
19. Spoerke, E.D.; Stupp, S.I. Synthesis of a poly(L-lysine)-calcium phosphate hybrid on titanium surfaces for enhanced bioactivity. *Biomaterials* **2005**, *26*, 5120–5129. [CrossRef]
20. Galli, D.; Benedetti, L.; Bongio, M.; Maliardi, V.; Silvani, G.; Ceccarelli, G.; Ronzoni, F.; Conte, S.; Benazzo, F.; Graziano, A.; et al. In vitro osteoblastic differentiation of human mesenchymal stem cells and human dental pulp stem cells on poly-L-lysine-treated titanium-6-aluminium-4-vanadium. *J. Biomed. Mater. Res. A* **2011**, *97*, 118–126. [CrossRef]
21. Varoni, E.; Canciani, E.; Palazzo, B.; Varasano, V.; Chevallier, P.; Petrizzi, L.; Dellavia, C.; Mantovani, D.; Rimondini, L. Effect of Poly-L-Lysine coating on titanium osseointegration: From characterization to in vivo studies. *J. Oral Implantol.* **2015**, *41*, 626–631. [CrossRef] [PubMed]
22. Ballini, A.; Di Benedetto, A.; De Vito, D.; Scarano, A.; Scacco, S.; Perillo, L.; Posa, F.; Dipalma, G.; Paduano, F.; Contaldo, M.; et al. Stemness genes expression in naïve vs. osteodifferentiated human dental-derived stem cells. *Eur. Rev. Med. Pharmacol. Sci.* **2019**, *23*, 2916–2923.
23. Ballini, A.; Cantore, S.; Scacco, S.; Perillo, L.; Scarano, A.; Aityan, S.K.; Contaldo, M.; Cd Nguyen, K.; Santacroce, L.; Syed, J.; et al. A comparative study on different stemness gene expression between dental pulp stem cells vs. dental bud stem cells. *Eur. Rev. Med. Pharmacol. Sci.* **2019**, *23*, 1626–1633.
24. Di Benedetto, A.; Brunetti, G.; Posa, F.; Ballini, A.; Grassi, F.R.; Colaianni, G.; Colucci, S.; Rossi, E.; Cavalcanti-Adam, E.A.; Lo Muzio, L.; et al. Osteogenic differentiation of mesenchymal stem cells from dental bud: Role of integrins and cadherins. *Stem Cell Res.* **2015**, *15*, 618–628. [CrossRef] [PubMed]
25. Naddeo, P.; Laino, L.; La Noce, M.; Piattelli, A.; De Rosa, A.; Iezzi, G.; Laino, G.; Paino, F.; Papaccio, G.; Tirino, V. Surface biocompatibility of differently textured titanium implants with mesenchymal stem cells. *Dent. Mater.* **2015**, *31*, 235–243. [CrossRef] [PubMed]
26. Paino, F.; La Noce, M.; Giuliani, A.; De Rosa, A.; Mazzoni, S.; Laino, L.; Amler, E.; Papaccio, G.; Desiderio, V.; Tirino, V. Human DPSCs fabricate vascularized woven bone tissue: A new tool in bone tissue engineering. *Clin. Sci.* **2017**, *131*, 699–713. [CrossRef] [PubMed]
27. Mangano, C.; Paino, F.; d'Aquino, R.; De Rosa, A.; Iezzi, G.; Piattelli, A.; Laino, L.; Mitsiadis, T.; Desiderio, V.; Mangano, F.; et al. Human dental pulp stem cells hook into biocoral scaffold forming an engineered biocomplex. *PLoS ONE* **2011**, *6*, e18721. [CrossRef] [PubMed]
28. Laino, L.; La Noce, M.; Fiorillo, L.; Cervino, G.; Nucci, L.; Russo, D.; Herford, A.S.; Crimi, S.; Bianchi, A.; Biondi, A.; et al. Dental Pulp Stem Cells on Implant Surface: An In Vitro Study. *BioMed Res. Int.* **2021**, *2021*, 3582342. [CrossRef]
29. Tetè, G.; D'Orto, B.; Nagni, M.; Agostinacchio, M.; Polizzi, E.; Agliardi, E. Role of induced pluripotent stem cells (IPSCS) in bone tissue regeneration in dentistry: A narrative review. *J. Biol. Regul. Homeost. Agents* **2020**, *34*, 1–10. [PubMed]
30. Ivanovski, S.; Lee, R. Comparison of peri-implant and periodontal marginal soft tissues in health and disease. *Periodontology 2000* **2018**, *76*, 116–130. [CrossRef]
31. Smeets, R.; Henningsen, A.; Jung, O.; Heiland, M.; Hammächer, C.; Stein, J.M. Definition, etiology, prevention and treatment of peri-implantitis—A review. *Head Face Med.* **2014**, *10*, 34. [CrossRef] [PubMed]
32. Thoma, D.S.; Naenni, N.; Figuero, E.; Hämmerle, C.H.F.; Schwarz, F.; Jung, R.E.; Sanz-Sánchez, I. Effects of soft tissue augmentation procedures on peri-implant health or disease: A systematic review and meta-analysis. *Clin. Oral Impl. Res.* **2018**, *29*, 32–49. [CrossRef] [PubMed]
33. Ballini, A.; Cantore, S.; Farronato, D.; Cirulli, N.; Inchingolo, F.; Papa, F.; Malcangi, G.; Inchingolo, A.D.; Dipalma, G.; Sardaro, N.; et al. Periodontal disease and bone pathogenesis: The crosstalk between cytokines and porphyromonas gingivalis. *J. Biol. Regul. Homeost. Agents* **2015**, *29*, 273–281.
34. Cantore, S.; Mirgaldi, R.; Ballini, A.; Coscia, M.F.; Scacco, S.; Papa, F.; Inchingolo, F.; Dipalma, G.; De Vito, D. Cytokine gene polymorphisms associate with microbiogical agents in periodontal disease: Our experience. *Int. J. Med. Sci.* **2014**, *11*, 674–679. [CrossRef] [PubMed]
35. Inchingolo, F.; Martelli, F.S.; Gargiulo Isacco, C.; Borsani, E.; Cantore, S.; Corcioli, F.; Boddi, A.; Nguyễn, K.C.D.; De Vito, D.; Aityan, S.K.; et al. Chronic Periodontitis and Immunity, Towards the Implementation of a Personalized Medicine: A Translational Research on Gene Single Nucleotide Polymorphisms (SNPs) Linked to Chronic Oral Dysbiosis in 96 Caucasian Patients. *Biomedicines* **2020**, *8*, 115. [CrossRef] [PubMed]
36. Di Domenico, M.; Feola, A.; Ambrosio, P.; Pinto, F.; Galasso, G.; Zarrelli, A.; Di Fabio, G.; Porcelli, M.; Scacco, S.; Inchingolo, F.; et al. Antioxidant Effect of Beer Polyphenols and Their Bioavailability in Dental-Derived Stem Cells (D-dSCs) and Human Intestinal Epithelial Lines (Caco-2) Cells. *Stem Cells Int.* **2020**, *2020*, 8835813. [CrossRef] [PubMed]

37. Cantore, S.; Ballini, A.; De Vito, D.; Martelli, F.S.; Georgakopoulos, I.; Almasri, M.; Dibello, V.; Altini, V.; Farronato, G.; Dipalma, G.; et al. Characterization of human apical papilla-derived stem cells. *J. Biol. Regul. Homeost. Agents* **2017**, *31*, 901–910. [PubMed]
38. Boccellino, M.; Di Stasio, D.; Dipalma, G.; Cantore, S.; Ambrosio, P.; Coppola, M.; Quagliuolo, L.; Scarano, A.; Malcangi, G.; Borsani, E.; et al. Steroids and growth factors in oral squamous cell carcinoma: Useful source of dental-derived stem cells to develop a steroidogenic model in new clinical strategies. *Eur. Rev. Med. Pharmacol. Sci.* **2019**, *23*, 8730–8740. [PubMed]
39. Cosentino, C.; Di Domenico, M.; Porcellini, A.; Cuozzo, C.; De Gregorio, G.; Santillo, M.R.; Agnese, S.; Di Stasio, R.; Feliciello, A.; Migliaccio, A.; et al. p85 regulatory subunit of PI3K mediates cAMP-PKA and estrogens biological effects on growth and survival. *Oncogene* **2007**, *26*, 2095–2103. [CrossRef]
40. Donini, C.F.; Di Zazzo, E.; Zuchegna, C.; Di Domenico, M.; D'Inzeo, S.; Nicolussi, A.; Avvedimento, E.V.; Coppa, A.; Porcellini, A. The p85α regulatory subunit of PI3K mediates cAMP-PKA and retinoic acid biological effects on MCF7 cell growth and migration. *Int. J. Oncol.* **2012**, *40*, 1627–1635.
41. Boccellino, M.; Di Domenico, M.; Donniacuo, M.; Bitti, G.; Gritti, G.; Ambrosio, P.; Quagliuolo, L.; Rinaldi, B. AT1-receptor blockade: Protective effects of irbesartan in cardiomyocytes under hypoxic stress. *PLoS ONE* **2018**, *13*, e0202297. [CrossRef]
42. Vanacore, D.; Messina, G.; Lama, S.; Bitti, G.; Ambrosio, P.; Tenore, G.; Messina, A.; Monda, V.; Zappavigna, S.; Boccellino, M.; et al. Effect of restriction vegan diet's on muscle mass, oxidative status, and myocytes differentiation: A pilot study. *J. Cell Physiol.* **2018**, *233*, 9345–9353. [CrossRef]
43. D'Angelo, S.; La Porta, R.; Napolitano, M.; Galletti, P.; Quagliuolo, L.; Boccellino, M. Effect of Annurca apple polyphenols on human HaCaT keratinocytes proliferation. *J. Med. Food* **2012**, *15*, 1024–1031. [CrossRef]
44. Tyagi, N.; Bhardwaj, A.; Srivastava, S.K.; Arora, S.; Marimuthu, S.; Deshmukh, S.K.; Singh, A.P.; Carter, J.E.; Singh, S. Development and Characterization of a Novel in vitro Progression Model for UVB-Induced Skin Carcinogenesis. *Sci. Rep.* **2015**, *5*, 13894. [CrossRef]
45. Renò, F.; Rizzi, M.; Cannas, M. Gelatin-based anionic hydrogel as biocompatible substrate for human keratinocyte growth. *J. Mater. Sci. Mater. Med.* **2012**, *23*, 565–571. [CrossRef] [PubMed]
46. Guo, C.; Cui, W.; Wang, X.; Lu, X.; Zhang, L.; Li, X.; Li, W.; Zhang, W.; Chen, J. Poly-l-lysine/Sodium Alginate Coating Loading Nanosilver for Improving the Antibacterial Effect and Inducing Mineralization of Dental Implants. *ACS Omega* **2020**, *5*, 10562–10571. [CrossRef] [PubMed]
47. Zhang, Y.; Wang, F.; Huang, Q.; Patil, A.B.; Hu, J.; Fan, L.; Yang, Y.; Duan, H.; Dong, X.; Lin, C. Layer-by-layer immobilizing of polydopamine-assisted ε-polylysine and gum Arabic on titanium: Tailoring of antibacterial and osteogenic properties. *Mater. Sci. Eng. C Mater. Biol. Appl.* **2020**, *110*, 110690. [CrossRef] [PubMed]
48. Wang, S.S.; Tang, Y.L.; Pang, X.; Zheng, M.; Tang, Y.J.; Liang, X.H. The maintenance of an oral epithelial barrier. *Life Sci.* **2019**, *227*, 129–136. [CrossRef]
49. Liu, J.; Mao, J.J.; Chen, L. Epithelial–Mesenchymal Interactions as a Working Concept for Oral Mucosa Regeneration. *Tissue Eng. Part B Rev.* **2011**, *17*, 25–31. [CrossRef]
50. Rizzo, A.; Di Domenico, M.; Carratelli, C.R.; Mazzola, N.; Paolillo, R. Induction of proinflammatory cytokines in human osteoblastic cells by Chlamydia pneumoniae. *Cytokine* **2011**, *56*, 450–457. [CrossRef]
51. Contaldo, M.; Itro, A.; Lajolo, C.; Gioco, G.; Inchingolo, F.; Serpico, R. Overview on Osteoporosis, Periodontitis and Oral Dysbiosis: The Emerging Role of Oral Microbiota. *Appl. Sci.* **2020**, *10*, 6000. [CrossRef]
52. Contaldo, M.; Lucchese, A.; Lajolo, C.; Rupe, C.; Di Stasio, D.; Romano, A.; Petruzzi, M.; Serpico, R. The Oral Microbiota Changes in Orthodontic Patients and Effects on Oral Health: An Overview. *J. Clin. Med.* **2021**, *10*, 780. [CrossRef]
53. Rizzo, A.; Losacco, A.; Carratelli, C.R.; Domenico, M.D.; Bevilacqua, N. Lactobacillus plantarum reduces Streptococcus pyogenes virulence by modulating the IL-17, IL-23 and Toll-like receptor 2/4 expressions in human epithelial cells. *Int. Immunopharmacol.* **2013**, *17*, 453–461. [CrossRef]
54. Ballini, A.; Dipalma, G.; Isacco, C.G.; Boccellino, M.; Di Domenico, M.; Santacroce, L.; Nguyễn, K.C.D.; Scacco, S.; Calvani, M.; Boddi, A.; et al. Oral Microbiota and Immune System Crosstalk: A Translational Research. *Biology* **2020**, *9*, 131. [CrossRef]
55. Inchingolo, F.; Dipalma, G.; Cirulli, N.; Cantore, S.; Saini, R.S.; Altini, V.; Santacroce, L.; Ballini, A.; Saini, R. Microbiological results of improvement in periodontal condition by administration of oral probiotics. *J. Biol. Regul. Homeost. Agents* **2018**, *32*, 1323–1328. [PubMed]
56. Cantore, S.; Ballini, A.; De Vito, D.; Abbinante, A.; Altini, V.; Dipalma, G.; Inchingolo, F.; Saini, R. Clinical results of improvement in periodontal condition by administration of oral probiotics. *J. Biol. Regul. Homeost. Agents* **2018**, *32*, 1329–1334. [PubMed]
57. Papi, P.; Letizia, C.; Pilloni, A.; Petramala, L.; Saracino, V.; Rosella, D.; Pompa, G. Peri-implant diseases and metabolic syndrome components: A systematic review. *Eur. Rev. Med. Pharmacol. Sci.* **2018**, *22*, 866–875.
58. Boccellino, M.; Giuberti, G.; Quagliuolo, L.; Marra, M.; D'Alessandro, A.M.; Fujita, H.; Giovane, A.; Abbruzzese, A.; Caraglia, M. Apoptosis induced by interferon-alpha and antagonized by EGF is regulated by caspase-3-mediated cleavage of gelsolin in human epidermoid cancer cells. *J. Cell Physiol.* **2004**, *201*, 71–83. [CrossRef] [PubMed]
59. Di Domenico, M.; Pinto, F.; Quagliuolo, L.; Contaldo, M.; Settembre, G.; Romano, A.; Coppola, M.; Ferati, K.; Bexheti-Ferati, A.; Sciarra, A.; et al. The Role of Oxidative Stress and Hormones in Controlling Obesity. *Front. Endocrinol.* **2019**, *10*, 540. [CrossRef] [PubMed]

60. Cattoni, F.; Teté, G.; Calloni, A.M.; Manazza, F.; Gastaldi, G.; Capparè, P. Milled versus moulded mock-ups based on the superimposition of 3D meshes from digital oral impressions: A comparative in vitro study in the aesthetic area. *BMC Oral Health* **2019**, *19*, 230. [CrossRef] [PubMed]
61. Contaldo, M.; Lucchese, A.; Gentile, E.; Zulli, C.; Petruzzi, M.; Lauritano, D.; Amato, M.R.; Esposito, P.; Riegler, G.; Serpico, R. Evaluation of the intraepithelial papillary capillary loops in benign and malignant oral lesions by in vivo Virtual Chromoendoscopic Magnification: A preliminary study. *J. Biol. Regul. Homeost. Agents* **2017**, *31*, 11–22.
62. Contaldo, M.; Lauritano, D.; Carinci, F.; Romano, A.; Di Stasio, D.; Lajolo, C.; Della Vella, F.; Serpico, R.; Lucchese, A. Intraoral confocal microscopy of suspicious oral lesions: A prospective case series. *Int. J. Dermatol.* **2020**, *59*, 82–90. [CrossRef]
63. Contaldo, M.; Di Stasio, D.; della Vella, F.; Lauritano, D.; Serpico, R.; Santoro, R.; Lucchese, A. Real Time In Vivo Confocal Microscopic Analysis of the Enamel Remineralization by Casein Phosphopeptide-Amorphous Calcium Phosphate (CPP-ACP): A Clinical Proof-of-Concept Study. *Appl. Sci.* **2020**, *10*, 4155. [CrossRef]
64. Gentile, E.; Di Stasio, D.; Santoro, R.; Contaldo, M.; Salerno, C.; Serpico, R.; Lucchese, A. In vivo microstructural analysis of enamel in permanent and deciduous teeth. *Ultrastruct. Pathol.* **2015**, *39*, 131–134. [CrossRef]
65. Camerlingo, C.; d'Apuzzo, F.; Grassia, V.; Perillo, L.; Lepore, M. Micro-Raman spectroscopy for monitoring changes in periodontal ligaments and gingival crevicular fluid. *Sensors* **2014**, *14*, 22552–22563. [CrossRef]
66. Boccellino, M.; Quagliuolo, L.; Verde, A.; La Porta, R.; Crispi, S.; Piccolo, M.T.; Vitiello, A.; Baldi, A.; Signorile, P.G. In vitro model of stromal and epithelial immortalized endometriotic cells. *J. Cell Biochem.* **2012**, *113*, 1292–1301. [CrossRef]
67. Pedata, P.; Boccellino, M.; La Porta, R.; Napolitano, M.; Minutolo, P.; Sgro, L.A.; Zei, F.; Sannolo, N.; Quagliuolo, L. Interaction between combustion-generated organic nanoparticles and biological systems: In vitro study of cell toxicity and apoptosis in human keratinocytes. *Nanotoxicology* **2012**, *6*, 338–352. [CrossRef] [PubMed]
68. Giannelli, G.; Milillo, L.; Marinosci, F.; Lo Muzio, L.; Serpico, R.; Antonaci, S. Altered expression of integrins and basement membrane proteins in malignant and pre-malignant lesions of oral mucosa. *J. Biol. Regul. Homeost. Agents* **2001**, *15*, 375–380. [PubMed]
69. Zannella, C.; Shinde, S.; Vitiello, M.; Falanga, A.; Galdiero, E.; Fahmi, A.; Santella, B.; Nucci, L.; Gasparro, R.; Galdiero, M.; et al. Antibacterial Activity of Indolicidin-Coated Silver Nanoparticles in Oral Disease. *Appl. Sci.* **2020**, *10*, 1837. [CrossRef]
70. Tetè, G.; Cisternino, L.; Giorgio, G.; Sacchi, L.; Montemezzi, P.; Sannino, G. Immediate versus delayed loading of post-extraction implants in the aesthetic zone: A prospective longitudinal study with 4-year follow-up. *J. Biol. Regul. Homeost. Agents* **2020**, *34*, 19–25. [PubMed]

Article

In Situ and Ex Situ Designed Hydroxyapatite: Bacterial Cellulose Materials with Biomedical Applications

Adrian Ionut Nicoara [1,2], Alexandra Elena Stoica [1,2,*], Denisa-Ionela Ene [3], Bogdan Stefan Vasile [1,2], Alina Maria Holban [4] and Ionela Andreea Neacsu [1,2]

1. Department of Science and Engineering of Oxide Materials and Nanomaterials, Faculty of Applied Chemistry and Materials Science, University Politehnica of Bucharest, 060042 Bucharest, Romania; adrian.nicoara@upb.ro (A.I.N.); bogdan.vasile@upb.ro (B.S.V.); ionela.neacsu@upb.ro (I.A.N.)
2. National Research Center for Micro and Nanomaterials, Faculty of Applied Chemistry and Materials Science, University Politehnica of Bucharest, 060042 Bucharest, Romania
3. Faculty of Engineering in Foreign Languages, University Politehnica of Bucharest, 060042 Bucharest, Romania; denisa_ionela.ene@stud.fim.upb.ro
4. Microbiology Department, Faculty of Biology, University of Bucharest, 060101 Bucharest, Romania; alina.m.holban@bio.unibuc.ro
* Correspondence: oprea.elena19@gmail.com; Tel.: +40-784069104

Received: 22 September 2020; Accepted: 23 October 2020; Published: 27 October 2020

Abstract: Hydroxyapatite (HAp) and bacterial cellulose (BC) composite materials represent a promising approach for tissue engineering due to their excellent biocompatibility and bioactivity. This paper presents the synthesis and characterization of two types of materials based on HAp and BC, with antibacterial properties provided by silver nanoparticles (AgNPs). The composite materials were obtained following two routes: (1) HAp was obtained in situ directly in the BC matrix containing different amounts of AgNPs by the coprecipitation method, and (2) HAp was first obtained separately using the coprecipitation method, then combined with BC containing different amounts of AgNPs by ultrasound exposure. The obtained materials were characterized by means of XRD, SEM, and FT-IR, while their antimicrobial effect was evaluated against Gram-negative bacteria (*Escherichia coli*), Gram-positive bacteria (*Staphylococcus aureus*), and yeast (*Candida albicans*). The results demonstrated that the obtained composite materials were characterized by a homogenous porous structure and high water absorption capacity (more than 1000% *w/w*). These materials also possessed low degradation rates (<5% in simulated body fluid (SBF) at 37 °C) and considerable antimicrobial effect due to silver nanoparticles (10–70 nm) embedded in the polymer matrix. These properties could be finetuned by adjusting the content of AgNPs and the synthesis route. The samples prepared using the in situ route had a wider porosity range and better homogeneity.

Keywords: bacterial cellulose; hydroxyapatite; nanoAg; tissue engineering; antimicrobial composite

1. Introduction

One of the most significant advances in the field of tissue engineering is the development of a porous three-dimensional matrix [1]. In order to act as an optimal bone support, the synthetic matrix must have a series of properties, including biocompatibility, biodegradability, appropriate porosity (similar to the replaced tissue), antimicrobial activity, and production reproducibility [2,3]. In addition to these requirements, it is also recommended that they have mechanical properties similar to natural bone, such as compressive strength, fatigue resistance, and high Young's modulus [4].

Such characteristics allow cell penetration, vascularization, and adequate nutrient and oxygen diffusion to cells and to the unformed extracellular matrix, which ensures cells viability. The pore size is, in fact, a key element for material efficiency. The pores must be large enough to allow cells to enter and move into the framework of the scaffold, while a small dimension allows the attachment of essential cell number at the same level [5]. Depending on the type of host tissue, all the support materials used in tissue engineering may have a macroporous structure with a particular pore size. For example, researchers suggest a pore size of 200–400 microns is optimal for bone tissue engineering [5,6].

The inorganic phase of the composites designed for bone replacement is usually hydroxyapatite (HAp) [7,8]. HAp is an essential element required for tissue regeneration, with the advantages of great biocompatibility, high plasticity, and remarkable mechanical properties because its chemical and crystalline structure is similar to natural bone apatite [2,9]. It also has an ultrafine structure and a large surface area that is advantageous for cell–biomaterial interactions and has been widely studied in applications for bone engineering [8,10].

For the organic phase of natural bone, replacement with bacterial cellulose (BC) has been attempted [11]. Even though the BC structure is chemically equivalent to plant cellulose (β-D-glucopyranose units linked by β-1,4 glycosidic bonds), it is free of by-products, such as lignin, pectin, hemicellulose, and other constituents of lignocellulosic materials. BC is a biodegradable polymer consisting of nanofibrillar structures, which determine a high specific surface area and a microporous structure. The unique 3D structure of BC is the main reason for its excellent retention and osteoinductivity, properties that make it a highly desirable substitute for collagen extracellular matrix in hard tissue engineering applications [12]. However, insufficient mechanical strength of the polymer restricts its direct application in vivo [13].

Many studies [14–17] have shown that BC could provide tissue regeneration and substitution, thus being used for bioengineering of hard, cartilaginous, and soft tissues. Bacterial cellulose is widely used as a wound dressing material, and nanomaterials obtained from BC show great antimicrobial properties [18–21]. BC can be combined with polymeric and nonpolymeric compounds to acquire or enhance antimicrobial, cell adhesion, and proliferation properties [13,22–25].

Scaffolds embedded with antimicrobial agents, antibiotics, or several forms of silver nanoparticles, which are known antimicrobial agents, are attracting great interest in biomedical research. Metallic silver and silver nanoparticles (AgNPs) have been reported to provide a wide variety of antimicrobial activities [12,26–28].

AgNPs are more toxic compared to bulk silver but they have a strong anti-inflammatory impact during tissue healing and can be integrated into composite materials to obtain antibacterial properties [29–31]. The human dietary intake of silver, owing to the widespread use of silver compounds, is estimated at 70–90 μg per day [32]. One of the main risk factors in tissue engineering and implant development is microbial infections. Bacterial colonization and the development of multicellular attached communities, called biofilms, are responsible for the high rate of failure in tissue engineering [33].

The purpose of this study was to develop a composite material based on hydroxyapatite, bacterial cellulose, and silver nanoparticles with biomedical applications. The material was obtained by the coprecipitation technique, which is a reliable, simple, economic, fairly rapid, and precise method that allows the synthesis of homogenous structures and favorable pore dimensions [10]. Studies have described the synthesis of bacterial cellulose/hydroxyapatite composites for bone healing applications using different methods [34–39]. In this work, AgNPs were integrated in the composite system in order to induce antibacterial properties.

2. Materials and Synthesis Methods

The chemical reagents were calcium nitrate tetrahydrate ($Ca(NO_3)_2 \bullet 4H_2O$, >99%), ammonium phosphate dibasic (($NH_4)_2HPO_4$, 99%), ammonium hydroxide (NH_4OH, 99%), sodium hydroxide (NaOH, 98%), silver nitrate ($AgNO_3$, >99%), sodium citrate ($C_6H_5O_7Na$, >99%), polyvinylpyrrolidone

((C$_6$H$_9$NO)$_n$), and sodium borohydride (NaBH$_4$, >99%), purchased from Sigma-Aldrich (St. Louis, MO, USA). The solvents were American Chemical Society (ACS, Washington, DC, USA) purity. Bacterial cellulose membrane was produced in the laboratory by *Gluconacetobacter* sp. strain isolated from traditionally fermented apple vinegar in the Microbiology Laboratory of the Chemical and Biochemical Engineering Department, University Politehnica of Bucharest, based on a protocol previously described [40].

In order to obtain 500 mL colloidal silver (100 ppm concentration), an aqueous silver nitrate solution (AgNO$_3$) was used as silver precursor, to which 30 mL sodium citrate (0.3 M) was added. After 12 min, 30 mL of Polyvinylpyrrolidone (PVP, 0.007 M) and 5 mL NaBH$_4$ (1 M) were added to reduce Ag$^+$ to Ag^0 nanoparticles. Finally, 5 mL of oxygenated water (30%) was added, and stirring was maintained for another 10 min approximately, until a light blue color (due to the size of the nanoparticles) was obtained (Figure 1) [41].

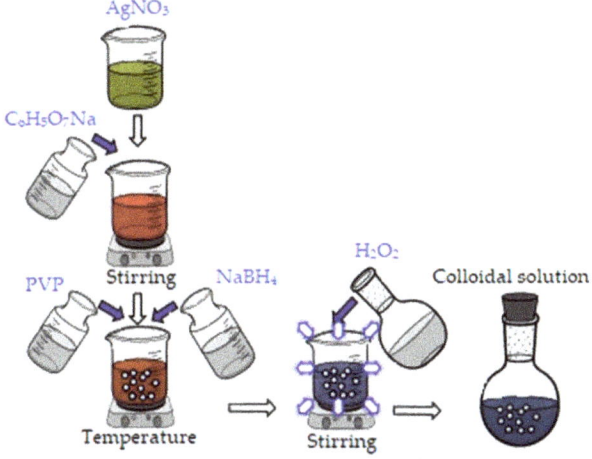

Figure 1. Synthesis of silver nanoparticles.

The bacterial cellulose synthesized by the Gram-negative bacteria (*Gluconacetobacter* sp.) was boiled at 80 °C in water alkalized with sodium hydroxide (pH 14, measured by colorimetric method). After purification, BC was washed in distilled water until it reached neutral pH. Afterward, it was minced using a blender (Silvercrest, Neckarsulm, Germany) and weighed according to the recipe. Previously, the amount of dry matter was determined on a quantity of BC by eliminating the humidity, and it was found that 0.25 g of dry BC can be obtained from 10.62 g of wet BC [17].

Two synthesis methods were used to obtain the bacterial cellulose and HAp-based composites. The first method involved obtaining in situ hydroxyapatite nanoparticles directly on cellulose fibers and subsequently adding the AgNPs solution, followed by homogenization using an ultrasound probe (composites further referenced as BC$_1$, BC$_2$, BC$_3$, and BC$_4$).

For the synthesis of 2 g HAp, the amount of precursors required to obtain the material with different concentrations of AgNPs (0, 1, 2, and 5 wt %) was calculated.

The Ca^{2+} and PO$_4$$^{3-}$ precursors, Ca(NO$_3$)$_2$•4H$_2$O, and (NH$_4$)$_2$HPO$_4$ were solubilized in distilled water, and bacterial cellulose was added in the calcium nitrate solution (see Figure 2). The mixtures were homogenized by magnetic stirring, and the ammonium phosphate solution was added dropwise. After homogenization, the pH was adjusted to 10.5 with an ammonium hydroxide solution. The obtained precipitates were aged for 24 h, then washed with distilled water until pH 7 was achieved. After washing, the appropriate amount of silver colloidal solution that had been previously obtained was added to each composition according to the centralizing table (Table 1). The obtained mixture was mixed for 3 min in the presence of ultrasound to ensure the best possible homogeneity and then poured into Petri dishes (d = 54 mm), frozen, and subsequently subjected to the freeze-drying process (freezing at −55 °C for 12 h, vacuum at 0.001 mbar for 12 h, and heating under vacuum for 24 h to 35 °C) in order to obtain porous composite materials [42].

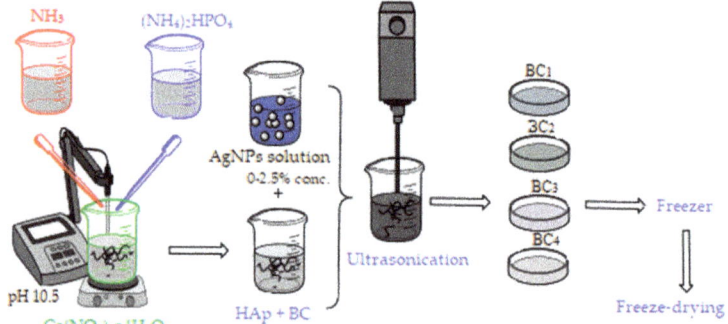

Figure 2. Synthesis of BC_1, BC_2, BC_3, and BC_4 composites; BC, bacterial cellulose.

Table 1. Bacterial cellulose, HAp, and AgNP content in the final composites.

Sample Name	Bacterial Cellulose Content (%)	HAp Content (%)	AgNP Content (%)
BC_1	50	50	0
BC_2	50	49	1
BC_3	50	48	2
BC_4	50	50	5
BC_5	50	50	0
BC_6	50	49	1
BC_7	50	48	2
BC_8	50	45	5

The second method of synthesis involved the separate synthesis of HAp by the coprecipitation method, followed by its addition to bacterial cellulose gel in the presence of ultrasound for 3 min, as described in Figure 3.

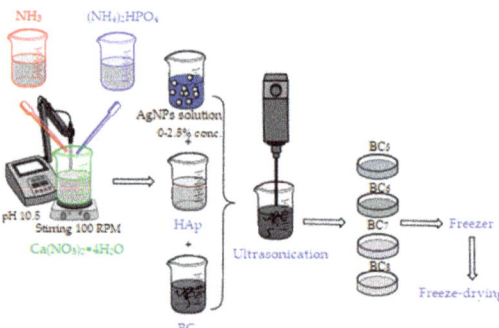

Figure 3. Synthesis of BC_5, BC_6, BC_7, and BC_8 composites.

After homogenization, the required amount of silver colloidal solution was added, followed by the steps previously described in the in situ method. The composites thus obtained by the ex situ method were noted as BC_5, BC_6, BC_7, and BC_8, and their composition is presented in Table 1.

3. Characterization Methods

3.1. Physicochemical Characterization

Investigation of the crystallinity of the powders was performed by means of X-ray diffraction (XRD) technique using the PANalytrical Empyrean (Malvern, Bruno, the Netherlands) equipment in Bragg–Brentano geometry equipped with a Cu anode (λCuKα = 1.541874 Å) X-ray tube. The spectra were acquired in the range of 10–80° 2θ angles (Bragg angle) with an acquisition step of 0.02° and an acquisition time of 100 s. The scanning electron microscopy (SEM) images were performed with a FEI Inspect F50 microscope coupled with an energy-dispersive spectrometer (EDS) (FEI, Eindhoven, the Netherlands). Both secondary electron and backscattered electron detectors were used at 30 kV accelerating voltage. The TEM images of AgNPs were obtained using the high-resolution transmission electron microscope TecnaiTM G2 F30 S-TWIN equipped with selected-area electron diffraction (SAED) detector, purchased from the company FEI. This microscope operates in transmission mode at a voltage of 300 kV with a resolution of 2 Å. Research conducted by Fourier transform infrared spectroscopy (FT-IR) involved the analysis of a small amount of samples using the Nicolet iS50R spectrometer (Thermo Fisher Waltham, MA, USA). The measurements were performed at room temperature utilizing the total reflection attenuation module (ATR), and 32 scans of the samples between 4000 and 400 cm^{-1} were performed using a resolution of 4 cm^{-1}. The differential thermal analysis (ATD-DSC) were performed using a Shimadzu DTG-TA-50H equipment (Shimadsu, Sanjo, Japan) at 25–700 °C with a heating rate of 10 °C/min.

The open porosity of the freeze-dried composite materials was calculated with Equation (1) for each material prepared in order to observe the porosity level according to the chosen manufacturing method, while the water absorption was calculated with Equation (2):

$$\text{Open porosity (\%)} = \frac{M_{we} - M_d}{M_{we} - M_w} \times 100 \qquad (1)$$

$$\text{Water absorption (\%)} = \frac{M_{we} - M_d}{M_d} \times 100 \qquad (2)$$

where M_{we} is the wet sample weight, M_d is the dry sample weight, and M_w is the sample weight in water.

3.2. Degradability

To test their biodegradability, the samples were placed in a 12-well plate in which phosphate-buffered saline (PBS) and simulated body fluid (SBF) were added, similar to the processes involved in the human body. After immersion of the samples in fluid, their integrity was monitored for 7 days. The degradation rate was calculated with Equation (3) for each material:

$$\text{Degradation (\%)} = \frac{M_{7day} - M_{initial}}{M_{initial}} \times 100 \qquad (3)$$

where M_{7days} is the sample weight in SBF after 7 days of immersion in SBF, and $M_{initial}$ is the sample weight after immersion in SBF. All the weight values were obtained at room temperature using a hydrostatic analytic balance.

3.3. Antimicrobial Efficiency

The antimicrobial behavior of the freeze-dried composite materials was qualitatively assessed by an adapted growth inhibition assay [43]. To cover a wide spectrum of clinically relevant model microbial species, one Gram-positive (*Staphylococcus aureus* ATCC 23235), one Gram-negative bacteria (*Escherichia coli* ATCC 25922), and one yeast (*Candida albicans* ATCC 10231) laboratory strain were used. The standard work protocol for the adapted version of the disc diffusion method implies the preparation of microbial suspensions of 0.5 McFarland standard density (1.5×10^8 colony forming units (CFU)/mL), prepared in sterile buffered saline solution. The obtained microbial suspensions were afterward used to swab inoculate the entire surface of the nutrient agar Petri dishes. After inoculation, identical size samples of the sterile coatings were aseptically placed on the inoculated agar surface, and the plates were incubated at 37 °C for 24 h to allow the growth of bacteria. After incubation, the growth inhibition zone diameter (mm) was measured. A wider inhibition zone suggests a higher antimicrobial effect of the fibrous dressing, reflecting the ability of AgNPs contained into the composite material to diffuse within the agar.

4. Results and Discussions

The thermal analysis corresponding to the composite samples are presented in Figure 4. The two minor weight losses that occurred at temperatures below 200 °C were probably related to the volatilization of solvents and physical water. The main mass loss was observed in the range 250–450 °C, with the corresponding exothermic effect being strong and intense and indicating burning of the organic component of the composite (bacterial cellulose). Regarding the compositional aspects, the thermal analysis allowed an accurate assessment of the loading degree depending on the material deposited on the surface or between the fibers of the bacterial cellulose. It was observed that certain changes associated with endothermic processes occurred in the thermogravimetric (TG) curve with the addition of silver nanoparticles. Hence, in the 450–600 °C interval, exothermic effects generated by the combustion of BC were observed (see Figure 4a), while in the 600–700 °C interval, it can be assumed that the oxidation of silver nanoparticles and dehydroxylation of HAp occurred [15].

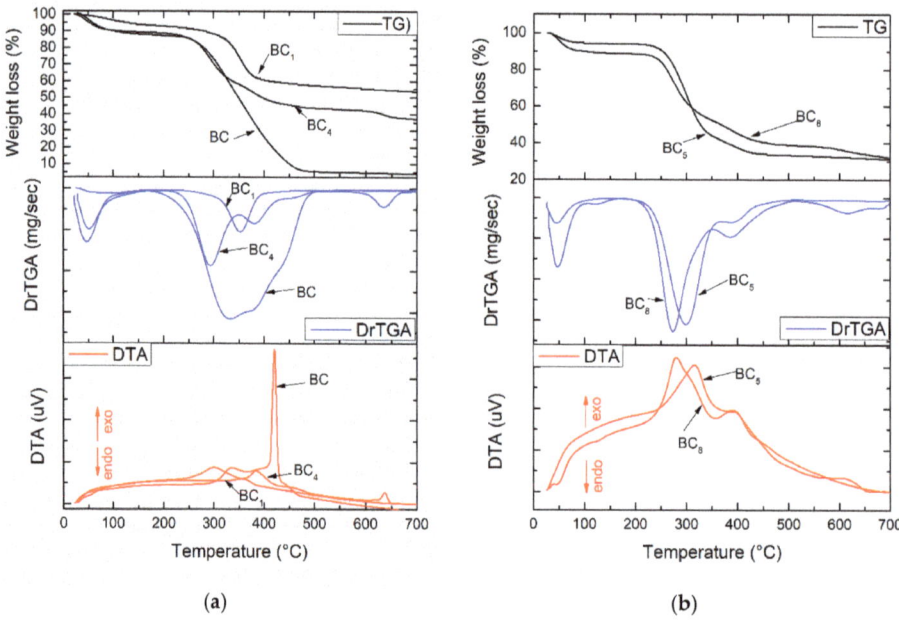

Figure 4. Thermal analysis corresponding to the composite samples: (a) BC_1 and BC_4; (b) BC_5 and BC_8.

We observed (Figure 4) significant differences regarding mass loss between the samples obtained by in situ vs. ex situ method. The total weight loss in the temperature range of 30–700 °C was 45% for BC_1, 65% for BC_4, 68% for BC_5, and 70% for BC_8. The composites obtained in situ had a lower weight loss, which suggests good loading of BC with calcium phosphate phases (HAp).

In order to demonstrate the composition, hydroxyapatite was analyzed by XRD technique. The diffractograms are presented in Figure 5.

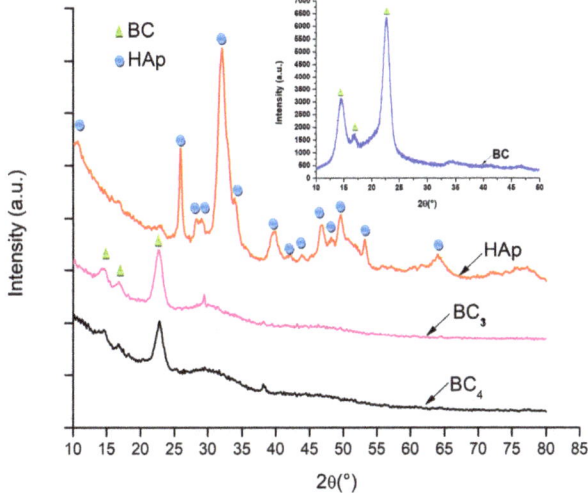

Figure 5. XRD analysis corresponding to the composites samples BC_3 and BC_4, and hydroxyapatite (HAp) obtained by the coprecipitation method.

Due to the fact that BC_5–BC_8 samples were made by direct mixing of bacterial cellulose with hydroxyapatite (ex situ), the composition of this sample was not expected to change; therefore, the XRD analysis was only performed for bacterial cellulose, simple hydroxyapatite, and BC_3 and BC_4 composites (in situ) [44,45].

It was observed that, in all the analyzed samples, the existence of bacterial cellulose and HAp was obvious. In addition, the low-intensity peak around $2\theta = 38°$, which can be assigned to the (111) crystalline plane, indicated the presence of silver nanoparticles in the composite structure. Investigation of the composites BC_3 and BC_4 revealed peaks located at 2θ values of 15, 16, and 23°, which can be attributed to bacterial cellulose according to ICDD 00-056-1718.

As the HAp peaks were poorly visible in XRD analysis, FT-IR analyses were used to better highlight hydroxyapatite formation. The results are presented in Figure 6.

The vibrational frequencies characteristic of bacterial cellulose were observed at 3500–3200 cm^{-1} (OH stretch vibrations) and 2958 cm^{-1} (CH_2 and CH_3 stretch vibrations). The wide band observed in the region of 3500–3200 cm^{-1}, attributed to the hydroxyl groups within the bacterial cellulose, increased in absorbance with higher silver content. This behavior suggested that the presence of HAp crystals affected cellulose hydroxyl groups, probably by covering them at the surface. Furthermore, the change observed for the band attributed to intramolecular hydrogen bonding (~3500 cm^{-1}) confirmed a strong interaction between the OH groups and calcium phosphate. The chemical interaction between HAp and BC stabilized the composite so that it could maintain its mechanical integrity, an aspect required for bone substituents [46].

The FT-IR bands observed at 1020 cm^{-1} and 570–600 cm^{-1} were attributed to the vibrational modes of PO_4^{3-}. Because the stretching vibration of CO_3^{2-} also appeared (at 1418 cm^{-1}), absorption of CO_2 from the air is suggested [47]. This is mainly a result of the affinity for carbonate of HAp as well as the lack of heat treatment during the in situ synthesis (which favors the release of CO_2). Carbonated hydroxyapatite contributes to the biomimetism increase of the obtained composites, which can promote the process of osteoregeneration. It was observed that the in situ method accelerated the nucleation of HAp crystals onto BC fibers instead of crystallization as higher absorbance values were registered for BC_5–BC_8 samples, which contained highly crystalized HAp. The bands observed at 1641 and 643 cm^{-1} correspond to the stretching and deformation vibrations of AgO, respectively, thus confirming the presence of silver in the obtained materials [48]. This result supports the idea that the composite material developed here possesses essential physicochemical properties and could be very useful for biomedical applications, especially hard tissue engineering.

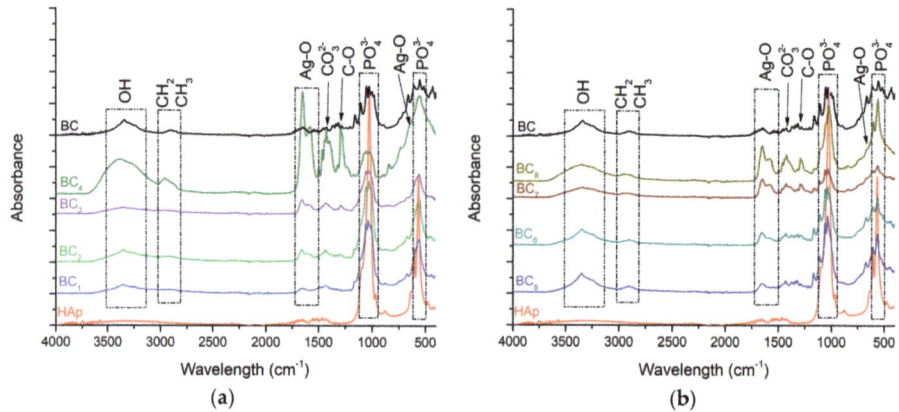

Figure 6. FT-IR analysis corresponding to the (**a**) in situ and (**b**) ex-situ composite samples.

Through the two analyses performed, it was possible to notice the elemental composition of materials (EDS) as well as the homogeneity of hydroxyapatite particle dispersion.

The SEM image highlighted the fibrous structure of BC (see green arrow), which were decorated with inorganic particles (see blue arrow). In the SEM images performed on the composites in which HAp was obtained in situ (Figure 7a–d), a better homogeneity was observed compared to the cases in which HAp was obtained separately and subsequently mixed with BC (ex situ) (Figure 8a–d. The interaction between HAp nanoparticles distributed in the 3D network of BC stabilized the composite so that it could maintain its mechanical integrity, an aspect required for bone substituents. In addition, EDS analysis confirmed the presence of the elements specific for hydroxyapatite (Ca, P, and O) as well as the presence of silver for the samples in which it was added (Figure 7(b_2–d_2) and Figure 8(b_2–d_2)).

Transmission electron microscopy images showed the silver nanostructure (Figure 9a,b), with the dimensions of the silver particles being in the range 3–60 nm. It could be observed that the quasi-spherical morphology of nanosilver and some areas were darker while others were brighter; the darker areas indicate a higher degree of crystallinity of the material.

Figure 9d shows a SAED image with information on the crystallinity of the analyzed material. The presence of diffraction rings with higher light intensity shows a high degree of crystallinity.

The calculated open porosity for each prepared material is presented in Figure 10a, and the calculated water absorption is presented in Figure 9b.

Figure 10a shows that the composites obtained by in situ approach had a large porosity compared to samples obtained by ex situ. This suggests that the in situ route will provide a biodegradable polymer with excellent water retention and, possibly, good osteoinductivity, which can be used as an artificial substitute for hard tissue.

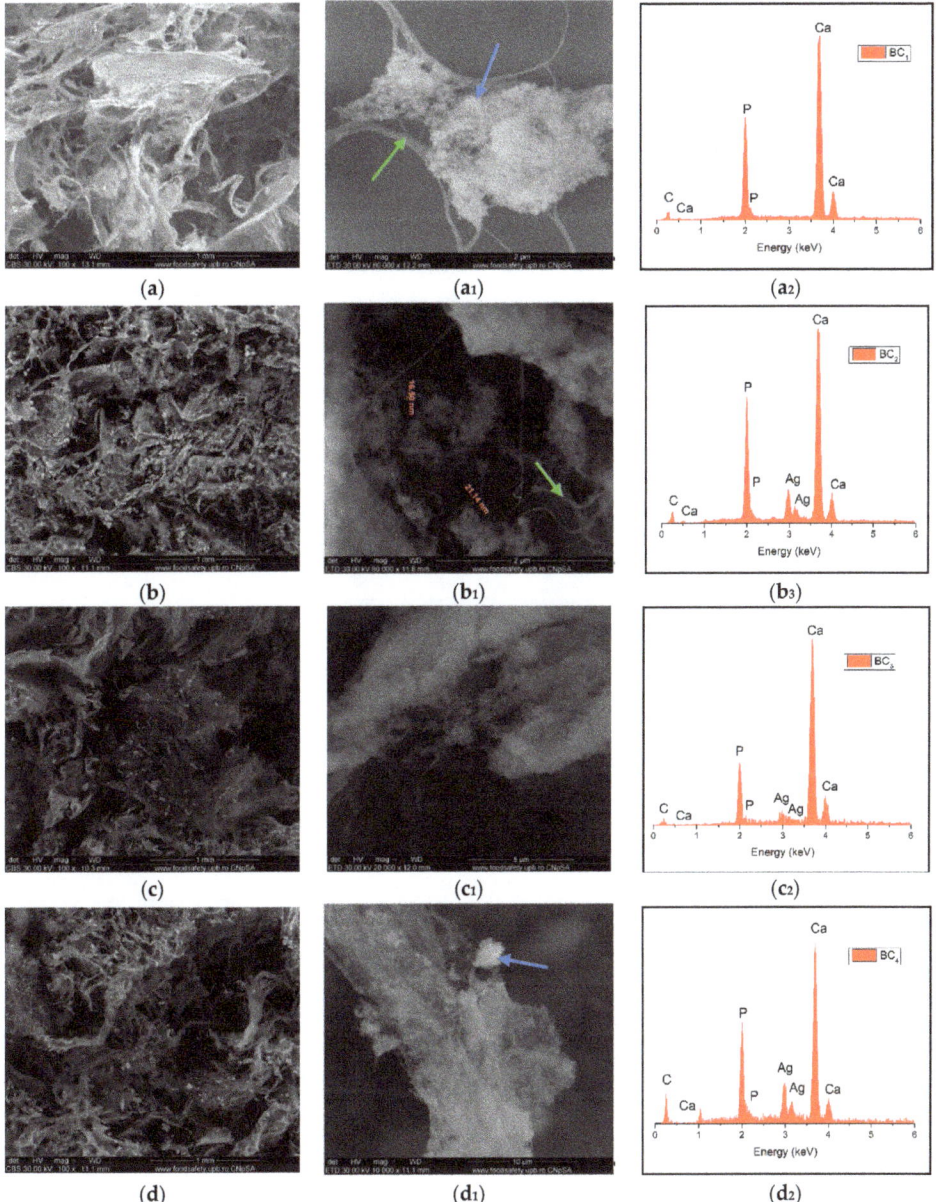

Figure 7. High-resolution backscattered-electron (BSE) images (and EDS spectra at 100× magnification for: BC_1 (**a**,**a₁**,**a₂**), BC_2 (**b**,**b₁**,**b₂**), BC_3 (**c**,**c₁**,**c₂**), BC_4 (**d**,**d₁**,**d₂**). (where **a₁**–**d₁** images represent a high magnification of a-d images area; green arrow indicates the fibrous structure of BC and blue arrow indicate inorganic particles).

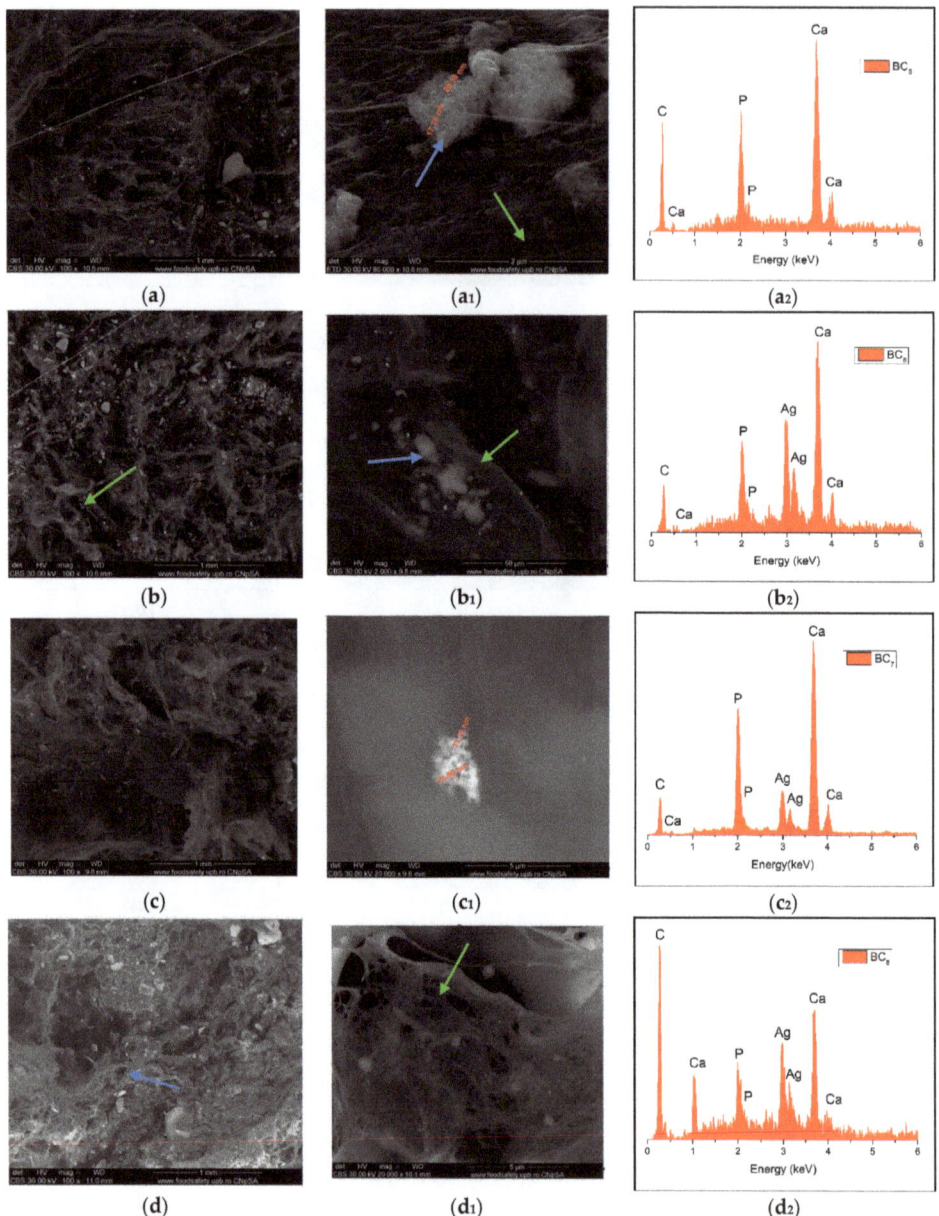

Figure 8. High-resolution backscattered electron (BSE) images and EDS spectra at 100× magnification for BC_5 (**a**,**a₁**,**a₂**), BC_6 (**b**,**b₁**,**b₂**), BC_7 (**c**,**c₁**,**c₂**), and BC_8 (**d**,**d₁**,**d₂**) (where **a₁**–**d₁** images represent a high magnification of a–d image area; green arrow indicates the fibrous structure of BC and blue arrow indicate inorganic particles).

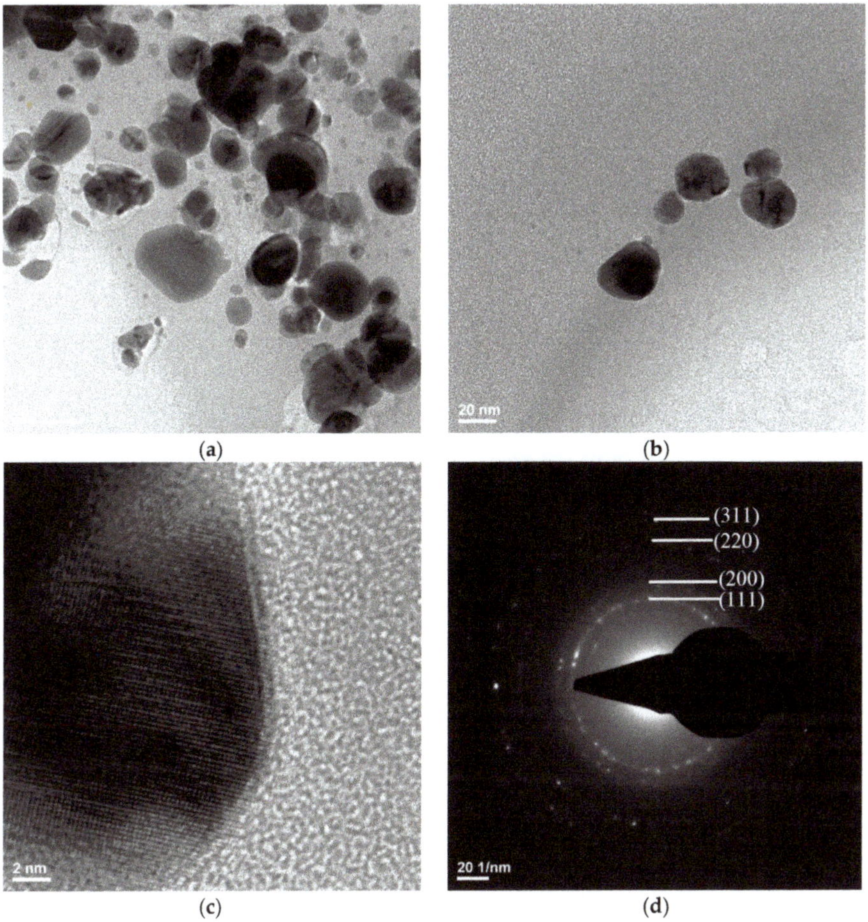

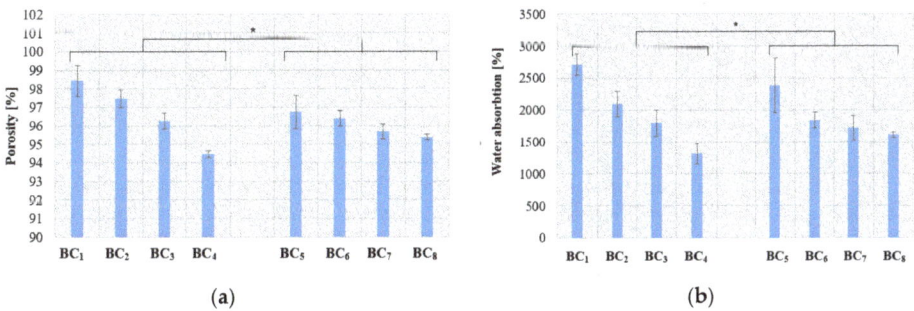

Figure 9. TEM (**a**,**b**), HR-TEM (**c**), and SAED (**d**) images for silver.

Figure 10. Porosity (**a**) and water absorption (**b**) results for the obtained composites (presented as mean ± S.D. of three replicates and * $p < 0.005$ obtained by single-factor ANOVA test.

As can be observed, when the silver nanoparticle concentration increased, the composite porosity decreased for susceptible types of composite, which resulted in lower absorption capacity. Even though high porosity, which is associated with increased absorption capacity, is an important structural parameter for bone substituents, the registered decrease due to Ag addition is not significant in this case as the water absorption was still greater than 1000–1500%.

Visual inspection (Figure 11) is an efficient technique for investigation of simulated in vitro degradation of BC samples, and it has also been used in recent literature [49] in order to investigate the degradation of cellulose-based materials. Generally, visual inspection implies macroscopic pictures of the immersed samples while observing, in a qualitative manner, the presence of detached fragments, the apparition of denser fragments that may provide additional mechanical integrity for cell growth, and so on. As expected, the composite materials did not show major changes after immersion in PBS and SBF.

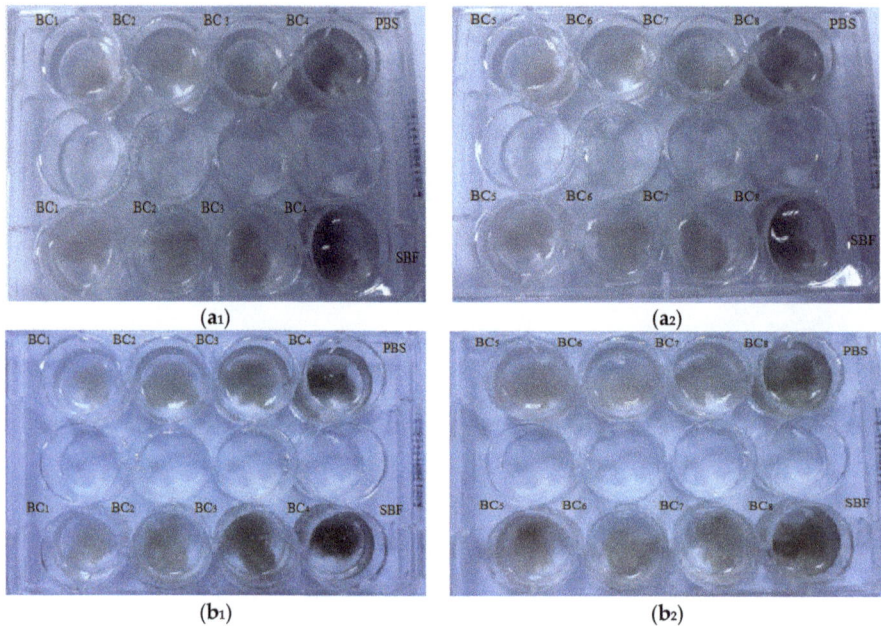

Figure 11. Macroscopic pictures of degradation of BC composites initially (a_1, a_2) and after seven days of immersion in phosphate-buffered saline (PBS) and simulated body fluid (SBF) (b_1, b_2).

After seven days, no major visual changes were observed, and the degradation was below 5% (see Figure 12), a sign that bacterial cellulose prevented the disintegration of the composite. A rapid biodegradation of the implanted material is not desired because it takes time for it to integrate better into the host tissue. Another problem is that by biodegradation, remnants/fragments of the material can reach the level of sensitive areas, which would be fatal.

Due to the higher homogeneity and better interactions between the phosphate phase (HAp) and BC, the composites obtained in situ had a lower degradation compared to those obtained using ex situ methods.

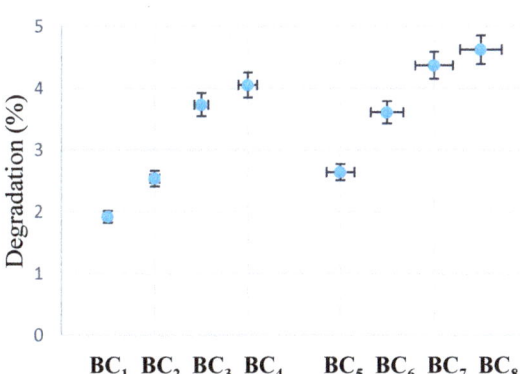

Figure 12. Degradability of composites after seven days of immersion in SBF at room temperature.

Antimicrobial Potential

The antimicrobial effect of the obtained composite materials was different among the tested samples, being influenced by the synthesis approach, AgNP content, and microbial species. Figure 13 shows the diameters of the inhibition area of the microorganisms grown in the presence of the tested materials.

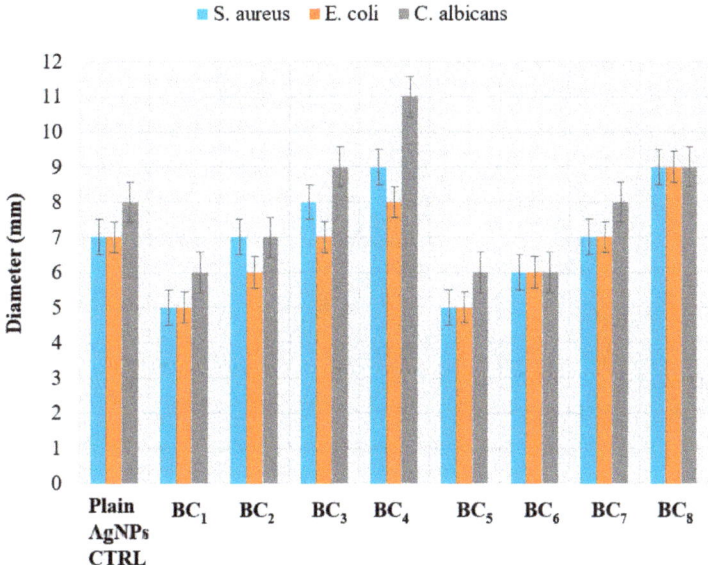

Figure 13. The diameter of the growth inhibition zone of the tested microorganisms grown in the presence of the obtained BC samples containing various amounts of AgNPs. The plain AgNP control is represented by 10 uL of AgNPs (used at maximum equivalent amount contained in BC samples), which was added to a commercial filter paper disc of a similar size as the obtained BC samples.

Because composites BC_1 and BC_5 did not contain antibacterial AgNPs, they were used as control samples for evaluation of the other samples.

The antimicrobial characteristic of the obtained composites was clearly influenced by the concentration of AgNPs, with samples with higher content of silver exhibiting the greatest microbial growth inhibition, regardless of the synthesis approach. It was observed that the materials obtained by the in situ method had a more pronounced overall antibacterial characteristic (growth inhibition zones ranging 5–11 mm) compared to the samples obtained by the ex situ method (diameter of inhibition zones ranging 5–9 mm).

The different antimicrobial effects of the materials obtained by in situ and ex situ routes correlated with their physicochemical properties. Samples obtained by the in situ route showed a larger porosity, suggesting that the bioactive compounds (i.e., AgNPs) may be absorbed more efficiently in pores. Moreover, a higher porosity degree can be directly associated with an easier release of the bioactive agent, therefore inducing an increase in the antimicrobial activity of the final composite. This idea is supported by the results obtained with the control AgNPs utilized in equivalent amounts and added to sterile commercial filter paper, which demonstrated lower inhibition zones.

The most efficient growth inhibition was observed against the yeast strain *C. albicans* ATCC 10231, with the result being relevant for samples obtained by both in situ and ex situ methods. However, the composites obtained in situ also showed increased antimicrobial activity against the Gram-positive *S. aureus* strain. This result suggests that the obtained composite materials may act differently on microbial cells, depending on the particularities of their cellular wall. Such differences were observed before with silver nanoparticles [50–53].

5. Conclusions

In this study, we report the synthesis of hydroxyapatite–bacterial cellulose silver nanocomposites obtained by two routes using coprecipitation, namely, in situ and ex situ assembly. These materials contained an organic part (bacterial cellulose), an inorganic part (hydroxyapatite), and an antimicrobial agent (AgNPs) contained in various amounts, thereby conferring new bioactive properties on the composite materials. Physicochemical and antimicrobial studies demonstrated that the most efficient in terms of potential biomedical applications were the samples obtained by the in situ approach. The porosity range of the in situ materials was greater than the porosity of ex situ composites, while the best antimicrobial activity was observed for the material coded BC_4, which had a content of 5 wt % AgNPs. Due to the physicochemical structure, together with the already demonstrated great antimicrobial properties and low biodegradability of these materials, they have potential applications as successful candidates for biomedical applications, especially in hard tissue engineering. Their current limitation relates to the fact that further tests performed on osteoblast differentiation and mineralization (e.g., alkaline phosphatase and alizarin red S) are needed.

Author Contributions: Conceptualization, A.I.N. and A.E.S.; methodology, B.S.V.; validation, A.I.N. and I.A.N.; formal analysis, A.I.N., D.-I.E., and B.S.V.; investigation, D.-I.E., I.A.N., and A.M.H.; writing—original draft preparation, A.E.S. and A.I.N.; writing—review and editing, A.E.S., I.A.N., B.S.V., and A.M.H.; supervision, B.S.V. All authors have read and agreed to the published version of the manuscript.

Funding: This work was financially supported by the project Smart Scaffolds Built on Biocellulose 3D Architecture or Artificial Electrospun Templates for Hard Tissue Engineering (ScaBiES), PN-III-P1-1.1-TE-2016-0871 (contract number 66/2018), financed by the Executive Unit for Financing Higher Education, Research, Development, and Innovation (UEFISCDI). The SEM and RAMAN analyses obtained on the samples were possible due to the EU-funded project POSCCE-A2-O2.2.1-2013-1/Priority Axe 2, project number 638/12.03.2014, ID 1970, SMIS-CSNR code 48652. The XRD analyses were financed by the European Social Fund and the Romanian Government under the contract number POSDRU/86/1.2/S/58146 (MASTERMAT).

Conflicts of Interest: The authors declare no conflict of interest.

References

1. Sarkar, C.; Chowdhuri, A.R.; Garai, S.; Chakraborty, J.; Sahu, S.K. Three-dimensional cellulose-hydroxyapatite nanocomposite enriched with dexamethasone loaded metal–organic framework: A local drug delivery system for bone tissue engineering. *Cellulose* **2019**, *26*, 7253–7269. [CrossRef]
2. Shkarina, S.; Shkarin, R.; Weinhardt, V.; Melnik, E.; Vacun, G.; Kluger, P.J.; Loza, K.; Epple, M.; Ivlev, S.I.; Baumbach, T.; et al. 3D biodegradable scaffolds of polycaprolactone with silicate-containing hydroxyapatite microparticles for bone tissue engineering: High-resolution tomography and in vitro study. *Sci. Rep.* **2018**, *8*, 8907. [CrossRef] [PubMed]
3. Atila, D.; Karataş, A.; Evcin, A.; Keskin, D.; Tezcaner, A. Bacterial cellulose-reinforced boron-doped hydroxyapatite/gelatin scaffolds for bone tissue engineering. *Cellulose* **2019**, *26*, 9765–9785. [CrossRef]
4. Blanco Parte, F.G.; Santoso, S.P.; Chou, C.-C.; Verma, V.; Wang, H.-T.; Ismadji, S.; Cheng, K.-C. Current progress on the production, modification, and applications of bacterial cellulose. *Crit. Rev. Biotechnol.* **2020**, *40*, 397–414. [CrossRef]
5. Woltje, M.; Ostermann, K.; Aibibu, D.; Rodel, G.; Cherif, C. Session 12: Biomaterials. *Biomed. Tech.* **2019**, *64*, 69–71. [CrossRef] [PubMed]
6. Lu, T.; Feng, S.; He, F.; Ye, J. Enhanced osteogenesis of honeycomb β-tricalcium phosphate scaffold by construction of interconnected pore structure: An in vivo study. *J. Biomed. Mater. Res. Part A* **2020**, *108*, 645–653. [CrossRef]
7. Hapuhinna, K.; Gunaratne, R.; Pitawala, J. Comparison between Differently Synthesized Hydroxyapatite Composites for Orthopedic Applications. *J. Mater. Sci. Chem. Eng.* **2019**, *7*, 16–28. [CrossRef]
8. Arcos, D.; Vallet-Regí, M. Substituted hydroxyapatite coatings of bone implants. *J. Mater. Chem. B* **2020**, *8*, 1781–1800. [CrossRef] [PubMed]
9. Chang, H.-H.; Yeh, C.-L.; Wang, Y.-L.; Fu, K.-K.; Tsai, S.-J.; Yang, J.-H.; Lin, C.-P. Neutralized Dicalcium Phosphate and Hydroxyapatite Biphasic Bioceramics Promote Bone Regeneration in Critical Peri-Implant Bone Defects. *Materials* **2020**, *13*, 823. [CrossRef]
10. Liu, Y.; Gu, J.; Fan, D. Fabrication of High-Strength and Porous Hybrid Scaffolds Based on Nano-Hydroxyapatite and Human-Like Collagen for Bone Tissue Regeneration. *Polymers* **2020**, *12*, 61. [CrossRef]
11. Nicomrat, D. Silver Nanoparticles Impregnated Biocellulose Produced by Sweet Glutinous Rice Fermentation with the Genus Acetobacter. *E3S Web Conf.* **2020**, *141*, 03003. [CrossRef]
12. Maneerung, T.; Tokura, S.; Rujiravanit, R. Impregnation of silver nanoparticles into bacterial cellulose for antimicrobial wound dressing. *Carbohydr. Polym.* **2008**, *72*, 43–51. [CrossRef]
13. Velu, R.; Calais, T.; Jayakumar, A.; Raspall, F. A Comprehensive Review on Bio-Nanomaterials for Medical Implants and Feasibility Studies on Fabrication of Such Implants by Additive Manufacturing Technique. *Materials* **2019**, *13*, 92. [CrossRef]
14. Busuioc, C.; Stroescu, M.; Stoica-Guzun, A.; Voicu, G.; Jinga, S.-I. Fabrication of 3D calcium phosphates based scaffolds using bacterial cellulose as template. *Ceram. Int.* **2016**, *42*, 15449–15458. [CrossRef]
15. Draghici, A.-D.; Busuioc, C.; Mocanu, A.; Nicoara, A.-I.; Iordache, F.; Jinga, S.-I. Composite scaffolds based on calcium phosphates and barium titanate obtained through bacterial cellulose templated synthesis. *Mater. Sci. Eng. C* **2020**, *110*, 110704. [CrossRef] [PubMed]
16. Khan, F.; Dahman, Y. A novel approach for the utilization of biocellulose nanofibres in polyurethane nanocomposites for potential applications in bone tissue implants. *Des. Monomers Polym.* **2012**, *15*, 1–29. [CrossRef]
17. Busuioc, C.; Ghitulica, C.D.; Stoica, A.; Stroescu, M.; Voicu, G.; Ionita, V.; Averous, L.; Jinga, S.I. Calcium phosphates grown on bacterial cellulose template. *Ceram. Int.* **2018**, *44*, 9433–9441. [CrossRef]
18. Halib, N.; Ahmad, I.; Grassi, M.; Grassi, G. The remarkable three-dimensional network structure of bacterial cellulose for tissue engineering applications. *Int. J. Pharm.* **2019**, *566*, 631–640. [CrossRef] [PubMed]
19. Frone, A.N.; Panaitescu, D.M.; Nicolae, C.A.; Gabor, A.R.; Trusca, R.; Casarica, A.; Stanescu, P.O.; Baciu, D.D.; Salageanu, A. Bacterial cellulose sponges obtained with green cross-linkers for tissue engineering. *Mater. Sci. Eng. C* **2020**, *110*, 110740. [CrossRef]

20. Liyaskina, E.; Revin, V.; Paramonova, E.; Nazarkina, M.; Pestov, N.; Revina, N.; Kolesnikova, S. Nanomaterials from bacterial cellulose for antimicrobial wound dressing. In *Journal of Physics: Conference Series*; IOP Publishing: Bristol, UK, 2017; p. 012034.
21. Mocanu, A.; Isopencu, G.; Busuioc, C.; Popa, O.-M.; Dietrich, P.; Socaciu-Siebert, L. Bacterial cellulose films with ZnO nanoparticles and propolis extracts: Synergistic antimicrobial effect. *Sci. Rep.* **2019**, *9*, 17687. [CrossRef]
22. Hodel, K.V.S.; Fonseca, L.M.d.S.; Santos, I.M.d.S.; Cerqueira, J.C.; Santos-Júnior, R.E.d.; Nunes, S.B.; Barbosa, J.D.V.; Machado, B.A.S. Evaluation of Different Methods for Cultivating Gluconacetobacter hansenii for Bacterial Cellulose and Montmorillonite Biocomposite Production: Wound-Dressing Applications. *Polymers* **2020**, *12*, 267. [CrossRef] [PubMed]
23. Basu, P.; Saha, N.; Alexandrova, R.; Saha, P. Calcium Phosphate Incorporated Bacterial Cellulose-Polyvinylpyrrolidone Based Hydrogel Scaffold: Structural Property and Cell Viability Study for Bone Regeneration Application. *Polymers* **2019**, *11*, 1821. [CrossRef]
24. Eslahi, N.; Mahmoodi, A.; Mahmoudi, N.; Zandi, N.; Simchi, A. Processing and properties of nanofibrous bacterial cellulose-containing polymer composites: A review of recent advances for biomedical applications. *Polym. Rev.* **2020**, *60*, 144–170. [CrossRef]
25. Chiaoprakobkij, N.; Seetabhawang, S.; Sanchavanakit, N.; Phisalaphong, M. Fabrication andcharacterization of novel bacterial cellulose/alginate/gelatin biocomposite film. *J. Biomater. Sci. Polym. Ed.* **2019**, *30*, 961–982. [CrossRef] [PubMed]
26. Bodea, I.M.; Cătunescu, G.M.; Stroe, T.F.; Dîrlea, S.A.; Beteg, F.I. Applications of bacterial-synthesized cellulose in veterinary medicine—A review. *Acta Vet. Brno* **2020**, *88*, 451–471. [CrossRef]
27. Ou, Q.; Huang, K.; Fu, C.; Huang, C.; Fang, Y.; Gu, Z.; Wu, J.; Wang, Y. Nanosilver-incorporated halloysite nanotubes/gelatin methacrylate hybrid hydrogel with osteoimmunomodulatory and antibacterial activity for bone regeneration. *Chem. Eng. J.* **2020**, *382*, 123019. [CrossRef]
28. Hussain, Z.; Abourehab, M.A.; Khan, S.; Thu, H.E. Silver nanoparticles: A promising nanoplatform for targeted delivery of therapeutics and optimized therapeutic efficacy. In *Metal Nanoparticles for Drug Delivery and Diagnostic Applications*; Elsevier: Amsterdam, The Netherlands, 2020; pp. 141–173.
29. Khurshid, Z.; Zafar, M.S.; Hussain, S.; Fareed, A.; Yousaf, S.; Sefat, F. Silver-substituted hydroxyapatite. In *Handbook of Ionic Substituted Hydroxyapatites*; Elsevier: Amsterdam, The Netherlands, 2020; pp. 237–257.
30. Preethi, G.U.; Unnikrishnan, B.S.; Sreekutty, J.; Archana, M.G.; Anupama, M.S.; Shiji, R.; Raveendran Pillai, K.; Joseph, M.M.; Syama, H.P.; Sreelekha, T.T. Semi-interpenetrating nanosilver doped polysaccharide hydrogel scaffolds for cutaneous wound healing. *Int. J. Biol. Macromol.* **2020**, *142*, 712–723. [CrossRef]
31. Janarthanan, P.; Sathasivam, T.; Li, T.H.; Dahlan, N.A.; Paramasivam, R. Silver Nanoparticles: Biological Synthesis and Applications. In *Biological Synthesis of Nanoparticles and Their Applications*; Taylor and Francis Group: Abington, UK, 2020; p. 93.
32. Ali, G.; Abd El-Moez, S.; Abdel-Fattah, W. Synthesis and characterization of nontoxic silvernano-particles with preferential bactericidal activity. *Biointerface Res. Appl. Chem.* **2019**, *9*, 4617–4623.
33. Wijnhoven, S.W.P.; Peijnenburg, W.J.G.M.; Herberts, C.A.; Hagens, W.I.; Oomen, A.G.; Heugens, E.H.W.; Roszek, B.; Bisschops, J.; Gosens, I.; Van De Meent, D.; et al. Nano-silver—A review of available data and knowledge gaps in human and environmental risk assessment. *Nanotoxicology* **2009**, *3*, 109–138. [CrossRef]
34. Khatoon, Z.; McTiernan, C.D.; Suuronen, E.J.; Mah, T.-F.; Alarcon, E.I. Bacterial biofilm formation on implantable devices and approaches to its treatment and prevention. *Heliyon* **2018**, *4*, e01067. [CrossRef]
35. Zimmermann, K.A.; LeBlanc, J.M.; Sheets, K.T.; Fox, R.W.; Gatenholm, P. Biomimetic design of a bacterial cellulose/hydroxyapatite nanocomposite for bone healing applications. *Mater. Sci. Eng. C* **2011**, *31*, 43–49. [CrossRef]
36. Saska, S.; Barud, H.; Gaspar, A.; Marchetto, R.; Ribeiro, S.J.L.; Messaddeq, Y. Bacterial cellulose-hydroxyapatite nanocomposites for bone regeneration. *Int. J. Biomater.* **2011**, *2011*. [CrossRef] [PubMed]
37. Wan, Y.; Huang, Y.; Yuan, C.; Raman, S.; Zhu, Y.; Jiang, H.; He, F.; Gao, C. Biomimetic synthesis of hydroxyapatite/bacterial cellulose nanocomposites for biomedical applications. *Mater. Sci. Eng. C* **2007**, *27*, 855–864. [CrossRef]
38. Hong, L.; Wang, Y.L.; Jia, S.R.; Huang, Y.; Gao, C.; Wan, Y.Z. Hydroxyapatite/bacterial cellulose composites synthesized via a biomimetic route. *Mater. Lett.* **2006**, *60*, 1710–1713. [CrossRef]

39. Wan, Y.Z.; Hong, L.; Jia, S.R.; Huang, Y.; Zhu, Y.; Wang, Y.L.; Jiang, H.J. Synthesis and characterization of hydroxyapatite–bacterial cellulose nanocomposites. *Compos. Sci. Technol.* **2006**, *66*, 1825–1832. [CrossRef]
40. Grande, C.J.; Torres, F.G.; Gomez, C.M.; Carmen Bañó, M. Nanocomposites of bacterial cellulose/hydroxyapatite for biomedical applications. *Acta Biomater.* **2009**, *5*, 1605–1615. [CrossRef] [PubMed]
41. Dincă, V.; Mocanu, A.; Isopencu, G.; Busuioc, C.; Brajnicov, S.; Vlad, A.; Icriverzi, M.; Roseanu, A.; Dinescu, M.; Stroescu, M. Biocompatible pure ZnO nanoparticles-3D bacterial cellulose biointerfaces with antibacterial properties. *Arab. J. Chem.* **2020**, *13*, 3521–3533. [CrossRef]
42. Leau, S.-A.; Marin, Ş.; Coară, G.; Albu, L.; Constantinescu, R.R.; Kaya, M.A.; Neacşu, I.-A. Study of wound-dressing materials based on collagen, sodium carboxymethylcellulose and silver nanoparticles used for their antibacterial activity in burn injuries. In Proceedings of the International Conference on Advanced Materials and Systems (ICAMS), Bucharest, Romania, 18–20 October 2018; pp. 123–128.
43. Ghorbani, F.; Li, D.; Ni, S.; Zhou, Y.; Yu, B. 3D printing of acellular scaffolds for bone defect regeneration: A review. *Mater. Today Commun.* **2020**, *22*, 100979. [CrossRef]
44. Anghel, I.; Grumezescu, A.M.; Holban, A.M.; Gheorghe, I.; Vlad, M.; Anghel, G.A.; Balaure, P.C.; Chifiriuc, C.M.; Ciuca, I.M. Improved activity of aminoglycosides entrapped in silica networks against microbial strains isolated from otolaryngological infections. *Farmacia* **2014**, *62*, 1.
45. Stumpf, T.R.; Yang, X.; Zhang, J.; Cao, X. In situ and ex situ modifications of bacterial cellulose for applications in tissue engineering. *Mater. Sci. Eng. C* **2018**, *82*, 372–383. [CrossRef]
46. Yin, N.; Chen, S.-Y.; Ouyang, Y.; Tang, L.; Yang, J.-X.; Wang, H.-P. Biomimetic mineralization synthesis of hydroxyapatite bacterial cellulose nanocomposites. *Progress Nat. Sci. Mater. Int.* **2011**, *21*, 472–477. [CrossRef]
47. Marković, S.; Veselinović, L.; Lukić, M.J.; Karanović, L.; Bračko, I.; Ignjatović, N.; Uskoković, D. Synthetical bone-like and biological hydroxyapatites: A comparative study of crystal structure and morphology. *Biomed. Mater.* **2011**, *6*, 045005. [CrossRef] [PubMed]
48. Markovic, M.; Fowler, B.O.; Tung, M.S. Preparation and comprehensive characterization of a calcium hydroxyapatite reference material. *J. Res. Natl. Inst. Stand. Technol.* **2004**, *109*, 553. [CrossRef] [PubMed]
49. Suk, J.S.; Xu, Q.; Kim, N.; Hanes, J.; Ensign, L.M. PEGylation as a strategy for improving nanoparticle-based drug and gene delivery. *Adv. Drug Deliv. Rev.* **2016**, *99*, 28–51. [CrossRef] [PubMed]
50. Hu, Y.; Catchmark, J.M. In vitro biodegradability and mechanical properties of bioabsorbable bacterial cellulose incorporating cellulases. *Acta Biomater.* **2011**, *7*, 2835–2845. [CrossRef]
51. Abbaszadegan, A.; Ghahramani, Y.; Gholami, A.; Hemmateenejad, B.; Dorostkar, S.; Nabavizadeh, M.; Sharghi, H. The effect of charge at the surface of silver nanoparticles on antimicrobial activity against gram-positive and gram-negative bacteria: A preliminary study. *J. Nanomater.* **2015**, *2015*. [CrossRef]
52. Ayodele, A.T.; Valizadeh, A.; Adabi, M.; Esnaashari, S.S.; Madani, F.; Khosravani, M.; Adabi, M. Ultrasound nanobubbles and their applications as theranostic agents in cancer therapy: A review. *Biointerface Res. Appl. Chem.* **2017**, *7*, 2253–2262.
53. Sabry, N.M.; Tolba, S.; Abdel-Gawad, F.K.; Bassem, S.M.; Nassar, H.F.; El-Taweel, G.E.; Okasha, A.; Ibrahim, M. Interaction between nano silver and bacteria: Modeling approach. *Biointerface Res. Appl. Chem.* **2018**, *8*, 3570–3574.

Publisher's Note: MDPI stays neutral with regard to jurisdictional claims in published maps and institutional affiliations.

© 2020 by the authors. Licensee MDPI, Basel, Switzerland. This article is an open access article distributed under the terms and conditions of the Creative Commons Attribution (CC BY) license (http://creativecommons.org/licenses/by/4.0/).

Article

Bioengineering Bone Tissue with 3D Printed Scaffolds in the Presence of Oligostilbenes

Francesca Posa [1,2,*,†], **Adriana Di Benedetto** [1,†], **Giampietro Ravagnan** [3], **Elisabetta Ada Cavalcanti-Adam** [2], **Lorenzo Lo Muzio** [1], **Gianluca Percoco** [4] and **Giorgio Mori** [1]

1. Department of Clinical and Experimental Medicine, University of Foggia, viale Pinto 1, 71122 Foggia, Italy; adriana.dibenedetto@unifg.it (A.D.B.); lorenzo.lomuzio@unifg.it (L.L.M.); giorgio.mori@unifg.it (G.M.)
2. Department of Biophysical Chemistry, Heidelberg University and Max Planck Institute for Medical Research, Jahnstraße 29, 69120 Heidelberg, Germany; eacavalcanti@mr.mpg.de
3. Glures srl. Unità Operativa di Napoli, Spin off Accademico dell'Università di Venezia Cà Foscari, Via delle Industrie 19b-30175 Venezia, Italy; ravagnangp@gmail.com
4. Department of Mechanics, Mathematics and Management, Polytechnic University of Bari, Via E. Orabona 4, 70125 Bari, Italy; gianluca.percoco@poliba.it
* Correspondence: francesca.posa@unifg.it
† These authors contributed equally to this paper.

Received: 13 August 2020; Accepted: 3 October 2020; Published: 9 October 2020

Abstract: Diseases determining bone tissue loss have a high impact on people of any age. Bone healing can be improved using a therapeutic approach based on tissue engineering. Scientific research is demonstrating that among bone regeneration techniques, interesting results, in filling of bone lesions and dehiscence have been obtained using adult mesenchymal stem cells (MSCs) integrated with biocompatible scaffolds. The geometry of the scaffold has critical effects on cell adhesion, proliferation and differentiation. Many cytokines and compounds have been demonstrated to be effective in promoting MSCs osteogenic differentiation. Oligostilbenes, such as Resveratrol (Res) and Polydatin (Pol), can increase MSCs osteoblastic features. 3D printing is an excellent technique to create scaffolds customized for the lesion and thus optimized for the patient. In this work we analyze osteoblastic features of adult MSCs integrated with 3D-printed polycarbonate scaffolds differentiated in the presence of oligostilbenes.

Keywords: bone; mesenchymal stem cells; biomaterial; polycarbonate; resveratrol; polydatin; osteogenic differentiation; focal adhesions; bone health

1. Introduction

The global increase in average life expectancy is leading to an escalation of age-related health problems which may affect organs or tissues. Although the bone tissue is capable of self-regeneration, there are several pathological conditions, determined by serious trauma or degenerative diseases, which require appropriate medical procedures in order to realize a complete recovery of the anatomical and functional properties of the tissue. These innovative therapeutic approaches are part of the so-called regenerative medicine [1,2]. The use of adult stem cells in bone reconstructive therapies offers significant benefits [3,4].

In fact, it is known that stem cells are capable of self-renewal and can differentiate into different cell types, ensuring the repair of most tissues, thus becoming highly useful for tissue engineering. Stem cells in human adults can be isolated from various tissues, including bone marrow, nervous tissue, peripheral blood, retina, liver, pancreas, tooth, and dental bud. In particular, among adult stem cells, mesenchymal stem cells (MSCs), originally identified in the bone marrow, which is still considered the best cell source for osteogenic differentiation, can be isolated also from several adult tissues such as adipose tissue, dental tissues, skin, brain, liver, and fetal tissues [5,6]. MSCs appropriately isolated and induced to differentiation, through integration with biocompatible scaffolds, could represent a valid therapeutic approach for the regeneration of connective tissues as bone and cartilage, healing defects of traumatic, degenerative or neoplastic origin. Unfortunately, several concerns may arise from autologous and allogeneic stem cell transplants which, in their turn, may also not be sufficient to accelerate the healing process in the case of large bone defects [7]. As a consequence, in the last decades an increasing pivotal role has been attributed to tissue-engineered bone grafts developed through a combined effort of engineering, biotechnology and biomaterial science [8]. In this perspective, especially for hard tissue regeneration, one of the most promising approaches is the use of customized scaffolds combined with factors allowing cell proliferation and osteogenic differentiation [9]. Matching tissue engineering with predictive medicine, based on mechanobiological computational models, would optimize healing processes [10]. The fabrication methods for tissue engineering are conventional methods, additive manufacturing techniques and 4D printing [11]. Among these methods, 3D printing is an encouraging technique, easily available to realize personalized scaffolds to be used for tissue regeneration [12].

Dental bud stem cells (DBSCs) are widely recognized as MSCs that can effectively differentiate into osteoblast-like cells [13] becoming a suitable candidate for bone regeneration. DBSCs express the typical mesenchymal stem markers and, as we have shown, their ability to acquire the osteoblastic features and to produce mineralized bone matrix in vitro, which is the crucial event of osteogenic differentiation [14,15], can be positively influenced by several molecules [16–19]. Polydatin (Pol) deserves a particular mention among the natural molecules capable of inducing DBSCs osteogenic differentiation [19] and opens new horizons to its possible use as a therapeutic agent, as we have exhaustively detailed in our new invention patent (patent n.16999PTIT entitled "Composizioni comprendenti o costituite da Polidatina per uso nel trattamento delle patologie ossee"—"Compositions comprising or consisting of Polydatin to be used in the treatment of bone pathologies", deposited with application number 102017000079581). Pol, that is a natural precursor of the polyphenolic compound Resveratrol (Res), is a glucoside we find in abundance in fruits and plants [20]. This glucoside shares some of the beneficial biological properties fully demonstrated for Res [21–23], but, in comparison, it also presents advantages: higher stability, significant abundance and better oral absorption [24–26]. We have previously shown that DBSCs positively responded to Res and Pol treatment increasing their osteogenic potential and, moreover, Pol appeared to be more effective than Res even when used at a very low concentration [19]. To induce bone regeneration, porous scaffolds with appropriate shape, pore size, porosity, degradability, biocompatibility, mechanical properties and desirable cellular responses are required. 3D printing has revealed to be very useful in this field, thanks to the capability to process complex shapes with a wide variety of biocompatible materials such as poly(l-lactic acid) (PLLA), poly(vinyl alcohol) (PVA), poly(lactic-co-glycolic acid) (PLGA) and polycaprolactone (PCL) starting from filaments and pellets [27]. In this study we investigate the combination of Res and Pol treatment and 3D-printed polycarbonate (PC) scaffolds to study the possible effects of this set-up on MSCs commitment into the osteoblastic lineage. The PC has been chosen to have artificial scaffolds with bio-compatible material that have strength and stiffness near to bone tissue [28].

2. Materials and Methods

2.1. Ethics

The study was conducted in compliance with the Declaration of Helsinki and the International Conference on Harmonization Principles of Good Clinical Practice. The research protocol was approved by the ethical committee, within the project BIADIDENT num. Rep 4159/2018, at the University of Foggia Ospedali Riuniti, and all participants gave informed consent allowing their anonymized information to be used for data analysis.

2.2. Scaffold Preparation

The scaffolds were manufactured using an Ultimaker S5, equipped with a AA nozzle diameter equal to 0.4 mm. At the best of author's knowledge, this printer can be considered as one of the best compromise between quality, price and system flexibility. The filament was the 3 mm 3DXMAX®, a high-heat 3D printing filament made using Lexan®. The print temperature was set to 285 °C and the bed temperature was kept to 110 °C, print speed 30 mm/s. The bed and nozzle temperature parameters are not inside the interval suggested by the 3D printer supplier, but since manufacturing time is lower than 5 min, the machine is able to complete the workpiece without damages. The different pore sizes were obtained setting on the Cura slicing software the distance between lines equal, respectively, to 0.75 mm for small pores, 0.9 mm for medium pores and 1.15 mm for large pores [29]. Figure 1 shows the Ultimaker 3D printer and samples of the manufactured scaffolds.

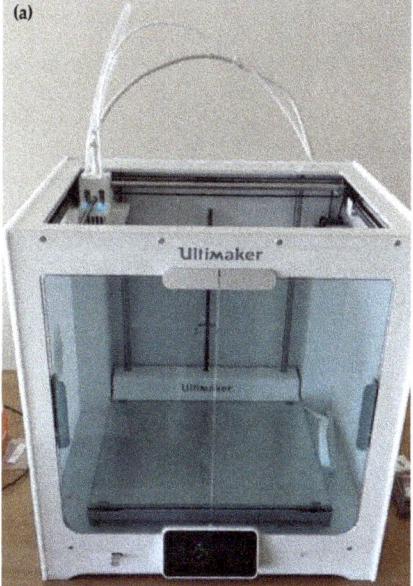
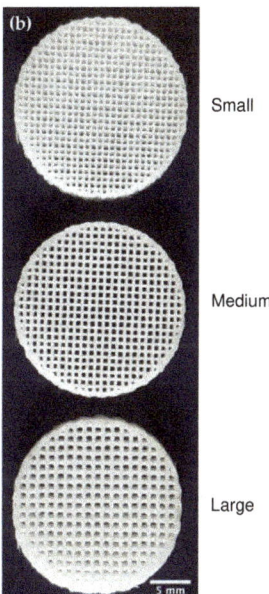

Figure 1. Printer and polycarbonate (PC) scaffolds. (**a**) Ultimaker S5 3D printer. (**b**) Printed scaffolds presenting pores of small (0.75 mm), medium (0.9 mm) and large (1.15 mm) dimensions. Images of representative scaffolds were chosen for the figure. Scale bar = 5 mm.

2.3. Patients and DBSCs Cultures

The dental buds were collected from ten healthy pediatric donors (eight-twelve years) who were subjected to the third molar extraction for orthodontic reasons; each patient's parents provided a written informed consent. The study was authorized by the Institutional Review Board of the Department of Clinical and Experimental Medicine, University of Foggia. The dental papilla, which corresponds to the internal section of the dental bud, and contains stem cells of mesodermal origin, was cut in small fragments and enzymatically digested. Single-cell suspensions were harvested by filtration, and seeded and expanded in vitro as already reported [30–32]. In the experiments aimed to examine Res and Pol effect on cell adhesion during the osteoblastic differentiation process, DBSCs were seeded at a density of $3 \times 10^3/cm^2$ and cultured in an osteogenic medium consisting of α-MEM supplemented with 2% heat inactivated fetal bovine serum (FBS), 10^{-8} M dexamethasone and 50 µg/mL ascorbic acid (Sigma Aldrich, Milan, Italy). DBSCs were maintained in the osteogenic medium supplemented also with 10 mM β-glycerophosphate (Sigma Aldrich, Milan, Italy), for the evaluation of Res and Pol effects on cell adhesion, proliferation, differentiation and examination of their ability in the induction of matrix mineralization in cultures on biomaterials.

2.4. Res and Pol Treatment

Res and Pol extracted from *Polygonum cuspidatum* (Japanese Knotweed), according to the procedure defined in the Patent EP1292320B1, were provided by Prof. Ravagnan. Res and Pol were dissolved in ethanol at 100 mM stock solutions [33] and then added to the culture media under low serum conditions (2% FBS) to the final concentration of 0.1 µM for both of them, as detailed in Di Benedetto et al. [19]. In the experiments control cells were not treated with Res or Pol and served as control group (Ctr), treated cells were exposed to Res or Pol (treatment group), that were added to the media at every renewal (every 3 days).

2.5. Immunofluorescence

For focal adhesion staining, DBSCs were cultured on glass coverslips for 4 days and then fixed in 4% (w/v) paraformaldehyde (PFA) in PBS. Cells were then washed with PBS and blocked in 1% BSA, 5% normal goat serum in PBS for 20 min, to avoid non-specific protein binding. The following antibodies were used: $\alpha_V\beta_3$ antibody 1:100 (clone LM609 antibody, cat. MAB1976, MerckMillipore, Merck KGaA, Darmstadt, Germany), followed by fluorescently labeled goat anti-mouse secondary antibody (Alexa Fluor 488, 2 µg/mL, Invitrogen ThermoFisher Scientific, Waltham, MA, USA). Samples were embedded in Mowiol containing 0.1% (v/v) DAPI for an additional staining of the nucleus. Cells were imaged by a multispectral confocal microscope Leica TCS SP5. The images were adjusted in brightness and color with ImageJ software (Research Services Branch, Image Analysis Software Version 1.52c, NIH, Bethesda, MD, USA).

2.6. Alizarin Red Staining (ARS)

DBSCs capacity to produce mineralized matrix nodules when cultured on the scaffolds was determined by performing ARS at 28 days of culture in osteogenic conditions. After removing the culture media, cells were rinsed with PBS, fixed in 10% formalin at RT for 10 min. Then cells were washed again with deionized water and stained using a 1% ARS solution for 10 min at RT. At the end of the incubation period the ARS solution was removed and cells were washed twice with deionized water and air dried. The quantification of ARS in the red stained monolayer was performed by extracting the dye and by reading the optical density (OD) in triplicate at 405 nm.

2.7. Statistical Analyses

Statistical analyses were performed by Student's *t*-test with the GraphPad Prism version 8.0.2 for MacOS software (San Diego, CA, USA). The results were considered statistically significant for $p < 0.05$ (indicated as § $p < 0.01$, * $p < 0.001$).

3. Results

3.1. Both Res and Pol Treatments Influence Cell Spreading and Focal Adhesion Assembly via $\alpha_V\beta_3$ Reorganization

To investigate the influence of Res and Pol on cell adhesion and spreading, which determine, as a consequence, DBSCs exhibition of osteoblastic features, the cells were cultured for 4 days on glass coverslips in absence of treatment (Ctr) or in presence of Res or Pol added to the media (Figure 2a–c). Such a short period of time was chosen because of DBSCs predisposition to proliferate and produce various cell layers when left in culture for a few days, a condition that would not have allowed a clear observation of focal adhesions. We examined $\alpha_V\beta_3$ integrin subcellular distribution by confocal immunofluorescence. This integrin receptor has already been shown to be crucial for the osteogenic differentiation process of MSCs [30] and Vitamin D or the supramolecular aggregate T-LysYal® (T-Lys) can enhance its expression and clusterization leading to the induction of the differentiation process [16,17]. As observable in Figure 1, in Ctr cells $\alpha_V\beta_3$ integrin clusters were hardly detectable and only few structures were present at the periphery of the cells (Figure 2a). On the other hand, the presence of the molecules in the osteogenic media clearly induced a higher expression and also a reorganization of $\alpha_V\beta_3$ integrin (Figure 2b,c). In particular, Pol treatment (Figure 2c) induced the strengthening of $\alpha_V\beta_3$ adhesion sites by forming more elongated and larger peripheral clusters in comparison to cells treated with Res (Figure 2b).

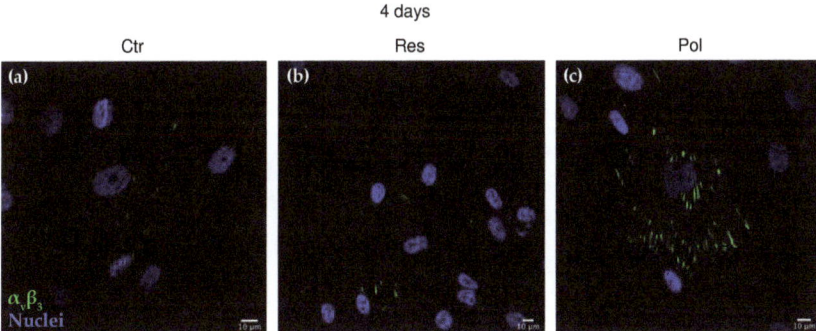

Figure 2. Polydatin (Pol) treatment induces clustering of $\alpha_V\beta_3$ integrin. Indirect immunofluorescence staining of $\alpha_V\beta_3$ integrin (green), detected by the antibody LM609, and nuclei (blue) in Dental bud stem cells (DBSCs). Midsection confocal microscopy images show the localization of integrin $\alpha_V\beta_3$ (green) in cells maintained for 4 days in osteogenic medium and treated with Resveratrol (Res) (**b**), Pol (**c**) and Control (Ctr) (**a**). Images of a representative experiment were chosen for the figure. Scale bar = 10 µm.

3.2. Res and Pol Treatments Prompt DBSCs Proliferation and Mineral Matrix Nodules Deposition on PC Scaffolds Presenting Pores of Medium Dimension (0.9 mm)

We analyzed the proliferation capacity of our cell model on PC scaffolds and their ability to differentiate into osteoblast-like cells producing mineralized matrix. Thus, we previously demonstrated also by FT-IR microscopic analysis that dental stem cells express osteoblastic features [34].

DBSCs were seeded on the biomaterials, which presented pores of medium dimensions (0.9 mm), and cultured in the osteogenic media without any treatment (Ctr) or exposed to Res and Pol treatments for a period of 4 weeks. Although in the first weeks of culture it was particularly difficult to find cells visible enough to be photographed using a phase contrast microscope (data not shown), after 3 weeks of differentiation, as shown in Figure 3a–c, cells appeared numerous and established strong contacts among them. In particular, in the control (Figure 3a), cells seemed to fill the corners of the scaffold pores, leaving a hole without any cell in the center of them. On the other hand, cells treated with Res and Pol (Figure 3b,c) were able to proliferate and interact with each other to cover the scaffold pores almost totally and worked to close them practically in a uniform way. Furthermore, long term cultures of DBSCs showed that the formation of calcium-rich deposits, evaluated by using the ARS after 28 days of osteogenic differentiation, was evident in the control (Figure 3d) and strengthened in the treatments (Figure 3e,f). Interestingly DBSCs capacity of mineralized matrix production was highly promoted when the scaffolds were used in combination with Pol treatment: the ARS quantification shown in the graph (Figure 3g) revealed that the production of mineral matrix nodules was greater in cells treated with 0.1 µM Res compared to the Ctr (19.65%), and remarkably enhanced when cells were exposed to 0.1 µM Pol (37.84%) if compared to the Ctr.

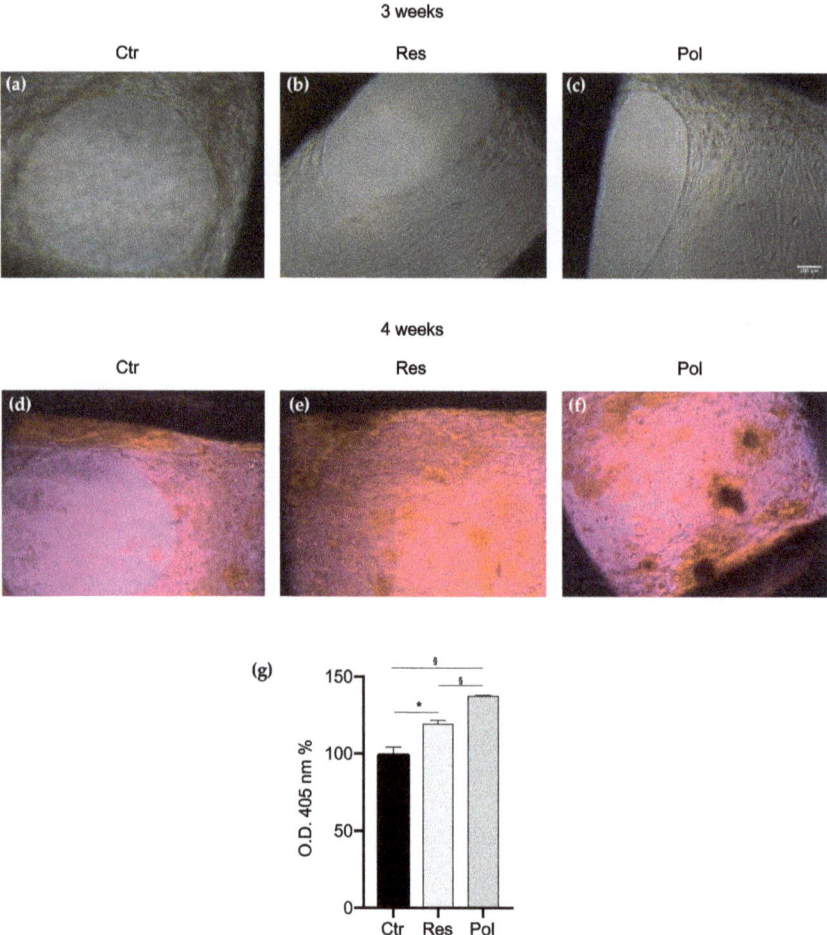

Figure 3. DBSCs proliferation and mineral matrix deposition on medium pore scaffolds. (**a–c**) Representative phase contrast pictures of DBSCs treated with Res, Pol or Ctr for 21 days in osteogenic conditions on scaffolds presenting pores of medium dimensions (0.9 mm). Scale bar = 100 μm. (**d–f**) ARS (Alizarin red staining) displayed mineral matrix deposition by DBSCs after 28 days of culture. (**g**) The graph shows ARS quantification using the optical density (OD) as mean percentage ± SD and is representative for three independent experiments performed in quadruplicates. § $p < 0.01$, * $p < 0.001$. Student's t-test was used for single comparisons. The biomaterial pores of a representative experiment were chosen for the figure.

3.3. Combined Effect of PC Scaffolds and Polydatin on DBSCs Proliferation and Mineralization

Since we observed a greater osteogenic potential when the treatment with Pol was present, to further explore the effect of this molecule on DBSCs osteoblastic differentiation, we cultured the cells on PC scaffolds presenting pores of two other different dimensions: small (0.75 mm) and large (1.15 mm). DBSCs were maintained in mineralizing conditions and stimulated with Pol for 28 days until the deposition of mineralized matrix. As shown in Figure 4, DBSCs proliferation on small pore scaffolds advanced with the progress of culture time (Figure 4a–d), and Pol treatment induced a substantial increase in the number of cells attached to the pores, an effect which was particularly evident after three weeks of osteogenic differentiation (Figure 4c,d). After 28 days of culture, we evaluated with

ARS how Pol stimulation influenced the mineralization capacity of our cell model and we observed that deposition of mineral matrix nodules was significantly higher in cells cultured with Pol, compared to the control (Figure 4e).

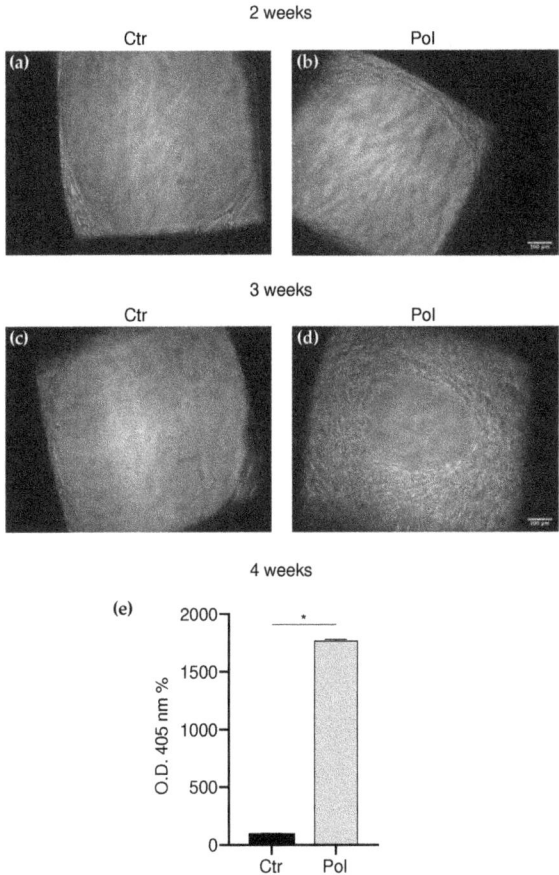

Figure 4. DBSCs proliferation and mineral matrix deposition on small pore scaffolds. (a–d) Representative phase contrast pictures of DBSCs treated with Pol or not (Ctr) for 14 days (**a**,**b**) and 21 days (**c**,**d**) in osteogenic conditions on scaffolds presenting pores of small dimensions (0.75 mm). Scale bar = 100 µm. (**e**) The graph shows ARS quantification using the OD as mean percentage ± SD and is representative for three independent experiments performed in quadruplicates. * $p < 0.001$. Student's *t*-test was used for single comparisons. The biomaterial pores of a representative experiment were chosen for the figure.

Interestingly, cells cultured on scaffolds presenting pores of large dimensions (Figure 5a–g) did not respond as well as those seeded on scaffolds with small pores. A very low number of cells were able to colonize the pores, the proliferation was not increased by the passing of time and the Pol treatment did not show any clear effect. Moreover, the ARS quantification evidenced that there was no significant difference between Ctr and Pol in the mineralization degree (Figure 5g).

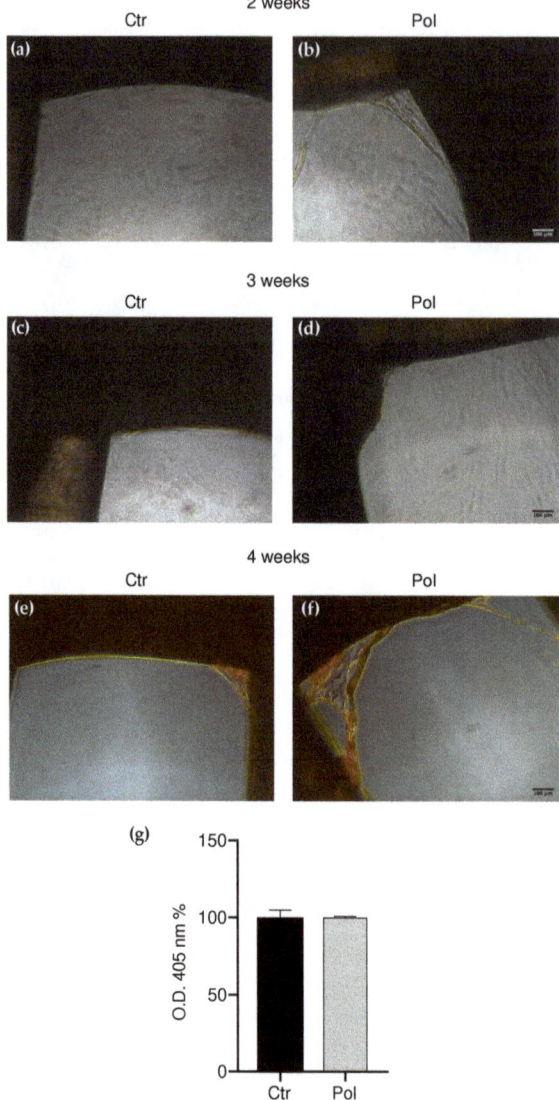

Figure 5. DBSCs proliferation and mineral matrix deposition on large pore scaffolds. (**a**–**d**) Representative phase contrast pictures of DBSCs treated with Pol or Ctr for for 14 days (**a**,**b**) and 21 days (**c**,**d**) in osteogenic conditions on scaffolds presenting pores of large dimensions (1.15 mm). Scale bar = 100 µm. (**e**,**f**) ARS (red staining) displayed mineral matrix deposition by DBSCs after 28 days of culture. (**g**) The graph shows ARS quantification using the OD as mean percentage ± SD and is representative for three independent experiments performed in quadruplicates. Student's *t*-test was used for single comparisons. The biomaterial pores of a representative experiment were chosen for the figure.

4. Discussion and Conclusions

It is well known that the tissue engineering market, which was globally worth about $4.7 billion in 2014, is estimated to reach a value close to $5.5 billion by 2022, considering only the US market. Adult stem cell research is today in an advanced phase of trialing and, in some diseases, cells are already part of therapeutic protocols for the treatment of illness and disabilities [35]. The involvement of precision medicine or even customized medicine proposes the personalization of health care with therapies, practices and/or "tailor-made" medical devices for the specific patient to be treated. The availability of optimized scaffolds, with shapes perfectly matching the lesion, would further reduce tissue regeneration times, especially after highly invasive surgical procedures. 3D printing is an excellent approach to design personalized scaffolds [36,37].

A correct regeneration process of hard tissues, as bone and cartilage, needs a biocompatible scaffold able to promote MSCs differentiation and transform a tissue repair in architectural and functional recovery. Bone lesions have multiple possible shapes and dimensions depending on the trauma or on the course of the chronic degenerative process [38].

In the case of MSCs bioengineering, the grafting site shape, its environment, morphology and dimension are basic for cell engraftment and differentiation [39].

Customized scaffolds, made rapidly and efficiently by 3D printers, could easily reproduce the perfect shape for the lesion and correctly create the ideal niche for MSCs engraftment [40], and their osteogenic differentiation would be optimized by the oligostilbenes Res and Pol.

In our study we found that both Res and Pol stimulated MSCs adhesion to the bone matrix protein Osteopontin via $\alpha_V\beta_3$ integrin and, specifically, Pol treatment prompted a greater reorganization of this integrin in focal adhesion sites. The elongated strings observed by immunofluorescence (Figure 2b,c) represent the classic arrangement of $\alpha_V\beta_3$ implicated in focal adhesion complexes. We can speculate that the already demonstrated osteogenic effect of Res and Pol on DBSCs [19] could be also related to the reorganization of $\alpha_V\beta_3$ integrin in focal adhesions. Moreover, as already known, the development of focal complexes on the surface of scaffolds is an essential event to trigger signals that stimulate MSCs proliferation and osteogenic differentiation [41,42]. Considering these two issues, we can state that oligostilbenes can be considered osteoinductive.

Furthermore, when we integrated MSCs on PC scaffolds, we found that both Res and Pol were able to induce the mineral matrix deposition. Gathering our observations, we can establish that the scaffolds were able to support the production of mineralized matrix, which is the final step and the main event of MSCs osteogenic differentiation, and, in addition, the treatment with the molecules object of our study positively assisted the mineralization process. In particular, in agreement with what we have recently demonstrated [19], Pol treatment induced an increase in the mineralization degree that was higher than the one observed in Res treatment.

Moreover, examining the structure of the scaffolds, we studied whether different pore sizes could affect MSCs acquisition of the osteogenic features. Thus we printed PC scaffolds with pores of 0.75, 0.9 and 1.15 mm; MSCs were cultured on them and induced to osteogenic differentiation. We focused on the use of Pol as treatment since we observed, in the initial experiments, that this molecule had a greater effect in the formation of calcium-rich deposits in differentiated MSCs when compared to Res treatment. The observed gathered data led us to conclude that the cell number tended to gradually decrease as the surface micropore was getting larger and subsequently also the mineralization capacity (Figures 3–5). We compared the results to detect the ideal pore size for cell proliferation and osteogenic differentiation and we deduced that the dimension of 0.75 mm represented the best size to be created with the 3D printer, among the different pore sizes analyzed; the smaller pores produce the optimal niche for MSCs to promote bone formation.

Thus, in conclusion, in this context we confirmed the osteogenic potential of Pol treatment on MSCs. Then we made a step forward by finding, in the combination of this treatment with PC scaffolds presenting small-sized pores, an optimal strategy to induce the osteogenic differentiation of MSCs and the subsequent deposition of mineralized matrix.

The results of this study suggest that the integration of the scaffolds, opportunely designed by 3D printing with MSCs, could optimize tissue regeneration; moreover Pol could be considered a promising approach to improve bone regeneration encouraging further studies for a deeper understanding of its biological mechanisms.

Author Contributions: Conceptualization, G.M., F.P., A.D.B. and G.P.; formal analysis, F.P., A.D.B. and G.M.; investigation, F.P. and A.D.B.; methodology, F.P. and A.D.B.; resources, G.M., G.R., L.L.M., E.A.C.-A.; writing—original draft preparation, F.P., G.M. and G.P.; writing—review and editing, F.P., G.M., G.P. and E.A.C.-A.; visualization, F.P. and A.D.B.; supervision, G.M.; funding acquisition, G.M. All authors have read and agreed to the published version of the manuscript.

Funding: This research was funded by Ministero dell'Istruzione, dell'Università e della Ricerca—PRIN 20098KM9RN, PI G.M. (Progetto di Ricerca d'Interesse Nazionale—Grant 2009). A.D.B. has received funding from the Fondo di Sviluppo e Coesione 2007–2013, APQ Ricerca Regione Puglia Programma regionale a sostegno della specializzazione intelligente e della sostenibilità sociale ed ambientale—Future In Research.

Conflicts of Interest: Francesca Posa, Adriana Di Benedetto, Giampietro Ravagnan, Lorenzo Lo Muzio and Giorgio Mori are name inventors of the Italian patent (16999PTIT deposited with application number 102017000079581) titled "Compositions comprising or consisting of Polydatin in the treatment of bone pathologies" related to the work described.

Abbreviations

MSCs	Mesenchymal Stem Cells
DBSCs	Dental Bud Stem Cells
Res	Resveratrol
Pol	Polydatin
ARS	Alizarin Red Staining
BMSCs	Bone Marrow Stem/Stromal Cells

References

1. Hutmacher, D.W.; Schantz, J.T.; Lam, C.X.; Tan, K.C.; Lim, T.C. State of the art and future directions of scaffold-based bone engineering from a biomaterials perspective. *J. Tissue Eng. Regen. Med.* **2007**, *1*, 245–260. [CrossRef] [PubMed]
2. Sakurada, K.; McDonald, F.M.; Shimada, F. Regenerative medicine and stem cell based drug discovery. *Angew. Chem. (Int. Engl.)* **2008**, *47*, 5718–5738. [CrossRef] [PubMed]
3. Giuliani, N.; Lisignoli, G.; Magnani, M.; Racano, C.; Bolzoni, M.; Dalla Palma, B.; Spolzino, A.; Manferdini, C.; Abati, C.; Toscani, D.; et al. New insights into osteogenic and chondrogenic differentiation of human bone marrow mesenchymal stem cells and their potential clinical applications for bone regeneration in pediatric orthopaedics. *Stem. Cells Int.* **2013**, *2013*, 312501. [CrossRef] [PubMed]
4. Iaquinta, M.R.; Mazzoni, E.; Bononi, I.; Rotondo, J.C.; Mazziotta, C.; Montesi, M.; Sprio, S.; Tampieri, A.; Tognon, M.; Martini, F. Adult Stem Cells for Bone Regeneration and Repair. *Front. Cell Dev. Biol.* **2019**, *7*, 268. [CrossRef]
5. Pittenger, M.F.; Mackay, A.M.; Beck, S.C.; Jaiswal, R.K.; Douglas, R.; Mosca, J.D.; Moorman, M.A.; Simonetti, D.W.; Craig, S.; Marshak, D.R. Multilineage potential of adult human mesenchymal stem cells. *Science* **1999**, *284*, 143–147. [CrossRef]
6. Zuk, P.A.; Zhu, M.; Mizuno, H.; Huang, J.; Futrell, J.W.; Katz, A.J.; Benhaim, P.; Lorenz, H.P.; Hedrick, M.H. Multilineage cells from human adipose tissue: Implications for cell-based therapies. *Tissue Eng.* **2001**, *7*, 211–228. [CrossRef]
7. Qin, Y.; Guan, J.; Zhang, C. Mesenchymal stem cells: Mechanisms and role in bone regeneration. *Postgrad. Med. J.* **2014**, *90*, 643–647. [CrossRef]
8. Rossi, F.; Santoro, M.; Perale, G. Polymeric scaffolds as stem cell carriers in bone repair. *J. Tissue Eng. Regen. Med.* **2015**, *9*, 1093–1119. [CrossRef]
9. Fahimipour, F.; Dashtimoghadam, E.; Hasani-Sadrabadi, M.M.; Vargas, J.; Vashaee, D.; Lobner, D.C.; Kashi, T.S.J.; Ghasemzadeh, B.; Tayebi, L. Enhancing cell seeding and osteogenesis of MSCs on 3D printed scaffolds through injectable BMP2 immobilized ECM-Mimetic gel. *Dent. Mater.* **2019**, *35*, 990–1006. [CrossRef]
10. Boccaccio, A.; Uva, A.E.; Fiorentino, M.; Mori, G.; Monno, G. Geometry Design Optimization of Functionally Graded Scaffolds for Bone Tissue Engineering: A Mechanobiological Approach. *PLoS ONE* **2016**, *11*, e0146935. [CrossRef]
11. Dogan, E.; Bhusal, A.; Cecen, B.; Miri, A.K. 3D Printing metamaterials towards tissue engineering. *Appl. Mater. Today* **2020**, *20*, 100752. [CrossRef] [PubMed]

12. Do, A.-V.; Khorsand, B.; Geary, S.M.; Salem, A.K. 3D Printing of Scaffolds for Tissue Regeneration Applications. *Adv. Healthc Mater.* **2015**, *4*, 1742–1762. [CrossRef] [PubMed]
13. Mori, G.; Ballini, A.; Carbone, C.; Oranger, A.; Brunetti, G.; Di Benedetto, A.; Rapone, B.; Cantore, S.; Di Comite, M.; Colucci, S. Osteogenic differentiation of dental follicle stem cells. *Int. J. Med. Sci.* **2012**, *9*, 480–487. [CrossRef] [PubMed]
14. Bajpayee, A.; Farahbakhsh, M.; Zakira, U.; Pandey, A.; Ennab, L.A.; Rybkowski, Z.; Dixit, M.K.; Schwab, P.A.; Kalantar, N.; Birgisson, B.; et al. In situ Resource Utilization and Reconfiguration of Soils Into Construction Materials for the Additive Manufacturing of Buildings. *Front. Mater.* **2020**, *7*, 52. [CrossRef]
15. Gentile, P.; Mattioli-Belmonte, M.; Chiono, V.; Ferretti, C.; Baino, F.; Tonda-Turo, C.; Vitale-Brovarone, C.; Pashkuleva, I.; Reis, R.L.; Ciardelli, G. Bioactive glass/polymer composite scaffolds mimicking bone tissue. *J. Biomed. Mater. Res. Part A* **2012**, *100*, 2654–2667. [CrossRef] [PubMed]
16. Di Benedetto, A.; Posa, F.; Marazzi, M.; Kalemaj, Z.; Grassi, R.; Lo Muzio, L.; Di Comite, M.; Cavalcanti-Adam, E.A.; Grassi, F.R.; Mori, G. Osteogenic and Chondrogenic Potential of the Supramolecular Aggregate T-LysYal®. *Front. Endocrinol (Lausanne)* **2020**, *11*, 285. [CrossRef] [PubMed]
17. Posa, F.; Di Benedetto, A.; Cavalcanti-Adam, E.A.; Colaianni, G.; Porro, C.; Trotta, T.; Brunetti, G.; Lo Muzio, L.; Grano, M.; Mori, G. Vitamin D promotes MSC osteogenic differentiation stimulating cell adhesion and αVβ3 expression. *Stem Cells Int.* **2018**, *2018*, 1–9. [CrossRef] [PubMed]
18. Posa, F.; Di Benedetto, A.; Colaianni, G.; Cavalcanti-Adam, E.A.; Brunetti, G.; Porro, C.; Trotta, T.; Grano, M.; Mori, G. Vitamin D effects on osteoblastic differentiation of mesenchymal stem cells from dental tissues. *Stem Cells Int.* **2016**, *2016*, 1–9. [CrossRef]
19. Di Benedetto, A.; Posa, F.; De Maria, S.; Ravagnan, G.; Ballini, A.; Porro, C.; Trotta, T.; Grano, M.; Muzio, L.L.; Mori, G. Polydatin, natural precursor of resveratrol, promotes osteogenic differentiation of mesenchymal stem cells. *Int. J. Med Sci.* **2018**, *15*, 944. [CrossRef]
20. Du, Q.H.; Peng, C.; Zhang, H. Polydatin: A review of pharmacology and pharmacokinetics. *Pharm. Biol.* **2013**, *51*, 1347–1354. [CrossRef]
21. Conti, V.; Izzo, V.; Corbi, G.; Russomanno, G.; Manzo, V.; De Lise, F.; Di Donato, A.; Filippelli, A. Antioxidant Supplementation in the Treatment of Aging-Associated Diseases. *Front. Pharmacol.* **2016**, *7*, 24. [CrossRef] [PubMed]
22. Salehi, B.; Mishra, A.P.; Nigam, M.; Sener, B.; Kilic, M.; Sharifi-Rad, M.; Fokou, P.V.T.; Martins, N.; Sharifi-Rad, J. Resveratrol: A Double-Edged Sword in Health Benefits. *Biomedicines* **2018**, *6*, 91. [CrossRef] [PubMed]
23. Gambini, J.; Inglés, M.; Olaso, G.; Lopez-Grueso, R.; Bonet-Costa, V.; Gimeno-Mallench, L.; Mas-Bargues, C.; Abdelaziz, K.M.; Gomez-Cabrera, M.C.; Vina, J.; et al. Properties of Resveratrol: In Vitro and In Vivo Studies about Metabolism, Bioavailability, and Biological Effects in Animal Models and Humans. *Oxidative Med. Cell. Longev.* **2015**, *2015*, 837042. [CrossRef] [PubMed]
24. Wang, H.L.; Gao, J.P.; Han, Y.L.; Xu, X.; Wu, R.; Gao, Y.; Cui, X.H. Comparative studies of polydatin and resveratrol on mutual transformation and antioxidative effect in vivo. *Phytomedicine* **2015**, *22*, 553–559. [CrossRef] [PubMed]
25. Chen, M.; Li, D.; Gao, Z.; Zhang, C. Enzymatic transformation of polydatin to resveratrol by piceid-β-D-glucosidase from Aspergillus oryzae. *Bioprocess Biosyst. Eng.* **2014**, *37*, 1411–1416. [CrossRef] [PubMed]
26. Xie, L.; Bolling, B.W. Characterisation of stilbenes in California almonds (Prunus dulcis) by UHPLC-MS. *Food Chem.* **2014**, *148*, 300–306. [CrossRef]
27. Wang, C.; Huang, W.; Zhou, Y.; He, L.; He, Z.; Chen, Z.; He, X.; Tian, S.; Liao, J.; Lu, B.; et al. 3D printing of bone tissue engineering scaffolds. *Bioact. Mater.* **2020**, *5*, 82–91. [CrossRef]
28. Alaboodi, A.S.; Sivasankaran, S. Experimental design and investigation on the mechanical behavior of novel 3D printed biocompatibility polycarbonate scaffolds for medical applications. *J. Manuf. Process.* **2018**, *35*, 479–491. [CrossRef]
29. Percoco, G.; Uva, A.E.; Fiorentino, M.; Gattullo, M.; Manghisi, V.M.; Boccaccio, A. Mechanobiological Approach to Design and Optimize Bone Tissue Scaffolds 3D Printed with Fused Deposition Modeling: A Feasibility Study. *Materials* **2020**, *13*, 648. [CrossRef]
30. Di Benedetto, A.; Brunetti, G.; Posa, F.; Ballini, A.; Grassi, F.R.; Colaianni, G.; Colucci, S.; Rossi, E.; Cavalcanti-Adam, E.A.; Muzio, L.L. Osteogenic differentiation of mesenchymal stem cells from dental bud: Role of integrins and cadherins. *Stem Cell Res.* **2015**, *15*, 618–628. [CrossRef]

31. Mori, G.; Brunetti, G.; Oranger, A.; Carbone, C.; Ballini, A.; Muzio, L.L.; Colucci, S.; Mori, C.; Grassi, F.R.; Grano, M. Dental pulp stem cells: Osteogenic differentiation and gene expression. *Ann. N. Y. Acad. Sci.* **2011**, *1237*, 47–52. [CrossRef] [PubMed]
32. Gronthos, S.; Mankani, M.; Brahim, J.; Robey, P.G.; Shi, S. Postnatal human dental pulp stem cells (DPSCs) in vitro and in vivo. *Proc. Natl. Acad. Sci. USA* **2000**, *97*, 13625–13630. [CrossRef] [PubMed]
33. Ravagnan, G.; De Filippis, A.; Cartenì, M.; De Maria, S.; Cozza, V.; Petrazzuolo, M.; Tufano, M.A.; Donnarumma, G. Polydatin, a natural precursor of resveratrol, induces β-defensin production and reduces inflammatory response. *Inflammation* **2013**, *36*, 26–34. [CrossRef] [PubMed]
34. Giorgini, E.; Conti, C.; Ferraris, P.; Sabbatini, S.; Tosi, G.; Centonze, M.; Grano, M.; Mori, G. FT-IR microscopic analysis on human dental pulp stem cells. *Vib. Spectrosc.* **2011**, *57*, 30–34. [CrossRef]
35. Moreno-Manzano, V.; García, E.O. Culturing Adult Stem Cells for Cell-Based Therapeutics: Neuroimmune Applications. In *Cell Culture*; IntechOpen: London, UK, 2018. [CrossRef]
36. Yan, Q.; Dong, H.; Su, J.; Han, J.; Song, B.; Wei, Q.; Shi, Y. A Review of 3D Printing Technology for Medical Applications. *Engineering* **2018**, *4*, 729–742. [CrossRef]
37. Jariwala, S.H.; Lewis, G.S.; Bushman, Z.J.; Adair, J.H.; Donahue, H.J. 3D Printing of Personalized Artificial Bone Scaffolds. *3D Print. Addit. Manuf.* **2015**, *2*, 56–64. [CrossRef]
38. Mori, G.; Di Benedetto, A.; Posa, F.; Muzio, L.L. Targeting MSCs for Hard Tissue Regeneration. In *MSCs and Innovative Biomaterials in Dentistry*; Springer: Berlin/Heidelberg, Germany, 2017; pp. 85–99. [CrossRef]
39. Cipitria, A.; Lange, C.; Schell, H.; Wagermaier, W.; Reichert, J.C.; Hutmacher, D.W.; Fratzl, P.; Duda, G.N. Porous scaffold architecture guides tissue formation. *J. Bone Miner. Res. Off. J. Am. Soc. Bone Miner. Res.* **2012**, *27*, 1275–1288. [CrossRef]
40. García-Sánchez, D.; Fernández, D.; Rodríguez-Rey, J.C.; Pérez-Campo, F.M. Enhancing survival, engraftment, and osteogenic potential of mesenchymal stem cells. *World J. Stem Cells* **2019**, *11*, 748–763. [CrossRef]
41. Biggs, M.J.P.; Dalby, M.J. Focal adhesions in osteoneogenesis. *Proc. Inst. Mech. Eng. H* **2010**, *224*, 1441–1453. [CrossRef]
42. Wang, Y.K.; Chen, C.S. Cell adhesion and mechanical stimulation in the regulation of mesenchymal stem cell differentiation. *J. Cell. Mol. Med.* **2013**, *17*, 823–832. [CrossRef]

© 2020 by the authors. Licensee MDPI, Basel, Switzerland. This article is an open access article distributed under the terms and conditions of the Creative Commons Attribution (CC BY) license (http://creativecommons.org/licenses/by/4.0/).

Article

Enhanced Osteogenic Differentiation of Human Primary Mesenchymal Stem and Progenitor Cultures on Graphene Oxide/Poly(methyl methacrylate) Composite Scaffolds

Katarzyna Krukiewicz [1,*], David Putzer [2], Nicole Stuendl [3], Birgit Lohberger [3] and Firas Awaja [4,5,*]

1. Department of Physical Chemistry and Technology of Polymers, Silesian University of Technology, 44-100 Gliwice, Poland
2. Experimental Orthopedics, Department of Orthopedic Surgery, Medical University Innsbruck, 6020 Innsbruck, Austria; david.putzer@i-med.ac.at
3. Department of Orthopedics and Trauma, Medical University of Graz, 8036 Graz, Austria; nicole.stuendl@medunigraz.at (N.S.); birgit.lohberger@medunigraz.at (B.L.)
4. School of Medicine, National University of Ireland, H91 CF50 Galway, Ireland
5. Engmat Ltd., Clybaun Road, H91 P96H Galway, Ireland
* Correspondence: katarzyna.krukiewicz@polsl.pl (K.K.); firas.awaja@gmail.com (F.A.)

Received: 4 June 2020; Accepted: 3 July 2020; Published: 5 July 2020

Abstract: Due to its versatility, small size, large surface area, and ability to interact with biological cells and tissues, graphene oxide (GO) is an excellent filler for various polymeric composites and is frequently used to expand their functionality. Even though the major advantage of the incorporation of GO is the enhancement of mechanical properties of the composite material, GO is also known to improve bioactivity during biomineralization and promote osteoblast adhesion. In this study, we described the fabrication of a composite bone cement made of GO and poly(methyl methacrylate) (PMMA), and we investigated its potential to enhance osteogenic differentiation of human primary mesenchymal stem and progenitor cells. Through the analysis of three differentiation markers, namely alkaline phosphatase, secreted protein acidic and rich in cysteine, and bone morphogenetic protein-2 in the presence and in the absence of an osteogenic differentiation medium, we were able to indicate a composite produced manually with a thick GO paper as the most effective among all investigated samples. This effect was related to its developed surface, possessing a significant number of voids and pores. In this way, GO/PMMA composites were shown as promising materials for the applications in bone tissue engineering.

Keywords: bone regeneration; graphene oxide; mesenchymal stem and progenitor cells; osteogenic differentiation; poly(methyl methacrylate)

1. Introduction

Since its discovery in 2004 [1], graphene has drawn immense attention of the scientific community and has become an object of intensive research. Due to its high planar surface area, superior mechanical strength, outstanding optical properties, as well as remarkable thermal and electrical conductivity [2–4], graphene has been widely used in a variety of applications, including transparent conductors, ultrafast transistors, precise biosensors, and tissue scaffolds [5]. The potential of graphene has been further expanded by introducing to its structure a variety of functional group, resulting in the fabrication of graphene oxide (GO). GO is usually produced by the oxidation of graphite, resulting in a partial breaking of sp^2 bonds present in its structure and subsequent increase in the distance between carbon layers [3]. Therefore, GO possesses both hydrophobic (due to the presence of pristine graphite structure) and hydrophilic (due to the presence of hydroxyl, epoxy, carbonyl, and carboxyl groups) parts, and is characterized by affinity for aromatic rings, excellent aqueous processability, amphiphilicity, ease of functionalization, and biocompatibility [3,5]. Consequently, GO has been marked as an excellent material for numerous biomedical applications, including the design of biosensors [5], drug delivery systems [6], antimicrobial coatings [7], cell imaging platforms [8], and in gene therapy [9].

Due to its versatility, small size, large surface area, and ability to interact with biological cells and tissues, GO is an excellent filler for various polymers and is frequently used to expand their functionality. For instance, Wan et al. [10] reported an increase in the tensile strength, modulus, and energy at break, as well as the improvement in bioactivity during biomineralization simultaneously with maintaining high porosity when an electrospun poly(ε-caprolactone) membrane was reinforced with GO nanoplatelets. Also Baradaran et al. [11] observed the increase in elastic modulus and fracture toughness, as well as the promotion in osteoblast adhesion and proliferation when GO was used as a filler for hydroxyapatite nanosheets. On the other hand, calcium phosphate mineralized graphene oxide/chitosan scaffolds were found to express biomimicry, providing a suitable environment for cell adhesion and growth, and maintaining high mechanical strength [12]. GO was also demonstrated to act as an excellent filler for such polymer matrices as poly(vinyl alcohol) [13], poly(carbonate urethane) [14], hyaluronic acid [15], and poly(acrylic acid) [16], resulting in the formation of robust composite materials with applicability in biomedical engineering.

Poly(methyl methacrylate) (PMMA) is a non-toxic thermoplastic polymer possessing a very good toxicological safety record in biomedicine [17]. PMMA is frequently used as a screw fixation in bone, bone cement, filler for bone cavities and skull defects, as well as vertebrae stabilization in osteoporotic patients [18]. The exceptional applicability of PMMA in orthopedic and dental applications is caused by its good processability, handling properties, biocompatibility, suitable mechanical strength, and Young's modulus [19]. Despite its numerous merits, the common complications of using PMMA is bone resorption observed after implantation, which is the effect of the formation of a weak-link zone derived from not sufficient interactions between cement and a bone [20]. Therefore, the modification of PMMA has become a very active area of research. A promising way to improve the biological performance of PMMA is to blend it with an antibiotic, e.g., gentamycin [21]. This modification approach is nowadays a well established strategy that allows prevention of periprosthetic infections and osteomyelitis. Another way to enhance the performance of PMMA is to incorporate in its structure a filler with a particular functionality. For instance, loading of PMMA with multiwalled carbon nanotubes could significantly improve the mechanical properties and reduce the exothermic polymerization reaction of the bone cement [22]. The use of tri-calcium phosphate and chitosan as inorganic/organic additives to PMMA decreased polymer curing temperature, extended setting time, and increased weight loss and porosity after degradation and, among all, promoted better osteointegration than pure PMMA [23]. Also graphene and graphene oxide have been used as fillers to PMMA [24], improving the mechanical properties of PMMA, particularly its fracture toughness and fatigue performance. What was interesting, GO was found to outperform graphene and provide greater enhancements due to its high functionalization that increased the interfacial adhesion between a filler and PMMA matrix. Simultaneously, the presence of graphene or GO was not found to have any negative effect on the biocompatibility of PMMA composites, potentially allowing their further clinical progression [25].

In this paper, the potential of GO/PMMA composites for the application in bone tissue engineering is assessed by the analysis of three differentiation markers expressed by human primary mesenchymal stem and progenitor cells (hMSPCs) cultured on the top of the composites. By performing the cell culture in the presence and absence of a specific induction medium, we were able to determine the efficiency of osteogenic differentiation of hMSPCs cultured on four types of GO/PMMA scaffolds, differing in the thickness of GO paper as well as the method of fabrication of the composites. Microscopic analysis of the surface of materials allowed investigation of the biological behavior of the materials with respect to their surface morphology.

2. Materials and Methods

2.1. Fabrication of GO/PMMA Composites

For the production of graphene oxide (GO) paper, 4 mg/mL suspension of GO flakes (GO monolayer content > 95%, oxygen content between 40% and 50%) in water was purchased from Graphenea (San Sebastián, Spain). GO suspension was inserted into Petri dishes with serological pipettes. The GO dispersion was dried in a shaking incubator with air fan for 48 h and inserted into an oven at 180 °C for 20 min. By changing the volume of GO suspension, two types of GO paper were fabricated, i.e., thin (8 mL of GO suspension) and thick (10 mL of GO suspension), designated as $GO_{(A)}$ paper and $GO_{(B)}$ paper, respectively.

For the fabrication of GO/PMMA composites, GO paper as well as SIMPLEX P (Stryker, Kalamazoo, MI, USA) radiopaque bone cement (prepared according to the manufacturer's instructions with a full dose of liquid monomer) were applied. The monomer was mixed to the polymer manually under laboratory conditions. The PMMA cement was then inserted in a metal casting form and covered with GO paper. The bone cement was kept in place for 30 min to guarantee sufficient hardening. Screws were closed after 15 min. Two different methods were used to prepare combined samples including GO paper and bone cement (Scheme 1).

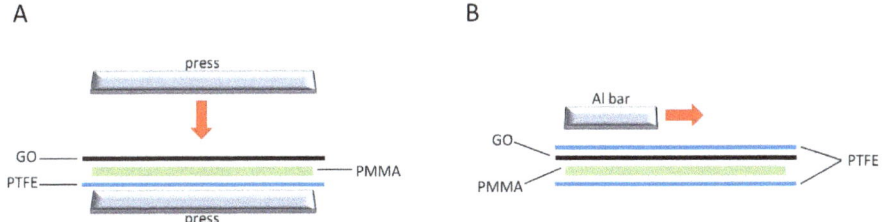

Scheme 1. Schematic representation of a fabrication process of $GO/PMMA_{(P)}$ (**A**) and $GO/PMMA_{(M)}$ (**B**) composite materials.

In the first method, GO paper was placed on the bottom of a steel flat press (covered with poly(tetrafluoroethylene), PTFE, sheet to simplify cement detachment). The cement was spread on the upper PTFE sheet and then placed in contact with the GO paper. Then, the press was closed after 15 min to reach minimum thickness until the cement was polymerized (30 min). The samples prepared in this way were designated as $GO/PMMA_{(P)}$. In the second method, a compound material was produced manually: GO paper was laid down on a PTFE sheet, and then the cement was spread on the upper PTFE sheet and then placed in contact with the GO paper. An aluminum bar was used to spread the cement on GO paper within the two PTFE sheets. In this case, spreading the cement was more difficult, and led to the formation of a non-homogeneous PMMA layer with GO paper broken up into little pieces, which may have been due to the shrinking and expanding behavior of PMMA during the cement hardening phase. The samples prepared in this way were designated as $GO/PMMA_{(M)}$.

Consequently, four types of samples were analyzed: $GO_{(A)}/PMMA_{(P)}$, $GO_{(B)}/PMMA_{(P)}$, $GO_{(A)}/PMMA_{(M)}$, and $GO_{(B)}/PMMA_{(M)}$, all with PMMA as the surface layer. For the scanning electron microscopic (SEM) investigations, a FEI Quanta 250 field emission gun (Thermo Fisher Scientific, Hillsboro, OR, USA) was used under high vacuum conditions and 20 kV high tension. The micrographs were recorded with the Everhart–Thornley–Detector in secondary electron (SE) mode. The surfaces were sputter coated with a 10 nm thin layer of gold in order to provide sufficient electrical conductivity.

2.2. Tissue Harvest and Cell Culture

Explant hMSPCs were established from tissue samples of spongiosa bone harvested during routine hip joint surgeries. The study protocol was approved by the local ethics committee (reference number 29-156ex16/17), and informed consent was obtained from each orthopedic surgery patient. The study included a total of three patients, aged between 25 and 35, excluding pregnant women and those suffering from local inflammatory processes, metabolic bone diseases, and impaired blood coagulation. The length of harvested bone samples was kept between 4 and 6 mm, and showed either cortical or cortical and cancellous structure. The samples were extensively rinsed with a phosphate-buffered saline (PBS; PAA Laboratory, Pasching, Austria) and transferred into 75 cm^2 culture flasks (TPP, Trasadingen, Switzerland) with an appropriate volume of culture medium. For cell isolation and expansion, the samples were incubated in a humidified atmosphere (5% CO_2, 37 °C).

2.3. Flow Cytometry

For a flow cytometric analysis, a total of 1×10^5 hMSPCs were resuspended in 200 µL PBS. The characterization of cells was achieved with the use of commercial monoclonal antibodies, namely CD73 PE, CD90 APC, CD105 PE, CD45 APC-Cy7, CD34 APC, CD14 FITC, CD19 APC, and HLA-DR APC (BD Bioscience, San Jose, CA, USA). Titration had previously been used to determine the optimal amount of each antibody. Subsequently, two-color staining panels were used to present a combination of antibodies with non overlapping spectra. Negative cell lines and matched fluorochrome-conjugated isotype controls were applied to perform a background staining for antibodies. FACS LSR II System (BD Bioscience), FACSDiva software (BD Bioscience), and FCS Express software (De Novo Software, Los Angeles, CA, USA) were employed to perform a flow cytometry analysis, to acquire and to analyze obtained data, respectively. Rainbow Beads (BD Bioscience) was used to check the day-to-day consistency of measurements. To exclude debris and cell aggregates, viable cells were gated on forward scatter (FSC) and side scatter (SSC). hMSPCs were defined by their phenotype and analyzed on a logarithmic scale. Data from all donors were analyzed by collecting 10,000 events under identical parameters.

2.4. Multilineage Differentiation Analysis

A seeding density of hMSPCs was established at 10^4 cells/cm^2, and the cells were seeded in an expansion medium composed of Dulbecco's modified Eagle's medium (DMEM-F12; GIBCO Invitrogen), 10% FBS (Lonza, Braine-l'Alleud, Belgium), 1% L-glutamine, 1% penicilline/streptomycine, and 0.1% amphotericine B. Additionally, 100 nM dexamethasone, 0.1 mM ascorbic acid-2-phosphate, and 10 mM β-glycerophosphate (all Sigma Aldrich, St. Louis, MO, USA) were added to the differentiation medium to induce osteogenesis. Histochemical staining (Alkaline Phosphatase kit No. 85; Sigma Aldrich) was used to assay alkaline phosphatase (ALP) activity after 7 and 14 days of culture. According to the instructions of the manufacturer, ALP enzyme activity was calculated basing on the absorbance of p-nitrophenol phosphate (405 nm) [26]. Adipogenic differentiation was performed in a medium containing 100 nM dexamethasone, 50 µM indomethacine (Sigma Aldrich), and 0.135 IE/mL insulin (Novo Nordisk, Bagsværd, Denmark), and was detected by Oil Red O staining of the adipocyte specific fat vacuoles after 21 days of culture. Chondrogenic differentiation was initiated by culturing cells in DMEM-F12 supplemented with 10% FBS, 100 µM L-ascorbic acid, and 1 ng/mL TGF-β3 (Lonza). Alcian blue staining was applied to verify the production of glycosaminoglycans and mucopolysaccharides after 21 days of culture. Cells were then fixed with 10% formaldehyde and stained with 1% Alcian blue in 3% acetic acid solution at pH 2.5.

2.5. Real-Time RT-PCR

RNeasy Mini Kit and DNase-I treatment (Qiagen, Hilden, Germany) were used to isolate total RNA from undifferentiated and osteogenic differentiated hMSPCs cultured on different GO surface modifications (on the GO-uppermost surface of the composites) on day 21. A total of 1 µg of RNA was reverse transcribed with iScriptcDNA Synthesis Kit, (BioRad Laboratories Inc., Hercules, CA, USA) using a blend of oligo(dT) and random hexamer primers. SsoAdvanced Universal SYBR Green Supermix (Bio-Rad) and CFX96 Touch (BioRad) were used for the amplification and measurements, respectively. A standard 3-step PCR temperature protocol (annealing temperature of 60 °C) was used for each qPCR, and was followed by a melting curve protocol both to confirm a single gene-specific peak and to detect primer dimerization. ΔΔCt method was used for the relative quantification of expression levels, and was based on the geometric mean of the internal controls TBP (TATA-box binding protein), RPLP0 (ribosomal protein, lateral stalk, subunit P0), and B2M (β-2 microglobulin), respectively. The expression levels (Ct) of the target genes were normalized to the reference genes (ΔCt), and the difference between the ΔCt value of the test sample and the ΔCt of the control sample gave the ΔΔCt value. Consequently, the expression ratio was calculated as $2^{\Delta\Delta Ct}$. Three QuantiTect primer assays (Qiagen) were selected for real time RT-PCR, namely ALPL, SPARC, and BMP2.

2.6. Statistical Analysis

Differences between groups were evaluated by means of a Student's unpaired t-test and the exact Wilcoxon test with the PASW statistics 18 software (IBM Corporation, Somers, NY, USA). Two-sided p-values ($p < 0.001$ ***; $p < 0.01$ **; $p < 0.05$ *) were considered statistically significant. SigmaPlot® 14.0 (Systat Software Inc., San Jose, CA, USA) was used to make graphical representations.

3. Results

3.1. Surface Characterization of GO/PMMA

The protocol of GO paper fabrication resulted in the formation of a self-supporting, uniform, and black material, with the average thickness of 5 ± 1 µm for $GO_{(A)}$ and 16 ± 1 µm for $GO_{(B)}$, and the average specific weight of 0.87 ± 0.08 mg/cm^2 for $GO_{(A)}$ and 2.90 ± 0.08 mg/cm^2 for $GO_{(B)}$. The morphology of $GO_{(A)}$ and $GO_{(B)}$, as presented in the SEM images (Figure 1), showed some wrinkles on the surfaces which were the most probably the edges of graphene oxide, revealing strong adhesion between GO platelets. Overall, the surface of GO paper was relatively smooth, and there were no obvious defects (pores or cracks) observed.

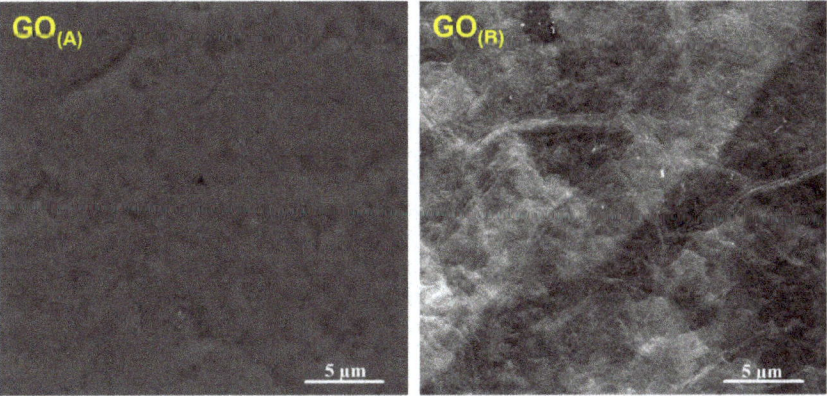

Figure 1. SEM micrographs presenting surface morphology of $GO_{(A)}$ and $GO_{(B)}$ paper.

As demonstrated in SEM micrographs of GO/PMMA composites (Figure 2), PMMA was covering the surface of GO paper, and more uniform surface was obtained when a thin $GO_{(A)}$ paper was used as a filler. The interface between GO and PMMA can be seen as a border region, particularly in the SEM image of $GO_{(A)}/PMMA_{(P)}$. $GO_{(A)}$ paper was thickly coated with PMMA, while the presence of a thick $GO_{(B)}$ paper was found to introduce wrinkles to the surface of polymer composite. Moreover, $GO_{(B)}$ paper seemed to protrude cleanly from the fracture site. The surface of $GO_{(B)}/PMMA_{(M)}$ was observed to exhibit a significant number of voids and pores.

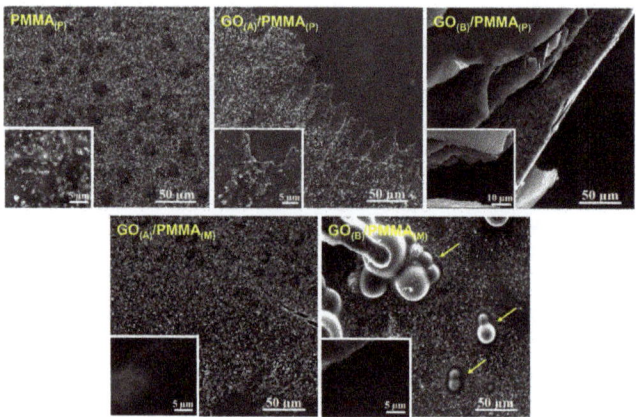

Figure 2. SEM micrographs presenting surface morphology of $GO_{(A)}/PMMA_{(P)}$, $GO_{(B)}/PMMA_{(P)}$, $GO_{(A)}/PMMA_{(M)}$, and $GO_{(B)}/PMMA_{(M)}$ (arrows indicate discussed surface structures), as well as untreated PMMA, with high magnification images as the insets.

3.2. hMSPC Characterization and Multilineage Differentiation Analysis

Cells providing morphologic characteristics of human primary MSPCs (mononuclear, fibroblast-like, spindle shaped, plastic-adherent) were isolated from all samples within 4–8 days. hMSPCs showed a positive expression of CD73 (99.8 ± 0.1%), CD90 (99.9 ± 0.1%), CD105 (69.1 ± 9.8%) of gated cells. The typical forward/side scatter characteristics of 71.5 ± 4.9% were gated. The negativity for CD14 (0.2 ± 0.2%), CD19 (0.6 ± 0.1%), CD34 (0.4 ± 0.3%), CD45 (23.9 ± 7.8%), and HLA-DR (0.5 ± 0.3%) confirmed the phenotype of MSPCs (Figure 3A).

ALP activity was measured of absorbance (optical dense, OD) of p-nitrophenol in supernatant at the wavelength of 405 nm over 14 days (Figure 3B). ALP expression was detected on day 7 and day 14, respectively, when the cells were osteogenically differentiated with a significant increase ($p < 0.001$). No expression of ALP was observed in any of the samples of undifferentiated negative controls. Due to the interaction of the cationic dye Alcian blue and acid glycosaminoglycans, augmented blue coloration was noticed for chondrogenic differentiated hMSPCs, and not for undifferentiated controls (Figure 3C). As a result of chondrogenic differentiation, a 4.7-fold increase ($p < 0.05$) was noticed for the expression of aggrecan. In order to demonstrate the multilineage ability, hMSPCs were also differentiated in the adipogenic lineage. The adipogenic cell differentiation was demonstrated with the formation of lipid vacuoles which were visualized by Oil Red O staining on day 21 (Figure 3D). These results explicitly characterized our primary cells as hMSCPs.

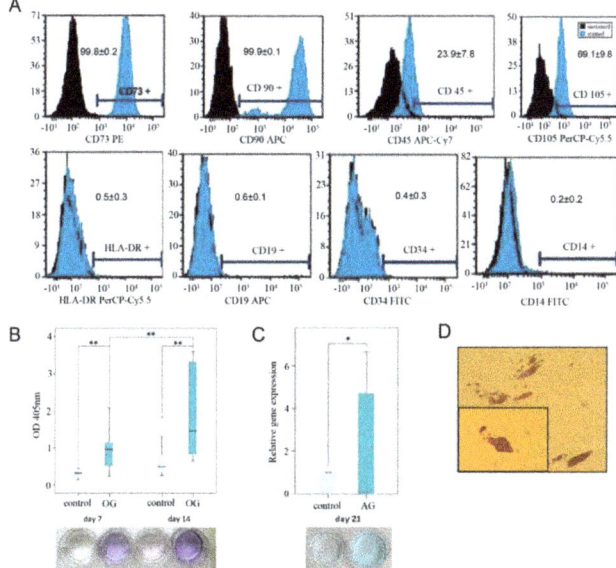

Figure 3. hMSPC characterization and multilineage differentiation analysis. The used hMSPCs were characterized according (**A**) the positive expression of CD73, CD90, CD105, and negative expression CD14, CD19, CD34, CD45, and HLA-DR using multicolor fluorescence-activated cell sorting analyses. The values indicated the percentage of positively stained cells. The capacity for multilineage differentiation potential was confirmed by (**B**) ALP staining for osteogenic differentiation, (**C**) Alcian blue staining and the expression of aggrecan for the chondrogenic differentiation, and (**D**) the Oil Red O staining of lipid droplets for the adipogenic lineage; $p < 0.01$ **; $p < 0.05$ *.

3.3. Efficiency of Osteogenic Differentiation

ALP, SPARC, and BMP-2 assays were performed to assess the mineralization of hMSPCs cultured on the surface of $GO_{(A)}/PMMA_{(P)}$, $GO_{(B)}/PMMA_{(P)}$, $GO_{(A)}/PMMA_{(M)}$ and $GO_{(B)}/PMMA_{(M)}$, as well as untreated PMMA. Consequently, the results shown in Figure 4 describe how strongly the expression of individual markers was increased by the osteogenic differentiation medium, with the undifferentiated hMSPCs as the control. As demonstrated by ALP assay, all investigated surfaces led to a significant increase in mineral deposition, with the most pronounced effect noticed for $GO_{(B)}/PMMA_{(M)}$ (8-fold increase when compared with a control). The same material was also found to lead to the significant increase in the expression of SPARC (2-fold increase with respect to control), even though the highest relative gene expression (3-fold increase with respect to control) was noticed for unmodified PMMA. All investigated surfaces, including GO/PMMA composites as well as unmodified PMMA samples, were shown to decrease the relative BMP-2 expression from 7 to 9 times when compared with a control.

The relative gene expression profiles of hMSPCs cultured in normal expansion medium were analyzed with respect to ALP, SPARC and BMP-2, and compared with an unmodified PMMA as the control. Consequently, the results presented in Figure 5 describe the expression of the individual markers by the osteogenic differentiation medium in relation to an unmodified PMMA control These results showed the unchanged SPARC expression, simultaneously with the decrease in ALP expression (approximately 2-fold) and a significant increase of BMP-2 expression (from 2-fold for $GO_{(A)}/PMMA_{(P)}$ to 2.5-fold for $GO_{(A)}/PMMA_{(M)}$) for GO/PMMA composites with respect to a PMMA control.

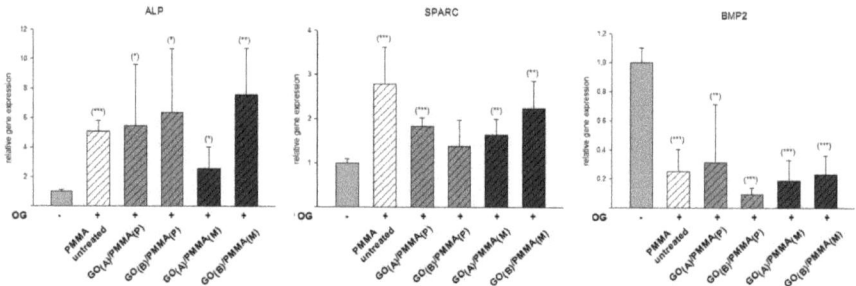

Figure 4. The expression of specific osteogenic markers (ALP, SPARC, and BMP-2) of hMSPCs cultured in the presence of osteogenic differentiation medium for 21 days on the surface of GO/PMMA composites as well as untreated PMMA, compared with undifferentiated hMSPCs cultivated in expansion medium as the negative control; $p < 0.001$ ***; $p < 0.01$ **; $p < 0.05$ *.

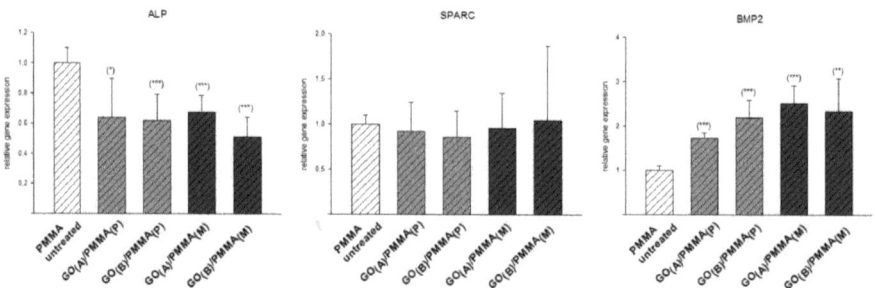

Figure 5. The expression of specific osteogenic markers (ALP, SPARC, and BMP-2) of hMSPCs cultured in the absence of a specific induction medium for 21 days on the surface of GO/PMMA composites as well as untreated PMMA; $p < 0.001$ ***; $p < 0.01$ **; $p < 0.05$ *.

To estimate the efficiency of osteogenic differentiation, GO/PMMA coatings were compared with an unmodified PMMA as a control (Figure 6). This kind of evaluation was chosen to show the efficiency of the osteogenic differentiation ability of each group in relation to the PMMA control. Particularly, $GO_{(B)}/PMMA_{(M)}$ composite was found to improve the expression of osteogenic differentiation markers (2-fold increase for ALP, 1.5-fold increase for SPARC, and 2-fold increase for BMP-2), while $GO_{(A)}/PMMA_{(M)}$ was found to decrease ALP expression (2-fold). All other effects were assessed as not significant.

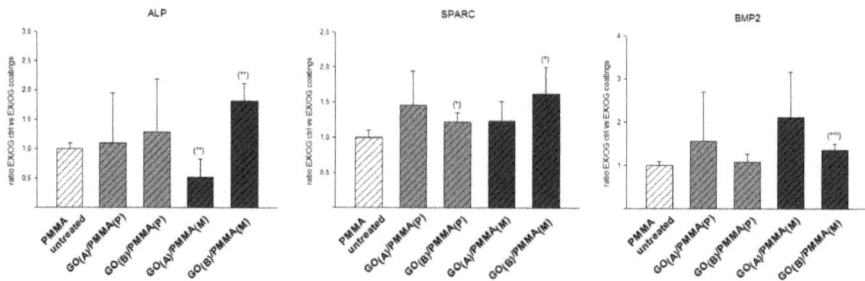

Figure 6. The expression of specific osteogenic markers (ALP, SPARC, and BMP-2) of hMSPCs cultured for 21 days on the surface of GO/PMMA composites with respect to a PMMA control; $p < 0.001$ ***; $p < 0.01$ **; $p < 0.05$ *.

4. Discussion

The greatest challenge of modern nanotechnology is to expand its scope from nanoscale into macroscale. Therefore, the possibility of fabricating free-standing, paper-like materials basing on nanoscale components is a subject of intensive research [27]. The protocol of GO paper fabrication, introduced by us, allowed formation of self-supporting, uniform films composed of stacked platelets of graphene oxide. The irregularity of surface of GO paper seemed to increase with its thickness, which is of special importance for further biological studies since higher roughness promotes cell adhesion [28].

GO paper is known to possess many functional groups, and could easily form hydrogen bonds with hydrophilic polymers. The presence of hydrophobic methacrylate groups in PMMA, however, discourages these interactions to occur. Therefore, since PMMA chains would rather remain in a coiled conformation, they are supposed to fit well into wavy structures of the nanosheets of GO paper leading to more efficient packing within intersheet gallery [29]. Consequently, SEM micrographs demonstrated a good interlocking between GO paper and PMMA matrix. Particularly $GO_{(A)}$ paper was found to be thickly coated with PMMA, suggesting strong adhesion between this polymer and GO. $GO_{(B)}$ paper, on the other hand, seemed to protrude from the fracture site, suggesting a weak interfacial bonding between $GO_{(B)}$ and PMMA. This is consistent with some previous studies indicating that small sizes of GO sheets promote the formation of homogeneous composites [30]. Moreover, the presence of voids and pores on the surface of $GO_{(B)}/PMMA_{(M)}$ could be associated with the presence of unreacted residual monomer, which is volatile and is supposed to be released after polymerization [31]. This would mean that some double bonds present in GO paper could be attacked by the radical species formed during MMA polymerization, retarding or inhibiting the reaction of polymerization [24].

As a biological model, cells providing morphologic characteristics of human primary MSPCs (mononuclear, fibroblast-like, spindle shaped, plastic-adherent) were isolated and cultured on the surface of GO/PMMA composites. The phenotype of hMSPCs was confirmed according to the criteria of the International Society for Cellular Therapy [32] for defining multipotent mesenchymal stromal cells. In addition, hMSPCs were successfully differentiated towards the osteogenic, chondrogenic, and adipogenic lineage, which was confirmed by ALP expression, Alcian blue staining, aggrecan expression, and Oil Red O staining. To assess the efficiency of osteogenic differentiation of hMSPCs cultured onto PMMA as well as GO/PMMA composites, three differentiation markers were analyzed, namely alkaline phosphatase (ALP), secreted protein acidic and rich in cysteine (SPARC), and bone morphogenetic protein-2 (BMP-2). ALP is a metalloenzyme playing an important role in the mineralization of tissue cells [33]. ALP is found to act as both a mineralization promoter by increasing the local concentration of phosphate, as well as an inhibitor of mineral formation by decreasing the concentration of extracellular pyrophosphate. Since ALP is observed to be highly expressed in mineralized cells, it can be used to predict their bone forming capacity under different conditions. Osteonectin (SPARC), on the other hand, is the most abundant non-collagenous extracellular matrix protein present in bone [34]. SPARC gene dosage has a dramatic effect on bone volume and is one of critical regulators of bone remodeling, calcium turnover and an initiator of mineralization [34,35]. Therefore, SPARC can be used for the examination of osteogenic differentiation. Another marker for osteogenic differentiation is bone morphogenetic protein-2 (BMP-2), which is known as a potent osteogenic factor with roles in both normal bone healing and pathological bone formation in soft tissues [36]. BMP-2 is found to facilitate osteogenic differentiation through inducing ALP activity, promoting mineralization and enhancing adherence of cells [37].

As demonstrated by ALP assay, all investigated surfaces led to a significant increase in mineral deposition. This effect was accompanied with a decrease in BMP-2 expression, which might suggest that the presence of a differentiation medium had a stronger effect on the osteogenic differentiation than BMP-2, resulting in down regulation of the latter. Nevertheless, the expression of specific osteogenic markers, such as ALP and SPARC, was found to be significantly increased by the osteogenic differentiation medium. In the absence of a specific induction medium, on the other hand, the cells cultured on GO/PMMA composites were found to be able to induce an increase in BMP-2. In this way,

GO/PMMA composites were shown to be able to drive cellular differentiation without any addition of osteogenic supplements, just as reported for hMSCs in collagen matrices subjected cyclic tensile strain [38], even though the process of osteogenic differentiation was found to be much more effective when cells were culture in the presence of a medium trigger.

The comparison of GO/PMMA coatings with an unmodified PMMA as a control suggested that among all investigated GO/PMMA composite materials, $GO_{(B)}/PMMA_{(M)}$ composite was the most efficient inducer of osteogenic differentiation, particularly basing on the relative expression of ALP (other changes were found to be statistically non-significant). This effect should be assigned to its surface morphology, as exhibiting a significant number of voids and pores. As presented by Abagnale et al. [39], specific patterns present on the surface can boost differentiation of MSCs towards specific cell types. These patterns should be in the micrometer range to be able to support the differentiation processes initiated by induction media, and this requirement is met particularly by $GO_{(B)}$ paper and $GO_{(B)}/PMMA_{(M)}$ composite. All these data suggest that GO/PMMA composites, particularly $GO_{(B)}/PMMA_{(M)}$ produced manually with a thick GO paper, may direct hMSPCs toward osteogenic differentiation and can serve as promising materials in bone tissue engineering. The properties of GO/PMMA make this material advantageous for potential applications as screws or implants fixation. Previous studies [25,40] showed that PMMA filled with GO conforms to physicochemical and mechanical demands of these clinical applications. In this study, we have shown that incorporation of GO into PMMA matrix may provide an additional functionality to the resulting composite material, enhancing its biocompatibility through facilitating osteogenesis. Still, before the introduced material could be considered as a bone cement, further studies should include a comprehensive biomechanical characterization of GO/PMMA.

5. Conclusions

In this paper, we present the potency of GO/PMMA composites to induce osteogenic differentiation of hMSPCs. Through the analysis of three differentiation markers, namely ALP, SPARC, and BMP-2 in the presence and in the absence of a specific induction medium, we were able to assess the efficiency of osteogenic differentiation of hMSPCs cultured on four types of GO/PMMA composites, differing in the thickness of GO paper as well as the method of fabrication of the composite. All investigated GO/PMMA composite materials were found to effectively induce osteogenic differentiation, and to outperform both unmodified PMMA and a negative control (undifferentiated hMSPCs). Among GO/PMMA composite materials, a composite produced manually with a thick GO paper ($GO_{(B)}/PMMA_{(M)}$) acted as the most efficient inducer of osteogenic differentiation, particularly basing on the relative expression of ALP (other changes were found to be statistically non-significant). Since $GO_{(B)}/PMMA_{(M)}$ was the composite surface possessing a significant number of voids and pores, its developed surface morphology was supposed to be responsible for directing hMSPCs toward osteogenic differentiation. In this way, GO/PMMA composites, and particularly $GO_{(B)}/PMMA_{(M)}$, were shown as promising materials in bone tissue engineering.

Author Contributions: Conceptualization, D.P., B.L., and F.A.; Investigation, D.P., N.S., B.L., and F.A.; Methodology, D.P., B.L., and F.A.; Supervision, F.A.; Writing—original draft, K.K.; Writing—review and editing, D.P., B.L., and F.A. All authors have read and agreed to the published version of the manuscript.

Funding: The cell studies described in this work were supported by the FFG Bridge program (Grant No. 861608). K.K. would like to thank the National Science Center, Poland (SONATA program, Grant No. 2016/23/D/ST5/01306) and the Silesian University of Technology, Poland (04/040/BK_20/0113).

Acknowledgments: The authors thank Sandra Ngo and Mariangela Fedel for their support in fabricating the GO, and Heike Kaltenegger for her support in isolating and characterizing the primary hMSPCs.

Conflicts of Interest: The authors declare no conflict of interest.

References

1. Novoselov, K.S.; Geim, A.K.; Morozov, S.V.; Jiang, D.; Zhang, Y.; Dubonos, S.V.; Grigorieva, I.V.; Firsov, A.A. Electric field in atomically thin carbon films. *Science* **2004**, *306*, 666–669. [CrossRef] [PubMed]
2. Ghosh, S.; Calizo, I.; Teweldebrhan, D.; Pokatilov, E.P.; Nika, D.L.; Balandin, A.A.; Bao, W.; Miao, F.; Lau, C.N. Extremely high thermal conductivity of graphene: Prospects for thermal management applications in nanoelectronic circuits. *Appl. Phys. Lett.* **2008**. [CrossRef]
3. Lee, J.; Kim, J.; Kim, S.; Min, D.H. Biosensors based on graphene oxide and its biomedical application. *Adv. Drug Deliv. Rev.* **2016**, *105*, 275–287. [CrossRef] [PubMed]
4. Chen, H.; Müller, M.B.; Gilmore, K.J.; Wallace, G.G.; Li, D. Mechanically strong, electrically conductive, and biocompatible graphene paper. *Adv. Mater.* **2008**, *20*, 3557–3561. [CrossRef]
5. Chung, C.; Kim, Y.K.; Shin, D.; Ryoo, S.R.; Hong, B.H.; Min, D.H. Biomedical applications of graphene and graphene oxide. *Acc. Chem. Res.* **2013**, *46*, 2211–2224. [CrossRef]
6. Liu, J.; Cui, L.; Losic, D. Graphene and graphene oxide as new nanocarriers for drug delivery applications. *Acta Biomater.* **2013**, *9*, 9243–9257. [CrossRef]
7. Choudhary, P.; Parandhaman, T.; Ramalingam, B.; Duraipandy, N.; Kiran, M.S.; Das, S.K. Fabrication of Nontoxic Reduced Graphene Oxide Protein Nanoframework as Sustained Antimicrobial Coating for Biomedical Application. *ACS Appl. Mater. Interfaces* **2017**, *9*, 38255–38269. [CrossRef]
8. Liu, Q.; Wei, L.; Wang, J.; Peng, F.; Luo, D.; Cui, R.; Niu, Y.; Qin, X.; Liu, Y.; Sun, H.; et al. Cell imaging by graphene oxide based on surface enhanced Raman scattering. *Nanoscale* **2012**, *4*, 7084–7089. [CrossRef]
9. Yin, F.; Hu, K.; Chen, Y.; Yu, M.; Wang, D.; Wang, Q.; Yong, K.T.; Lu, F.; Liang, Y.; Li, Z. SiRNA delivery with PEGylated graphene oxide nan osheets for combined photothermal and genetherapy for pancreatic cancer. *Theranostics* **2017**, *7*, 1133–1148. [CrossRef]
10. Wan, C.; Chen, B. Poly(ε-caprolactone)/graphene oxide biocomposites: Mechanical properties and bioactivity. *Biomed. Mater.* **2011**, *6*, 055010. [CrossRef]
11. Baradaran, S.; Moghaddam, E.; Basirun, W.J.; Mehrali, M.; Sookhakian, M.; Hamdi, M.; Moghaddam, M.R.N.; Alias, Y. Mechanical properties and biomedical applications of a nanotube hydroxyapatite-reduced graphene oxide composite. *Carbon N. Y.* **2014**, *69*, 32–45. [CrossRef]
12. Xie, C.; Lu, X.; Han, L.; Xu, J.; Wang, Z.; Jiang, L.; Wang, K.; Zhang, H.; Ren, F.; Tang, Y. Biomimetic Mineralized Hierarchical Graphene Oxide/Chitosan Scaffolds with Adsorbability for Immobilization of Nanoparticles for Biomedical Applications. *ACS Appl. Mater. Interfaces* **2016**, *8*, 1707–1717. [CrossRef] [PubMed]
13. Bai, H.; Li, C.; Wang, X.; Shi, G. A pH-sensitive graphene oxide composite hydrogel. *Chem. Commun.* **2010**, *46*, 2376–2378. [CrossRef] [PubMed]
14. Thampi, S.; Muthuvijayan, V.; Parameswaran, R. Mechanical characterization of high-performance graphene oxide incorporated aligned fibroporous poly(carbonate urethane) membrane for potential biomedical applications. *J. Appl. Polym. Sci.* **2015**, *132*, 41809. [CrossRef]
15. Byun, E.; Lee, H. Enhanced loading efficiency and sustained release of doxorubicin from hyaluronic acid/graphene oxide composite hydrogels by a mussel-inspired catecholamine. *J. Nanosci. Nanotechnol.* **2014**, *16*, 7395–7401. [CrossRef]
16. Shen, J.; Yan, B.; Li, T.; Long, Y.; Li, N.; Ye, M. Mechanical, thermal and swelling properties of poly(acrylic acid)-graphene oxide composite hydrogels. *Soft Matter* **2012**, *8*, 1831–1836. [CrossRef]
17. Feuser, P.E.; Gaspar, P.C.; Ricci-Júnior, E.; Da Silva, M.C.S.; Nele, M.; Sayer, C.; De Araújo, P.H.H. Synthesis and characterization of poly(methyl methacrylate) pmma and evaluation of cytotoxicity for biomedical application. *Macromol. Symp.* **2014**, *343*, 65–69. [CrossRef]
18. Frazer, R.Q.; Byron, R.T.; Osborne, P.B.; West, K.P. PMMA: An essential material in medicine and dentistry. *J. Long-Term. Eff. Med. Implants* **2005**, *15*, 629–639. [CrossRef]
19. Gohil, S.V.; Suhail, S.; Rose, J.; Vella, T.; Nair, L.S. Polymers and Composites for Orthopedic Applications. In *Materials and Devices for Bone Disorders*; Bose, S., Bandyopadhyay, A., Eds.; Elsevier/Academic Press: Amsterdam, The Netherlands, 2017; ISBN 9780128028032.
20. He, Q.; Chen, H.; Huang, L.; Dong, J.; Guo, D.; Mao, M.; Kong, L.; Li, Y.; Wu, Z.; Lei, W. Porous surface modified bioactive bone cement for enhanced bone bonding. *PLoS ONE* **2012**, *7*, e42525. [CrossRef]

21. Miola, M.; Bistolfi, A.; Valsania, M.C.; Bianco, C.; Fucale, G.; Verné, E. Antibiotic-loaded acrylic bone cements: An in vitro study on the release mechanism and its efficacy. *Mater. Sci. Eng. C* **2013**, *33*, 3025–3032. [CrossRef]
22. Ormsby, R.W.; Modreanu, M.; Mitchell, C.A.; Dunne, N.J. Carboxyl functionalised MWCNT/polymethyl methacrylate bone cement for orthopaedic applications. *J. Biomater. Appl.* **2014**, *29*, 209–221. [CrossRef] [PubMed]
23. Fang, C.H.; Lin, Y.W.; Sun, J.S.; Lin, F.H. The chitosan/tri-calcium phosphate bio-composite bone cement promotes better osteo-integration: An in vitro and in vivo study. *J. Orthop. Surg. Res.* **2019**, *14*, 162. [CrossRef] [PubMed]
24. Paz, E.; Forriol, F.; Del Real, J.C.; Dunne, N. Graphene oxide versus graphene for optimisation of PMMA bone cement for orthopaedic applications. *Mater. Sci. Eng. C* **2017**, *77*, 1003–1011. [CrossRef]
25. Paz, E.; Ballesteros, Y.; Abenojar, J.; Del Real, J.C.; Dunne, N.J. Graphene oxide and graphene reinforced PMMA bone cements: Evaluation of thermal properties and biocompatibility. *Materials* **2019**, *12*, 3146. [CrossRef] [PubMed]
26. Sabokbar, A.; Millett, P.J.; Myer, B.; Rushton, N. A rapid, quantitative assay for measuring alkaline phosphatase activity in osteoblastic cells in vitro. *Bone Miner.* **1994**, *27*, 57–67. [CrossRef]
27. Dikin, D.A.; Stankovich, S.; Zimney, E.J.; Piner, R.D.; Dommett, G.H.B.; Evmenenko, G.; Nguyen, S.T.; Ruoff, R.S. Preparation and characterization of graphene oxide paper. *Nature* **2007**, *448*, 457–460. [CrossRef]
28. Ku, S.H.; Park, C.B. Myoblast differentiation on graphene oxide. *Biomaterials* **2013**, *34*, 2017–2023. [CrossRef]
29. Compton, O.C.; Putz, K.W.; Brinson, L.C.; Nguyen, S.T. Composite Graphene Oxide-Polymer Laminate and Method. U.S. Patent Application No. 8,709,213, 20 October 2011.
30. Gudarzi, M.M.; Sharif, F. Self assembly of graphene oxide at the liquid-liquid interface: A new route to the fabrication of graphene based composites. *Soft Matter* **2011**, *7*, 3432–3440. [CrossRef]
31. Gonalves, G.; Cruz, S.M.A.; Ramalho, A.; Grácio, J.; Marques, P.A.A.P. Graphene oxide versus functionalized carbon nanotubes as a reinforcing agent in a PMMA/HA bone cement. *Nanoscale* **2012**, *4*, 2937–2945. [CrossRef]
32. Dominici, M.; Le Blanc, K.; Mueller, I.; Slaper-Cortenbach, I.; Marini, F.C.; Krause, D.S.; Deans, R.J.; Keating, A.; Prockop, D.J.; Horwitz, E.M. Minimal criteria for defining multipotent mesenchymal stromal cells. The International Society for Cellular Therapy position statement. *Cytotherapy* **2006**, *8*, 315–317. [CrossRef]
33. Golub, E.E.; Boesze-Battaglia, K. The role of alkaline phosphatase in mineralization. *Curr. Opin. Orthop.* **2007**, *18*, 444–448. [CrossRef]
34. Delany, A.M.; Hankenson, K.D. Thrombospondin-2 and SPARC/osteonectin are critical regulators of bone remodeling. *J. Cell Commun. Signal.* **2009**, *3*, 227–238. [CrossRef] [PubMed]
35. Zong, S.; Zeng, G.; Zou, B.; Li, K.; Fang, Y.; Lu, L.; Xiao, D.; Zhang, Z. Effects of Polygonatum sibiricum polysaccharide on the osteogenic differentiation of bone mesenchymal stem cells in mice. *Int. J. Clin. Exp. Pathol.* **2015**, *8*, 6169–6180. [PubMed]
36. Rui, Y.F.; Lui, P.P.Y.; Lee, W.Y.W.; Chan, K.M. Higher BMP receptor expression and BMP-2-induced osteogenic differentiation in tendon-derived stem cells compared with bone-marrow-derived mesenchymal stem cells. *Int. Orthop.* **2012**, *36*, 1099–1107. [CrossRef] [PubMed]
37. Sun, J.; Li, J.; Li, C.; Yu, Y. Role of bone morphogenetic protein-2 in osteogenic differentiation of mesenchymal stem cells. *Mol. Med. Rep.* **2015**, *12*, 4230–4237. [CrossRef]
38. Sumanasinghe, R.D.; Bernacki, S.H.; Loboa, E.G. Osteogenic differentiation of human mesenchymal stem cells in collagen matrices: Effect of uniaxial cyclic tensile strain on bone morphogenetic protein (BMP-2) mRNA expression. *Tissue Eng.* **2006**, *12*, 3459–3465. [CrossRef] [PubMed]
39. Abagnale, G.; Steger, M.; Nguyen, V.H.; Hersch, N.; Sechi, A.; Joussen, S.; Denecke, B.; Merkel, R.; Hoffmann, B.; Dreser, A.; et al. Surface topography enhances differentiation of mesenchymal stem cells towards osteogenic and adipogenic lineages. *Biomaterials* **2015**, *61*, 316–326. [CrossRef]
40. Park, C.; Park, S.; Lee, D.; Choi, K.S.; Lim, H.P.; Kim, J. Graphene as an Enabling Strategy for Dental Implant and Tissue Regeneration. *Tissue Eng. Regen. Med.* **2017**, *14*, 481–493. [CrossRef]

© 2020 by the authors. Licensee MDPI, Basel, Switzerland. This article is an open access article distributed under the terms and conditions of the Creative Commons Attribution (CC BY) license (http://creativecommons.org/licenses/by/4.0/).

Article

Lactoferrin-Hydroxyapatite Containing Spongy-Like Hydrogels for Bone Tissue Engineering

Ana R. Bastos [1,2], Lucília P. da Silva [1,2], F. Raquel Maia [1,2,3,*], Sandra Pina [1,2], Tânia Rodrigues [1,2], Filipa Sousa [1,2], Joaquim M. Oliveira [1,2,3], Jillian Cornish [4], Vitor M. Correlo [1,2,3] and Rui L. Reis [1,2,3]

1. 3B's Research Group, I3Bs—Research Institute on Biomaterials, Biodegradables and Biomimetics, University of Minho, Headquarters of the European Institute of Excellence on Tissue Engineering and Regenerative Medicine, AvePark, Parque de Ciência e Tecnologia, Zona Industrial da Gandra, 4805-017 Barco, Guimarães, Portugal
2. ICVS/3B's—PT Government Associated Laboratory, 4710-057 Braga, Portugal
3. The Discoveries Centre for Regenerative and Precision Medicine, Headquarters at University of Minho, Avepark, 4805-017 Barco, Guimarães, Portugal
4. Department of Medicine, University of Auckland, Auckland 1023, New Zealand
* Correspondence: raquel.maia@i3bs.uminho.pt

Received: 24 May 2019; Accepted: 25 June 2019; Published: 27 June 2019

Abstract: The development of bioactive and cell-responsive materials has fastened the field of bone tissue engineering. Gellan gum (GG) spongy-like hydrogels present high attractive properties for the tissue engineering field, especially due to their wide microarchitecture and tunable mechanical properties, as well as their ability to entrap the responsive cells. Lactoferrin (Lf) and Hydroxyapatite (HAp) are bioactive factors that are known to potentiate faster bone regeneration. Thus, we developed an advanced three-dimensional (3D) biomaterial by integrating these bioactive factors within GG spongy-like hydrogels. Lf-HAp spongy-like hydrogels were characterized in terms of microstructure, water uptake, degradation, and concomitant release of Lf along the time. Human adipose-derived stem cells (hASCs) were seeded and the capacity of these materials to support hASCs in culture for 21 days was assessed. Lf addition within GG spongy-like hydrogels did not change the main features of GG spongy-like hydrogels in terms of porosity, pore size, degradation, and water uptake commitment. Nevertheless, HAp addition promoted an increase of the pore wall thickness (from ~13 to 28 µm) and a decrease on porosity (from ~87% to 64%) and mean pore size (from ~12 to 20 µm), as well as on the degradability and water retention capabilities. A sustained release of Lf was observed for all the formulations up to 30 days. Cell viability assays showed that hASCs were viable during the culture period regarding cell-laden spongy-like hydrogels. Altogether, we demonstrate that GG spongy-like hydrogels containing HAp and Lf in high concentrations gathered favorable 3D bone-like microenvironment with an increased hASCs viability with the presented results.

Keywords: gellan gum; hydroxyapatite; lactoferrin; bone biomaterials

1. Introduction

Over recent years, tissue engineering (TE) has been pushing forward bone tissue regeneration field, especially in what concerns biomaterials' improvement. Further, the main struggle of bone tissue engineering is to reach a synergistic effect through the combination of biomaterials and cells [1]. Researchers have been demonstrating the importance of biochemical and microstructure cues in the success of designed biomaterials [2]. An optimal biomaterial should not only provide an adequate structural support for bone tissue engineering, but also promote tissue re-growth [3]. The essential prerequisites include: (i) a three dimensional (3D) porous structure, presenting (ii) optimized surface properties, which will potentiate cell attachment, migration, proliferation, and the retention of normal cell functions; (iii) sustained biodegradability; (iv) biomechanical properties that mimic the native tissue; (v) reproducibility; and, (vi) bioactivity, to replace large cortical bone defects and enable load transmission [4–7].

Lf is an iron-binding glycoprotein from the transferrin family [8]. Lf is well known by its antimicrobial, antibacterial, antiviral, antiparasitic anti-neoplastic, anti-inflammatory, and immunomodulatory activities on the immune system, which is a plus for every approach [9–11]. Nowadays, despite the molecular mechanism and action of this protein not being completely understood, Lf alone or combined with others systems has been explored for bone tissue engineering [3,12]. Lf alone has been showed to promote the proliferation of rat osteoblasts in a dose and time dependent manner [13]. The integration of recombinant human Lf and MC3T3-E1 osteoblast-like cells in an injectable hydrogel has been shown to enable MC3T3-E1 cells viability, proliferation, and differentiation supporting proteins phosphorylation/dephosphorylation [14]. In vivo systemic administration of Lf alone [15,16] or in a gelatin hydrogel [17] has also been shown to enhance the new bone formation on a bone defect site.

Hydroxyapatite (HAp) ($Ca_{10}(PO_4)_6(OH)_2$) presents a huge similarity to the inorganic component of the bone matrix and it has been widely used in bone tissue engineering as bone graft materials, coatings for implants, and bone fillers due to its ability to directly bond to the deficient apatite layer of the bone through the carbonated calcium [18–21]. HAp also presents osteoconductive and osteoinductive properties, as it facilitates the migration, adhesion, proliferation, and differentiation of progenitor cells, and the cell-mediated release of growth factors that stimulate bone formation in vivo [22]. We have previously developed Gellan Gum (GG) spongy-like hydrogels that contain HAp for bone tissue engineering [23,24]. GG spongy-like hydrogels show optimal conditions for Tissue Engineering and Regenerative Medicine (TERM) due to their porous microstructure arrangement, mechanical stability, as evidenced by their high flexibility and resilience to deformation, and high water content that altogether potentiate cell adhesion and spreading [25–27]. The addition of HAp (10 and 20% w/v) and $CaCl_2$, as a crosslinker, to the GG spongy-like hydrogels, improved the bioactivity of the materials, as evidenced by the formation of uniformly distributed apatite-like crystal phases on materials surface in a osteogenic medium, being similar in terms of composition and structure to bone-apatite [24]. In osteogenic culture conditions, HAp-containing spongy-like hydrogels prompted hASCs adhesion and spreading [24], as well as the differentiation of bone marrow cells that were isolated from mice long bones towards pre-osteoclasts [23] in the presence of vitamin D3.

The combination of Lf and HAp has been scarcely explored for bone tissue engineering. From the few existing in vitro studies, Lf-Hap have shown promising results, as evidenced by enhanced osteoblasts proliferation [3,28] and rADSCs differentiation towards osteoblasts [29]. Hence, in this work, we combined the unique properties of Lf and HAp in spongy-like hydrogels to reinforce the bone-like microenvironment. Different Lf-HAp spongy-like hydrogels formulations were developed and characterized, envisaging a material with appropriate physical-chemical properties and Lf temporal release for future bone tissue engineering approaches. Human adipose-derived stem cells (hASCs) were seeded within the Lf-Hap spongy-like hydrogels due to their promising features, such as their regenerative potential, self-renewal properties, capacity to differentiate into the osteogenic lineage, as well as their high and easy availability from the subcutaneous liposuction from adipose/fat tissue [30,31]. The ability of these biomaterials to behave as a platform to support hASCs cell adhesion and growth was then studied.

2. Materials and Methods

2.1. Preparation of Hydrogels

Gelzan CM (Sigma, St. Louis, MO, USA) was dissolved in distilled water at 90 °C for 30 min. with stirring. Lactoferrin (Lf, Bovine origin, New Zealand) was dissolved in distilled water. Subsequently, to the solution of Gellan Gum (GG), Lactoferrin (Lf), and Hydroxyapatite (HAp, Plasma Biotal, Buxton, UK) were added at 50 °C, as described in Table 1. After complete homogenization, the crosslinking solution $CaCl_2$ (Sigma, USA) was added and it was allowed to stabilize at room temperature for 30 min. Posteriorly, the hydrogels were cut in discs of 5 mm of diameter and 2 mm of thickness and replenished with phosphate-buffered saline solution (PBS, Sigma, USA) for 30 h to enable complete crosslinking, frozen at −80 °C for 18–20 h, and then freeze-dried (CryoDos -80, Telstar, Terrassa, Barcelona, Spain) for at least three days. Materials were sterilized by ethylene oxide.

Table 1. Different formulations of Gellan Gum (GG) spongy-like hydrogels with or without Lactoferrin (Lf), and/or Hydroxyapatite (HAp).

Name	GG % (w/v)	Lf % (w/v)	HAp % (w/v)
GG	1.25	-	-
GG/Low Lf	1.25	0.05	-
GG/High Lf	1.25	0.15	-
GG/Low HAp	1.25	-	1.00
GG/High HAp	1.25	-	10.00
GG/Low Lf/ Low HAp	1.25	0.05	1.00
GG/High Lf/Low HAp	1.25	0.15	1.00
GG/Low Lf/High HAp	1.25	0.05	10.00
GG/High Lf/High HAp	1.25	0.15	10.00

2.2. Physico-Chemical Characterization of Spongy-Like Hydrogels'

2.2.1. Fourier Transformed Infrared Spectroscopy (FTIR)

Fourier Transformed Infrared (FTIR) analysis was performed while using attenuated total reflectance (ATR) (IRPrestige-21, Shimadzu Corporation, Kyoto, Japan) in transmittance mode and in the region of 550–4000 cm^{-1} for spectroscopic study.

2.2.2. X-ray Diffraction (XRD) Analysis

The qualitative analyses of crystalline phases that were presented on the GG, GG/Lf, GG/HAp, and GG/Lf/HAp spongy-like hydrogels were obtained by XRD while using a conventional Bragg–Brentano diffractometer (Bruker D8 Advance DaVinci, Rheinstetten, Germany) that was equipped with CuKα radiation, produced at 40 kV and 40 mA. The data sets were collected in the 2θ range of 5–70° with a step size of 0.04° and 1 s for each step.

2.2.3. Scanning Electron Microscopy

Scanning electron microscopy (SEM, JSM-6010 LV, JEOL, Tokyo, Japan) and micro computed tomography (micro-CT) were used to assess the microstructure of the dried polymeric networks of the different formulations. For SEM analysis, the samples were immersed in liquid nitrogen to cut the sample into cross-sections, and the internal surface of the dried polymeric networks was observed.

2.2.4. Micro Computed Tomography

The microstructure of dried polymeric networks was analyzed while using a high resolution X-Ray microtomography SkyScan 1272 System (SkyScan, Kontich, Belgium). The pixel size used was 2.5 µm, the exposure time was 1s, and the source conditions used were 50 kV of energy and 200 µA of current. A binary picture was created using 150 slides while using a thresholding between 20 and 255 in a grey scale. The morphometric analysis, which includes porosity, pore size, interconnectivity, and wall thickness, were assessed by CT-analyzer program (CTAn, v1.17.0.0., SkyScan, Belgium).

2.2.5. Degradation Studies: Mass Loss

Lf-HAp containing spongy-like hydrogels were immersed in PBS at 37 °C for 30 days, with stirring. The initial weight (w_i) of the samples was recorded and the samples were then immersed in PBS and weighed again (final weight, w_f). The percentage of mass loss along the time was calculated according to the equation:

$$Mass\ loss(\%) = \frac{(W_f - W_i)}{W_i} \times 100$$

Every three days, the PBS solution was replaced and the supernatant was collected and then kept at −80 °C for further analysis of the amount of Lf released.

2.2.6. Water Uptake

Lf-HAp containing spongy-like hydrogels were weighed in the dried state (w_d) and were then immersed in PBS at 37 °C during three days. Along this period of time, at different time points (30 min., 1, 2, 4, 6, 24, 48, and 72 h), the samples were weighed in the wet state (w_w). The percentage of water uptake along the time was calculated according to the equation:

$$Water\ Uptake(\%) = \frac{(W_w - W_d)}{W_d} \times 100$$

2.2.7. Quantification of Lf Release: BCA Assay

A micro-Bicinchoninic Acid Assay (micro-BCA) was performed according to manufacturer instructions In order to evaluate the amount of Lf released from spongy-like hydrogels. Briefly, 75 µL of bicinchoninic acid was placed in each well of a 96 well-plate and 75 µL of the supernatant, previously collected during mass loss assessment, was added to each one. These well-plates were incubated for 2 h at 37 °C. After this time, the absorbance at 562 nm was measured while using a Microplate Reader (SYNERGY HT, BIO-TEK, USA). The supernatants from GG and GG with HAp spongy-like hydrogels were used as Blank. The corrected absorbance readings were converted into protein concentrations while using a calibration curve of Lf. The theoretical value according to the final volume and concentration of Lf into GG spongy-like hydrogels was calculated. GG/Low Lf containing spongy-like hydrogels, according to the theoretical calculations, can release a maximum of 20 µg, while G/High Lf can release a maximum of 60 µg.

2.3. In Vitro Studies

2.3.1. Cell Isolation and Culture

Human adipose-derived stem cells (hASCs) were isolated from the lipoaspirate samples following a protocol previously established with the Department of Plastic Surgery of Hospital da Prelada (Porto, Portugal). All subjects gave their informed consent for inclusion before they participated in the study. The study was conducted in accordance with the Declaration of Helsinki, and the protocol was approved by the Ethics Committee of Hospital da Prelada (P.I. N° 005/2019) and 3B's Research Group. The isolated cells were used for subsequent studies. The cells were routinely cultured in α-MEM medium (Alfagene, Carcavelos, Portugal) that was supplemented with 10% fetal bovine serum (FBS, Australia origin, Alfagene, Portugal), 1% antibiotic/antimycotic (Alfagene, Portugal), and maintained at 37 °C under a humidified atmosphere of 5% v/v CO_2 in air. The medium was changed twice a week and cells, at maximum passage 4, were seeded into spongy-like hydrogels, as described in the next section.

2.3.2. Cell Entrapment

A cell suspension (40×10^3, 20 µL) was dispensed dropwise on the top of the polymeric networks, which were previously hydrated for 15 min. with culture medium. Posteriorly, these constructs were incubated at 37 °C during 3 h, with 5% CO_2 to allow for maximum cell entrapment within the structures. Fresh medium (500 µL) was added after this period of time.

2.3.3. Cell Viability

Cell viability was assessed at 21 days while using 20% (v/v) of Alamar Blue reagent (ALAMAR BLUE®, AbD, Kidlington, Oxford, UK) in α-MEM culture medium, followed by 3 h of incubation at 37 °C with 5% CO_2 and protected from light. Afterwards, 100 µL of supernatant was transferred from each well in triplicate to a new 96-well cell culture plate. Fluorescence intensity was read at 530/20 nm (excitation) and 590/35 nm (emission) using a microplate reader (Synergy HT, Bio-Tek, USA). Alamar Blue in medium was used as a blank. DNA normalized the corrected absorbance readings at day 21. Three specimens of each formulation were used to assess cell viability along the time and three independent experiments were performed.

2.3.4. DNA Content

DNA content was assessed after 21 days of culture and it was used to normalize the cell viability values of the same time point. Constructs were washed with PBS and transferred to 1.5 mL tubes and heated at 70 °C for 30 min. Next, 1 mL of ultra-pure water was added into each tube that were placed at 37 °C for 1 h, and then frozen at −80 °C until analysis. Before DNA quantification, the samples were placed in an ultrasound bath for 1 h at 37 °C. Finally, Quant-IT PicoGreen dsDNA Assay Kit 2000 assays (Alfagene®, Portugal) was used according to the manufacturer's instructions. The fluorescence intensity was read at 485/20 nm (excitation) and 530/20 nm (emission) while using a microplate reader (Synergy HT, Bio-Tek, Winooski, VT, USA) and the readings were converted while using a standard curve that was produced with standard dsDNA solutions at different concentrations.

2.3.5. Cytoskeleton Morphology (Phalloidin/DAPI)

After three and 21 days of culture, the constructs were washed with PBS during 5 min., fixed with 10% formalin for one hour, and then washed again. Next, Phalloidin-TRITC (Sigma, USA) (1:80) was added and placed at room temperature during 1 h, protected from light. The constructs were counterstained with 4,6-Diamidino-2-phenylindole, Dilactate (DAPI, Sigma, USA) (1:5000) for 5 min. in the dark. Finally, they were visualized by Confocal Laser Scanning Microscope (TCS SP8, Leica, Mannheim, Germany).

2.4. Statistics

Statistical analysis was performed while using the GraphPad Prism 5.0 software. The data were analyzed while using the Shapiro–Wilk normality test. The results that did not present a normal distribution were analyzed using the Kruskal–Wallis test with Dunn's multiple comparison post-test for statistical analysis. The results that presented a normal distribution were analyzed while using a One-way ANOVA and Tukey's multiple comparisons T-test. The significance level between the groups were set for $* p < 0.05$, $** p < 0.01$, and $*** p < 0.001$. The data were presented as mean ± standard deviation (SD). Data in each figure are representative experiment of three experiments (n = 3).

3. Results

3.1. FTIR and XRD Analysis

The ATR-FTIR spectra of the spongy-like hydrogels that are presented in Figure 1 show the stretching vibrations of C–O–C bonds and the bending mode of mehtyl vC–H, respectively, at 1022 and

1593 cm^{-1} [32]. Regarding the Lf addition to the GG formulations, a signal of tyrosine was registered at 1138 cm^{-1} [33]. Additionally, an absorption peak appears at 1010 cm^{-1}, which can be assigned to N≡C or C=C stretch and C–H deformation vibrations of tryptophan. The presence of HAp was ascertained by the detection of the peak around 600 cm^{-1}, in the region of the v4 bending mode of $PO_4{}^{3-}$ and the peak around 1000 cm^{-1} that corresponded to the region of the v1 stretching mode of $PO_4{}^{3-}$ [24]. All of the GG/Lf/HAp spongy-like hydrogels show the broad band between 3500 and 3700 cm^{-1} that is attributed to the hydroxyl groups (vOH) stretching vibration due to the medium hydrogen bond of intramolecular and intermolecular type.

Figure 2 displays the XRD patterns of the spongy-like hydrogels. It can be observed that all of the compositions show the intensity peaks corresponding to the GG crystalline structure with the main peak being located at 32° [32]. The presence of a higher content of Lf led to a slight crystallinity decrease without additional peaks being observed (Figure 2iii). The crystallographic phases identification of the hydrogels containing HAP was accomplished by comparing the XRD patterns with the standard ICDD PDF 01-074-0565 of HAp (Figure 2iv–ix). As expected, with increasing the HAp content in the hydrogels, the crystallinity became more evident, as shown in Figure 2v,viii,ix.

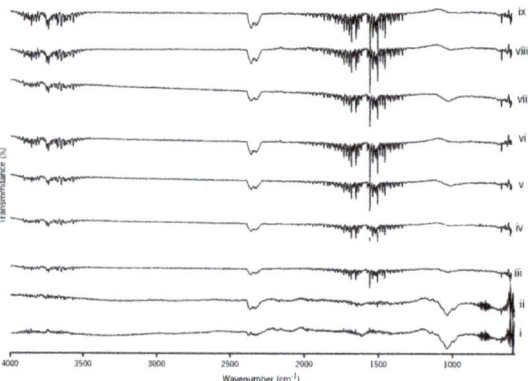

Figure 1. Fourier Transformed Infrared (FTIR) spectra of GG/Lf/HAp spongy-like hydrogels: (i) GG, (ii) GG/Low Lf, (iii) GG/High Lf, (iv) GG/Low HAp, (v) GG/High HAp, (vi) GG/Low Lf/Low HAp, (vii) GG/High Lf/Low HAp, (viii) GG/Low Lf/High HAp, and (ix) GG/High Lf/High HAp.

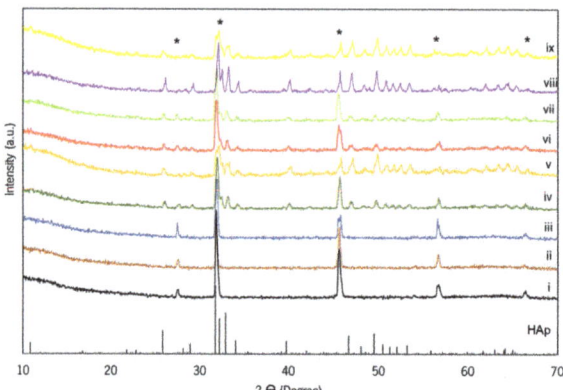

Figure 2. X-ray Diffraction (XRD) patterns of the GG/Lf/HAp spongy-like hydrogels: (i) GG, (ii) GG/Low Lf, (iii) GG/High Lf, (iv) GG/Low HAp, (v) GG/High HAp, (vi) GG/Low Lf/Low HAp, (vii) GG/High Lf/Low HAp, (viii) GG/Low Lf/High HAp and (ix) GG/High Lf/High HAp. HAp (ICDD PDF 01-074-0565), and * GG.

3.2. SEM and Micro-CT Analysis

Figure 3 shows the representative images of the cross section surface of the different GG/Lf/HAp dried polymeric networks (DPN) that were obtained by SEM. GG DPN containing just Lf (Low or High) showed a porous structure with smooth surfaces, which indicated that the presence of Lf had no effect in the microstructure when compared with the control (GG) (Figure 3A–C). Regarding GG DPN containing HAp (Figure 3D,E), smaller pores and rough surfaces were observed. This effect was more evident in the formulations containing higher HAp concentrations (1% in relation to 10%). The microstructure of GG DPN containing Lf/HAp was similar to the respective GG/HAp DPN (Figure 3F,G in relation to Figure 3D; Figure 3H,I in relation to Figure 3E). These results indicate that no evident effect was observed by adding Lf, while the addition of HAp affected the microstructure.

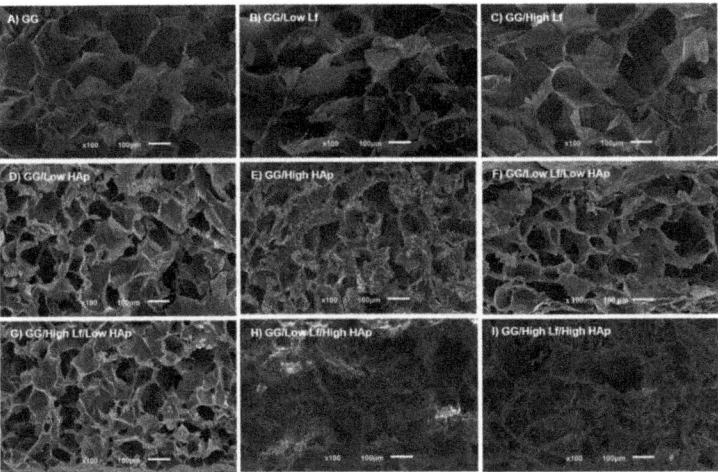

Figure 3. SEM representative images of GG dried polymeric networks (DPN) with or without Lactoferrin (Lf) and/or Hydroxyapatite (HAp): (**A**) GG, (**B**) GG/Low Lf, (**C**) GG/High Lf, (**D**) GG/Low HAp, (**E**) GG/High HAp, (**F**) GG/Low Lf/Low HAp, (**G**) GG/High Lf/Low HAp, (**H**) GG/Low Lf/High HAp, and (**I**) GG/High Lf/High HAp.

Porosity, pore wall thickness, and average pore size were quantified by micro-CT analysis (Table 2). The obtained results revealed that the addition of HAp has an inverse effect on the porosity of the DPN: the porosity tended to decrease with the increase of HAp concentration from ~87% to 64%, respectively. Despite this tendency, no other statistically significant differences on the average porosity were observed. Nevertheless, the same trend was verified on the mean pore size for all of the conditions: a decrease of pore size was observed when HAp is added. For these formulations, the decrease on pore size was followed by an increase of the pore wall thickness, although no significant differences were observed. In accordance to SEM results, the HAp addition had a higher impact in the microstructure of GG polymeric networks than Lf addition. In addition, the 2D microarrangement of the dried polymeric networks can be observed in Figure 4.

Table 2. Micro computed tomography (Micro-CT) analysis of (A) mean porosity, (B) pore size, and (C) wall thickness of GG dried polymeric networks (DPN) with or without Lf and/or HAp. Data was presented as mean ± stdev, the statistical analysis was performed using a Kruskal–Wallis test followed by Dunn's test, in which all formulations were compared between them.

Name	Porosity (%)	Pore Size (µm)	Wall Thickness (µm)
GG	86.5 ± 1.34	73.18 ± 12.92	12,83 ± 0.87
GG/Low Lf	86.01 ± 1.81	73.14 ± 5.37	13.21 ± 0.14
GG/High Lf	86.98 ± 1.97	68.85 ± 0.93	13.45 ± 1.31
GG/Low HAp	78.44 ± 1.37	55.23 ± 5.22	17.43 ± 0.21
GG/High HAp	64.45 ± 3.79	42.80 ± 5.73	20.17 ± 0.66
GG/Low Lf/ Low HAp	81.30 ± 0.64	63.30 ± 3.66	16.07 ± 0.42
GG/High Lf/Low HAp	84.14 ± 3.21	79.06 ± 15.31	16.48 ± 0.62
GG/Low Lf/High HAp	67.98 ± 0.55	59.96 ± 3.53	27.95 ± 1.66
GG/High Lf/High HAp	71.22 ± 2.96	59.74 ± 2.00	25.66 ± 1.72

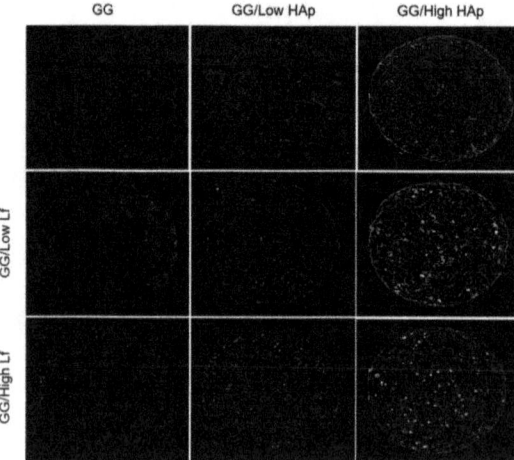

Figure 4. Two-dimensional (2D) images of the dried polymer networks of GG/Lf/HAp spongy-like hydrogels obtained by Micro-CT.

3.3. Water Uptake Analysis

The water uptake of the different GG DNP formulations was followed up to three days of immersion into PBS (Figure 5). All of the formulations have shown a burst of water uptake in the first hours of immersion, followed by an equilibrium phase corresponding to the maximum of water content. GG showed a water content of 2385% ± 357 after three days of immersion (Figure 5B). Similar values were observed for the GG/High Lf formulation. However, GG/Low Lf formulations have shown lower water content (1780% ± 143). All of the formulations containing HAp showed lower water content values (~1500%) when comparing to GG and both GG/Lf formulations. GG spongy-like hydrogels with 1 (GG/Low HAp) and 10% (GG/High HAp) of HAp showed a water content of 886% ± 160 and 603% ± 62, respectively. GG spongy-like hydrogels with 1% of HAp and Lf, independent of the concentration, showed a water content between 1277% ± 109 and 1438% ± 438, while the formulations containing 10% of HAp and Lf, the water content was lower (~300/500%). These results have shown that the addition of HAp at higher concentration (10%) to the GG formulation significantly decreased ($p < 0.05$) their water uptake ability (from 2385% ± 357 to 603% ± 63). Likewise, a significant decrease on water content between GG/High Lf (~2385% ± 683) and GG/High Lf/High HAp (~375% ± 35) ($p < 0.001$) was also observed due to HAp addition.

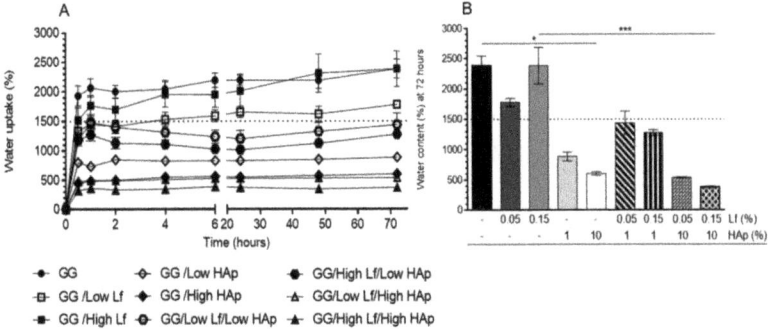

Figure 5. (**A**) Water uptake profile of dried polymeric networks (DPN) along 72 h (**B**) water content after 72 h. Data was presented as mean ± stdev and statistical analysis was performed while using a Kruskal-Wallis test followed by Dunn's test.

3.4. Degradation Tests and Lf Release Analysis

The degradation of different formulations was followed up to 30 days by quantifying the weight loss (Figure 6A). GG/Low Lf and GG/High Lf spongy-like hydrogels have shown higher mass loss in comparison with the other formulations (Figure 6A). In accordance, the GG/High Lf spongy-like hydrogels showed a significant mass loss when comparing to GG ($p < 0.05$), GG/Low Lf ($p < 0.01$), and Lf/HAp ($p < 0.001$) containing spongy-like hydrogels.

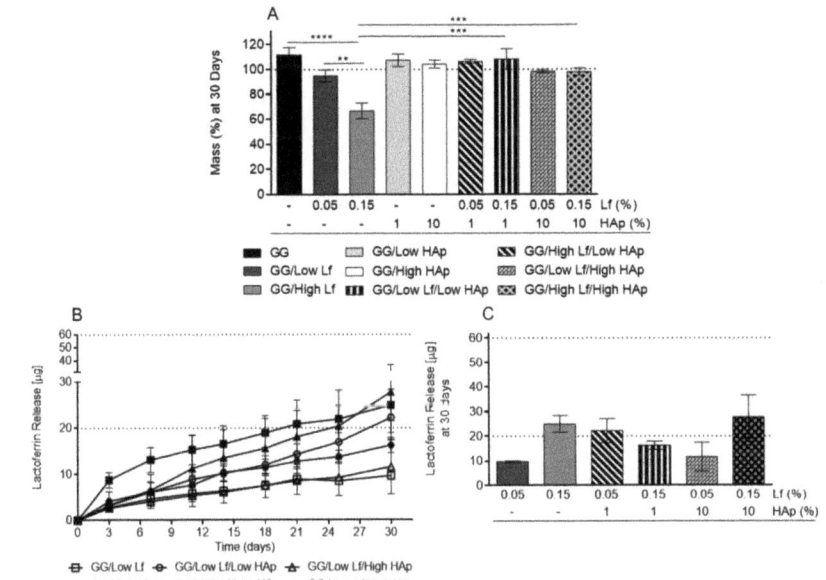

Figure 6. Degradation and Lf release analysis. Final mass of dried polymeric networks after 30 days (**A**). Lf release profile of Lf-HAp containing spongy-like hydrogels along 30 days (**B**) and at day 30 (**C**). The statistical analysis of the final mass was performed using a One-way ANOVA and Tukey's multiple comparisons T-test, while the Kruskal–Wallis test followed by Dunn's test was used for Lf release and all the formulations were compared between them.

3.5. Cell Viability and Cell Cytoskeleton Analysis

hASCs' cytoskeleton within spongy-like hydrogels was analyzed by Phalloidin staining (Figure 7). Independent of the spongy-like hydrogel formulation, cells attached and showed a spread morphology within the polymeric structure from three days of culture onward 21 days. Figure 4 shows cell cytoskeleton organization within GG/High Lf/High HAp, which is representative of the other spongy-like hydrogel formulations.

Cell viability within the spongy-like hydrogels was assessed through Alamar Blue®assay that was normalized by DNA at that time point (21 days) (Figure 8A). It was possible to verify that hASCs remained metabolically active after 21 days of culture. Noteworthy, the formulations containing Lf in combination or not with HAp showed the highest metabolic activity, while GG and GG/Low HAp spongy-like hydrogels showed the lowest metabolic activity. GG/High Lf, GG/High Lf/Low HAp, GG/Low Lf/High HAp, and GG/High Lf/High HAp showed significantly higher metabolic activity when comparing to the GG spongy-like hydrogels. A significant increase in metabolic activity was also verified between GG/Low HAp and GG/High Lf/Low HAp. Regarding the DNA concentration depicted in Figure 5B, it was possible to observe that, after 21 days, the cells were present in all the spongy-like hydrogels tested. Moreover, the results showed that spongy-like hydrogels with Lf combined or not with HAp have higher amounts of DNA, while GG spongy-like hydrogels have the lowest. However, statistical differences were not observed.

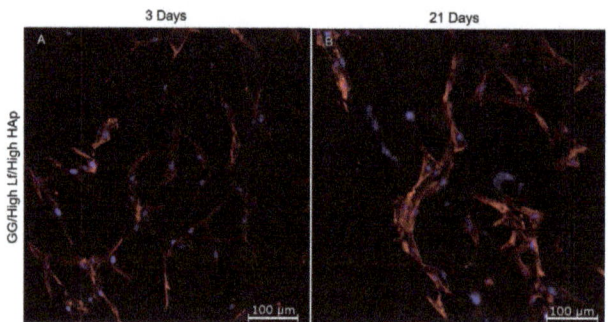

Figure 7. Representative images of cytoskeleton morphology (Phalloidin-TRITC, red) and nuclei (DAPI,blue) of hASCs within GG/High Lf/High HAp spongy-like hydrogels after (**A**) three and (**B**) 21 days of culture.

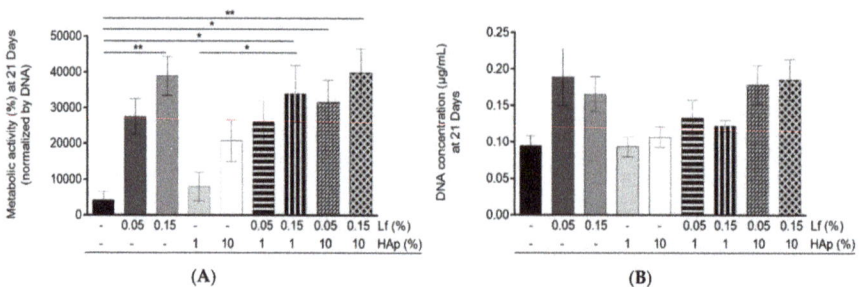

Figure 8. Cell viability assessment. (**A**) Metabolic activity of hASCs' normalized by Day 1 after 21 days of culture. (**B**) DNA concentration of hASCs' within spongy-like hydrogels after 21 days. The statistical analysis was performed using a One-way ANOVA and Tukey's multiple comparisons T-test and results were presented as mean ± standard deviation (SD) and the significance level between groups was set for: * $p < 0.05$, and ** $p < 0.01$.

4. Discussion

New biomaterials, cells, and growth factors are being combined to reach a synergistic effect and produce functional tissue engineered bone substitutes. With this in mind, we developed a batch of novel bioactive GG spongy-like hydrogels that contain different concentrations of Lf and/or HAp, and tested its capacity to support hASCs culture, envisioning its use as improved scaffolds for bone tissue engineering. In fact, hASCs can differentiate along the osteogenic lineage when submitted to specific growth factors, such as Lf [15,16,34].

The biomaterials microarchitecture, including pore size and porosity, are crucial parameters to match the native tissue characteristics and the required integrity [35]. This work showed that the presence of HAp significantly reduced the porosity and pore size. The HAp functioned as a nucleation agent for ice crystals during the freezing step of spongy-like hydrogels preparation, leading to the formation of pores in a higher amount, but with lower sizes [25,27]. Nevertheless, the results that were obtained regarding porosity and pore size, for all of the formulations, are in agreement with the literature that indicate that porosity between 35 and 75% and the pore size in the range of 50 to 400 µm are appropriated to be used in bone tissue engineering applications [36]. In our study, all of the developed spongy-like hydrogels presented pore sizes that were between ~40 to 80 µm, which are within the range that was described to allow for the ingrowth of capillaries and facilitate the exchange of nutrients and discharge of metabolites [1]. In fact, we could verify that hASCs that were cultured within all spongy-like hydrogels were metabolically active. Furthermore, the cells cultured within spongy-like hydrogels with Lf and/or HAp were more metabolic active than cells that were cultured within GG spongy-like hydrogels. This was an expected result, since Gellan Gum hydrogels do not present specific attachment sites for anchorage-dependent cells. When considering the pore size, it was also observed that the addition of HAp lead to a decrease in the porosity while thicker and rough pore walls were promoted. The commitment of this combined effect together with the intrinsic properties of the GG spongy-like hydrogels can overcome the problems that are associated with stress shielding [37]. It is well known that the thicker pore walls and roughness might be crucial for cells biological response since it directly influences cell migration and proliferation [38]. Moreover, it promotes structural support and adhesion sites, facilitates cell movement, regulates cell behavior, and sustains cell-to-cell recognition.

On the other hand, the microarchitecture of spongy-like hydrogels was not altered with the addition of Lf. In fact, this small protein was predicted to adsorb to the GG polymeric networks and it did not act as a nucleation agent for ice crystals in the freezing stage. GG spongy-like hydrogels Lf containing have shown higher porosity and water retention capacity, when compared with HAp containing spongy-like hydrogels. Similarly, in other studies, it has been verified that, when HAp is incorporated into a matrix, a lower water content is adsorbed/expected [23,39,40]. The achievement of the maximum water content occurs earlier due to the existence of lower microvoids between the interface of the GG matrix and the HAp crystals. Since HAp has very low water absorption, when it is added to the GG matrix, its natural/intrinsic hydrophilicity decreases. Overall, the microstructure of the developed spongy-like hydrogels was proper for the diffusion of oxygen and nutrients, since the cells showed their normal phenotype after 21 days of culture. Furthermore, it was also observed that the pore size did not negatively affect cell spreading and morphology, as demonstrated by cytoskeleton staining.

When considering the bioactivity potential of HAp and/or Lf, the release of these bioactive components to the surrounding environment is of high importance in an in vivo scenario. Hence, the release of these bioactive factors was studied in vitro along 31 days. The serum level of Lf on the human body is around 2 to 7 µg/mL, and it has been applied to achieve the desired effect from the contact of Lf with different cell types (e.g. proliferation and/or the inhibition of bone resorption) [17]. Having this in mind, we applied a protein concentration within our biomaterials that, when released, was similar to the protein concentration that is found in human body. Lf containing spongy-like hydrogels showed a significant mass loss, in contrast to the HAp containing spongy-like hydrogels. This can be explained by the higher release of non-crosslinked Lf from the semi-interpenetrating networks, which causes the disintegration of the polymeric networks. Interestingly, when they were combined, the HAp presence masked the Lf presence in all aspects that were previously mentioned. Perhaps it happens due to the high affinity between HAp and Lf, which is mainly provided by the electrostatic interactions between Lf and HAp surface [3,41]. Montesi et al., also reported that the protein quantity is a crucial parameter, since a lower amount of Lf only interacted with the negative surface ionic groups of HAp, while higher amount of Lf probably favored the protein–protein interactions (through hydrogen or hydrophobic/hydrophilic interactions) and formation of multiple protein layers with different molecular orientations [41]. Furthermore, these interactions were sustained by Lf incorporation during hydrogel preparation, which affected the hydrogel chains formation. It was also reported that the swelling ratio was correlated with the quantity of protein released [42], since a faster desorption of the Lf molecules being non-directly bounded on the particles surface is verified, when the scaffolds are immersed in a hydrated medium. However, in spite of non-existent significant differences, the formulations with 10% of HAp and Lf in both concentrations, showed lower swelling ratio and, consequently, a lower release. As expected, when Lf was present in a high amount, it had more protein to release. Montesi et al. demonstrated that, in addition to the HAp biocompatibility, osteoconductivity, and biodegradability, this bioceramic has the ability to bind several biomolecules without affecting their biological function [41]. Overall, some gaps have been identified based on the existing studies. In fact, most of the existing studies are performed while using animal cells/models hindering the correlation of these models with future clinical outcomes. Moreover, the results that were obtained can be conditioned, depending of the type and size of the defect, the degradation process, and the protein release, which resulted in different bone regeneration rates.

Our study demonstrates the synergetic effect from the combination of HAp with Lf and hASCs and their ability for use in spongy-like hydrogels for bone tissue engineering applications.

5. Conclusions

Bone tissue engineering has been explored by combining new biomaterials and cells to reach a synergistic effect. With this in mind, we studied the effect from the combination of bioactive compounds, Lf and HAp, within GG spongy-like hydrogels to be used in the bone tissue regeneration approaches.

The parameters studied, such as porosity, pore size, pore wall thickness, weight loss, water uptake, and Lf release, were tailored through the addition and/or combination of each compound. Lf containing spongy-like hydrogels showed similar features with GG spongy-like hydrogels. In contrast, differences in terms of all parameters studied were observed in GG/HAp and Lf-HAp containing spongy-like hydrogels. The HAp addition promoted an increase of the wall thickness and a decrease on porosity and pore size as well as, on degradability and water retention capabilities. Related to the Lf release, along the 30 days all formulations released at least half of the existent lactoferrin in both (0.05 and 0.15%) of the formulations containing Lf. Moreover, the capacity of these biomaterials to support hASCs in culture for 21 days was assessed. The microstructure of Lf/HAp-containing spongy-like hydrogels seemed to be proper for the diffusion of oxygen and nutrients, since the cells spread and showed their normal phenotype after 21 days. In conclusion, Lf and HAp in high concentrations assembled the required conditions to conduce these biomaterials to promote bone regeneration.

Author Contributions: All authors above mentioned have contributed to the preparation of the proposed manuscript, attested the authenticity of the data and agreed to its submission to the previous referred interest journal. For this work, no other persons have in any way contributed.

Funding: This research was funded by FROnTHERA (NORTE-01-0145-FEDER-0000232).

Acknowledgments: F.R.M. acknowledges FCT (SFRH/BPD/117492/2016), L.P.d.S. thanks FCT (SFRH/BD/78025/2011), J.M.O. thanks FCT Investigator program (IF/01285/2015) and A.R.B. thanks the funds provided by FCT under the doctoral program in Tissue Engineering, Regenerative Medicine and Stem Cells (PD/BD/143043/2018).

Conflicts of Interest: The authors declare no conflict of interest.

References

1. Gao, C.; Peng, S.; Feng, P.; Shuai, C. Bone biomaterials and interactions with stem cells. *Bone Res.* **2017**, *5*, 17059. [CrossRef]
2. Black, C.R.M.; Goriainov, V.; Gibbs, D.; Kanczler, J.; Tare, R.S.; Oreffo, R.O.C. Bone Tissue Engineering. *Curr. Mol. Biol. Rep.* **2015**, *1*, 132–140. [CrossRef]
3. Shi, P.; Wang, Q.; Yu, C.; Fan, F.; Liu, M.; Tu, M.; Lu, W.; Du, M. Hydroxyapatite nanorod and microsphere functionalized with bioactive lactoferrin as a new biomaterial for enhancement bone regeneration. *Colloids Surf. B Biointerfaces* **2017**, *155*, 477–486. [CrossRef]
4. Henkel, J.; Woodruff, M.A.; Epari, D.R.; Steck, R.; Glatt, V.; Dickinson, I.C.; Choong, P.F.M.; Schuetz, M.A.; Hutmacher, D.W. Bone Regeneration Based on Tissue Engineering Conceptions—A 21st Century Perspective. *Bone Res.* **2013**, *1*, 216–248. [CrossRef]
5. Kumar, S.D. Study of Development and Applications of Bioactive Materials and Methods In Bone Tissue Engineering. *Biomed. Res.* **2015**, *26*, S55–S61.
6. Silva, S.S.; Fernandes, E.M.; Silva-Correia, J.; Pina, S.C.A.; Vieira, S.; Oliveira, J.M.; Reis, R.L. Natural-Origin Materials for Tissue Engineering and Regenerative Medicine. In *Comprehensive Biomaterials II*, 2nd ed.; Ducheyne, P., Healy, K., Hutmacher, D.W., Grainger, D.W., Kirkpatrick, C.J., Eds.; Elsevier: Amsterdam, The Netherlands, 2017.
7. Matassi, F.; Nistri, L.; Paez, D.C.; Innocenti, M. New biomaterials for bone regeneration. *Clin. Cases Miner. Bone Metab.* **2011**, *8*, 21–24.
8. Lönnerdal, B. Lactoferrin: Molecular Structure and Biological Function. *Annu. Rev. Nutr.* **1995**, *15*, 93–110. [CrossRef]
9. Farnaud, S.; Evans, R.W. Lactoferrin—A multifunctional protein with antimicrobial properties. *Mol. Immunol.* **2003**, *40*, 395–405. [CrossRef]
10. Gao, R.; Watson, M.; Callon, K.E.; Tuari, D.; Dray, M.; Naot, D.; Amirapu, S.; Munro, J.T.; Cornish, J.; Musson, D.S. Local application of lactoferrin promotes bone regeneration in a rat critical-sized calvarial defect model as demonstrated by micro-CT and histological analysis. *J. Tissue Eng. Regen. Med.* **2018**, *12*, e620–e626. [CrossRef]
11. Bruni, N.; Capucchio, M.T.; Biasibetti, E.; Pessione, E.; Cirrincione, S.; Giraudo, L.; Corona, A.; Dosio, F. Antimicrobial Activity of Lactoferrin-Related Peptides and Applications in Human and Veterinary Medicine. *Molecules* **2016**, *21*, 752. [CrossRef]
12. Montesi, M.; Panseri, S.; Iafisco, M.; Adamiano, A.; Tampieri, A. Coupling Hydroxyapatite Nanocrystals with Lactoferrin as a Promising Strategy to Fine Regulate Bone Homeostasis. *PLoS ONE* **2015**, *10*, e0132633. [CrossRef]
13. Jun, W.; Bi-Yu, Z.; Chuang-Yue, Z. Effect of lactoferrin on rat osteoblast proliferation. *J. Hainan Med. Univ.* **2016**, *22*, 5–8.
14. Amini, A.A.; Nair, L.S. Recombinant human lactoferrin as a biomaterial for bone tissue engineering: Mechanism of antiapoptotic and osteogenic activity. *Adv. Healthc. Mater.* **2014**, *3*, 897–905. [CrossRef]
15. Yoshimaki, T.; Sato, S.; Tsunori, K.; Shino, H.; Iguchi, S.; Arai, Y.; Ito, K.; Ogiso, B. Bone regeneration with systemic administration of lactoferrin in non-critical-sized rat calvarial bone defects. *J. Oral Sci.* **2013**, *55*, 343–348. [CrossRef]
16. Li, W.; Zhu, S.; Hu, J. Bone Regeneration Is Promoted by Orally Administered Bovine Lactoferrin in a Rabbit Tibial Distraction Osteogenesis Model. *Clin. Orthop. Relat. Res.* **2015**, *473*, 2383–2393. [CrossRef]
17. Takaoka, R.; Hikasa, Y.; Hayashi, K.; Tabata, Y. Bone Regeneration by Lactoferrin Released from a Gelatin Hydrogel. *J. Biomater. Sci. Polym. Ed.* **2011**, *22*, 1581–1589. [CrossRef]
18. Prakasam, M.; Locs, J.; Salma-Ancane, K.; Loca, D.; Largeteau, A.; Berzina-Cimdina, L. Fabrication, Properties and Applications of Dense Hydroxyapatite: A Review. *J. Funct. Biomater.* **2015**, *6*, 1099–1140. [CrossRef]
19. Tayyebi, S.; Mirjalili, F.; Samadi, H.; Nemati, A. A Review of Synthesis and Properties of Hydroxyapatite/Alumina Nano Composite Powder. *Chem. J.* **2015**, *5*, 20–28.
20. Haider, A.; Haider, S.; Han, S.S.; Kang, I.-K. Recent advances in the synthesis, functionalization and biomedical applications of hydroxyapatite: A review. *RSC Adv.* **2017**, *7*, 7442–7458. [CrossRef]

21. Kattimani, V.S.; Kondaka, S.; Lingamaneni, K.P. Hydroxyapatite—Past, Present, and Future in Bone Regeneration. *Bone Tissue Regen. Insights* **2016**, *7*. [CrossRef]
22. Wang, W.; Yeung, K.W. Bone grafts and biomaterials substitutes for bone defect repair: A review. *Bioact. Mater.* **2017**, *2*, 224–247. [CrossRef]
23. Maia, F.R.; Musson, D.S.; Naot, D.; Da Silva, L.P.; Bastos, A.R.; Costa, J.B.; Oliveira, J.M.; Correlo, V.M.; Reis, R.L.; Cornish, J. Differentiation of osteoclast precursors on gellan gum-based spongy-like hydrogels for bone tissue engineering. *Biomed. Mater.* **2018**, *13*, 035012. [CrossRef]
24. Manda, M.G.; da Silva, L.P.; Cerqueira, M.T.; Pereira, D.R.; Oliveira, M.B.; Mano, J.F.; Marques, A.P.; Oliveira, J.M.; Correlo, V.M.; Reis, R.L. Gellan gum-hydroxyapatite composite spongy-like hydrogels for bone tissue engineering. *J. Biomed. Mater. Res. Part A* **2018**, *106*, 479–490. [CrossRef]
25. Gantar, A.; Da Silva, L.P.; Oliveira, J.M.; Marques, A.P.; Correlo, V.M.; Novak, S.; Reis, R.L. Nanoparticulate bioactive-glass-reinforced gellan-gum hydrogels for bone-tissue engineering. *Mater. Sci. Eng. C* **2014**, *43*, 27–36. [CrossRef]
26. Cerqueira, M.T.; da Silva, L.P.; Correlo, V.M.; Reis, R.L.; Marques, A.P. Epidermis recreation in spongy-like hydrogels: New opportunities to explore epidermis-like analogues. *Mater. Today* **2015**, *18*, 468–469. [CrossRef]
27. Da Silva, L.P.; Cerqueira, M.T.; Sousa, R.A.; Reis, R.L.; Correlo, V.M.; Marques, A.P. Engineering cell-adhesive gellan gum spongy-like hydrogels for regenerative medicine purposes. *Acta Biomater.* **2014**, *10*, 4787–4797. [CrossRef]
28. Ohsugi, H.; Habuto, Y.; Honda, M.; Aizawa, M.; Kanzawa, N. Evaluation of the Anti-Bacterial Activity of a Novel Chelate-Setting Apatite Cement Containing Lactoferrin. *Key Eng. Mater.* **2013**, *529*, 187–191. [CrossRef]
29. Kim, S.E.; Lee, D.-W.; Yun, Y.-P.; Shim, K.-S.; Jeon, D.I.; Rhee, J.-K.; Park, K. Heparin-immobilized hydroxyapatite nanoparticles as a lactoferrin delivery system for improving osteogenic differentiation of adipose-derived stem cells. *Biomed. Mater.* **2016**, *11*, 25004. [CrossRef]
30. Dai, R.; Wang, Z.; Samanipour, R.; Koo, K.-I.; Kim, K. Adipose-Derived Stem Cells for Tissue Engineering and Regenerative Medicine Applications. *Stem Cells Int.* **2016**, *2016*, 6737345. [CrossRef]
31. Calabrese, G.; Giuffrida, R.; Forte, S.; Fabbi, C.; Figallo, E.; Salvatorelli, L.; Memeo, L.; Parenti, R.; Gulisano, M.; Gulino, R. Human adipose-derived mesenchymal stem cells seeded into a collagen-hydroxyapatite scaffold promote bone augmentation after implantation in the mouse. *Sci. Rep.* **2017**, *7*, 7110. [CrossRef]
32. Krishna, K.A.; Vishalakshi, B. Gellan gum-based novel composite hydrogel: Evaluation as adsorbent for cationic dyes. *J. Appl. Polym. Sci.* **2017**, *134*, 45527. [CrossRef]
33. Anghel, L.; Erhan, R. Structural Aspects of Lactoferrin and Serum Transferrin Observed by FTIR Spectroscopy. *Chem. J. Mold.* **2018**, *13*, 111–116.
34. Olimpio, R.M.C.; De Oliveira, M.; De Síbio, M.T.; Moretto, F.C.F.; Deprá, I.C.; Mathias, L.S.; Gonçalves, B.M.; Rodrigues, B.M.; Tilli, H.P.; Coscrato, V.E.; et al. Cell viability assessed in a reproducible model of human osteoblasts derived from human adipose-derived stem cells. *PLoS ONE* **2018**, *13*, e0194847. [CrossRef]
35. Chang, H.-I.; Wang, Y. Cell Responses to Surface and Architecture of Tissue Engineering Scaffolds. In *Regenerative Medicine and Tissue Engineering—Cells and Biomaterials*; Eberli, P.D., Ed.; InTechOpen: London, UK, 2011.
36. Deville, S.; Saiz, E.; Tomsia, A.P. Freeze casting of hydroxyapatite scaffolds for bone tissue engineering. *Biomaterials* **2006**, *27*, 5480–5489. [CrossRef]
37. Bobbert, F.S.L.; Zadpoor, A.A. Effects of bone substitute architecture and surface properties on cell response, angiogenesis, and structure of new bone. *J. Mater. Chem. B* **2017**, *5*, 6175–6192. [CrossRef]
38. Ryan, A.J.; Gleeson, J.P.; Matsiko, A.; Thompson, E.M.; O'Brien, F.J. Effect of different hydroxyapatite incorporation methods on the structural and biological properties of porous collagen scaffolds for bone repair. *J. Anat.* **2015**, *227*, 732–745. [CrossRef]
39. Dey, S.; Pal, S. Evaluation of Collagen-hydroxyapatite Scaffold for Bone Tissue Engineering. In *IFMBE Proceedings, Proceedings of the 13th International Conference on Biomedical Engineering, Singapore, 3–6 December 2008*; Springer: Berlin/Heidelberg, Germany, 2009.
40. Correlo, V.M.; Pinho, E.D.; Pashkuleva, I.; Bhattacharya, M.; Neves, N.M.; Reis, R.L. Water Absorption and Degradation Characteristics of Chitosan-Based Polyesters and Hydroxyapatite Composites. *Macromol. Biosci.* **2007**, *7*, 354–363. [CrossRef]

41. Montesi, M.; Panseri, S.; Iafisco, M.; Adamiano, A.; Tampieri, A. Effect of hydroxyapatite nanocrystals functionalized with lactoferrin in osteogenic differentiation of mesenchymal stem cells. *J. Biomed. Mater. Res. Part A* **2015**, *103*, 224–234. [CrossRef]
42. Cabañas, M.V.; Peña, J.; Román, J.; Ramírez-Santillán, C.; Matesanz, M.C.; Feito, M.J.; Portolés, M.T.; Vallet-Regí, M. Design of tunable protein-releasing nanoapatite/hydrogel scaffolds for hard tissue engineering. *Mater. Chem. Phys.* **2014**, *144*, 409–417. [CrossRef]

 © 2019 by the authors. Licensee MDPI, Basel, Switzerland. This article is an open access article distributed under the terms and conditions of the Creative Commons Attribution (CC BY) license (http://creativecommons.org/licenses/by/4.0/).

Article

In Vitro and In Vivo Evaluation of Starfish Bone-Derived β-Tricalcium Phosphate as a Bone Substitute Material

Haruka Ishida [1,2], Hisao Haniu [1,2,3,*], Akari Takeuchi [2,3], Katsuya Ueda [1,2], Mahoko Sano [1,3], Manabu Tanaka [4], Takashi Takizawa [4], Atsushi Sobajima [4], Takayuki Kamanaka [4] and Naoto Saito [1,2,3]

1. Institute for Biomedical Sciences, Interdisciplinary Cluster for Cutting Edge Research, Shinshu University, 3-1-1 Asahi, Matsumoto, Nagano 390-8621, Japan; 18hb401g@shinshu-u.ac.jp (H.I.); 19hb402j@shinshu-u.ac.jp (K.U.); 18bs212e@shinshu-u.ac.jp (M.S.); saitoko@shinshu-u.ac.jp (N.S.)
2. Department of Biomedical Engineering, Graduate School of Medicine, Science and Technology, Shinshu University, 3-1-1 Asahi, Matsumoto, Nagano 390-8621, Japan; taakari@shinshu-u.ac.jp
3. Department of Biomedical Engineering, Graduate School of Science and Technology, Shinshu University, 3-1-1 Asahi, Matsumoto, Nagano 390-8621, Japan
4. Department of Orthopaedic Surgery, Shinshu University School of Medicine, 3-1-1 Asahi, Matsumoto, Nagano 390-8621, Japan; m990054e@gmail.com (M.T.); takashitak@shinshu-u.ac.jp (T.T.); soba@shinshu-u.ac.jp (A.S.); kam17@shinshu-u.ac.jp (T.K.)
* Correspondence: hhaniu@shinshu-u.ac.jp; Tel.: +81-263-37-3555

Received: 18 April 2019; Accepted: 10 June 2019; Published: 11 June 2019

Abstract: We evaluated starfish-derived β-tricalcium phosphate (Sf-TCP) obtained by phosphatization of starfish-bone-derived porous calcium carbonate as a potential bone substitute material. The Sf-TCP had a communicating pore structure with a pore size of approximately 10 μm. Although the porosity of Sf-TCP was similar to that of Cerasorb M (CM)—a commercially available β-TCP bone filler—the specific surface area was roughly three times larger than that of CM. Observation by scanning electron microscopy showed that pores communicated to the inside of the Sf-TCP. Cell growth tests showed that Sf-TCP improved cell proliferation compared with CM. Cells grown on Sf-TCP showed stretched filopodia and adhered; cells migrated both to the surface and into pores. In vivo, vigorous tissue invasion into pores was observed in Sf-TCP, and more fibrous tissue was observed for Sf-TCP than CM. Moreover, capillary formation into pores was observed for Sf-TCP. Thus, Sf-TCP showed excellent biocompatibility in vitro and more vigorous bone formation in vivo, indicating the possible applications of this material as a bone substitute. In addition, our findings suggested that mimicking the microstructure derived from whole organisms may facilitate the development of superior artificial bone.

Keywords: starfish; calcium carbonate; porous calcium phosphate; β-tricalcium phosphate; bone substitute; angiogenesis

1. Introduction

Bone is a tissue with excellent regenerative ability; however, supplementation is necessary for reconstruction of bone defects that cannot be naturally healed due to fracture or tumor resection. Autologous bone grafting can be used to compensate for defects. In this method, grafts are harvested from a healthy part of the patient and transplanted into the defective part of the bone. Because the graft itself has bone-forming ability, the process after transplantation is usually effective. However, it may not be possible to obtain bone of the necessary amount, shape, and size, and there are other disadvantages, such as pain and deformation at the collection site. Additionally, problems such as infection and immune response arise when allogeneic bone grafting and heterogeneous bone grafting are used [1].

In order to overcome these problems, researchers have been attempting to develop artificial bone prosthetic materials showing good biocompatibility [2]. For example, a glass material that bonds to bones without showing foreign body reactions was reported by Hench et al. in 1971 [3]. Since then, the development of artificial bone has progressed due to the discovery of physiologically active materials, even inorganic materials, and the synthesis of hydroxyapatite (HAp) as a main component of bone inorganic matter and direct bonding to bone were described by Jarcho et al. [4] and Akao et al. [5]. In addition to HAp, β-tricalcium phosphate (β-TCP) has been extensively studied as a calcium phosphate-based biomaterial. β-TCP has higher solubility at neutral pH than HAp and is not only hydrolyzed in body fluids but is also biologically absorbed by osteoclasts, representing an effective bioabsorbable material that can be used as a bone substitute. Studies on β-TCP have been reported by Driskell et al. [6], and SynthoGraft, which was first developed in 1981, has been approved by the US Food and Drug Administration as an absorbable synthetic bone grafting material. Such bioceramics account for approximately 40% of total bone grafts [7].

Regardless of material, porous bodies with continuous pores are required for supporting osteoinductive factors and for binding, after implantation, with strong bone, cellular tissues, and blood vessels for invasion [8]. It is necessary to control pore size distribution and pore structure in order to achieve both secured pore structures that promote bone formation and strength when used as a filler material. Porous bodies have been produced by various methods [9,10]. However, no artificial bone with performance comparable to bone has been developed.

Roy and Linnehan have reported methods for converting coral-derived calcium carbonate into HAp as an artificial bone using biological materials [11]. Coral-derived HAp is a porous HAp with a pore size of 150 to 500 µm and has high biocompatibility [12]. This material has been commercialized as ProOsteon in the United States of America. However, coral harvesting is costly and associated with environmental problems.

The inorganic matter forming on starfish is composed of calcite granules containing Mg^{2+} and has a fine porous structure with a pore diameter of several tens of microns. Takeuchi et al. reported that this porous structure is converted into β-TCP containing Mg^{2+} by phosphatization of starfish [13].

Therefore, in this study, we evaluated the physical properties of starfish bone-derived β-TCP as a potential bone substitute material using MC3T3-E1 mouse calvaria-derived osteoblast-like cells and in vivo implantation into rat calvaria bone defects.

2. Materials and Methods

2.1. Ethics Statement

All experiments were carried out according to institutional guidelines for animal experimentation at Shinshu University School of Medicine. All protocols used in this study were reviewed and approved by the Division of Laboratory Animal Research (#280054). All surgery was performed under general anesthesia (intraperitoneal injection of sodium pentobarbital), and all efforts were made to minimize suffering. All rats were euthanized using isoflurane inhalation at the end of the study.

2.2. Preparation of Porous Calcium Phosphate Derived from Starfish Bone

Collection of starfish-bone-derived calcite and its phosphatization to β-TCP were carried out as previously described [13]. Organic substances were dissolved by immersing starfish (*Patiria pectinifera*) in a commercial bleach (sodium hypochlorite aqueous solution with a volume ratio of about 6%; Hiter®; Kao Corp., Tokyo, Japan). After dissolving organic matter, starfish bone was obtained by filtering, washing with ion-exchanged water, and drying at 60 °C for 24 h. Briefly, 1.5 g of starfish bone granules were phosphatized in a reaction vessel for hydrothermal treatment (inner volume 25 mL) with 20 mL of a 0.5 mol/L diammonium hydrogen phosphate aqueous solution [$(NH_4)_2HPO_4$, FUJIFILM Wako Pure Chemical Corp., Osaka, Japan], at 200 °C for 72 h. Phosphatization treatment was performed under more severe conditions than autoclave sterilization, so samples were considered sterile and treated accordingly. The product was then washed with ion-exchanged water and dried at 60 °C for 24 h.

The obtained phosphate-treated starfish-bone-derived β-TCP (Sf-TCP) was sieved at 150–500 μm to obtain materials with the same size as the β-TCP used in the clinical setting (Cerasorb M [CM]; curasan AG, Kleinostheim, Germany).

2.3. Observation of Surface Structures by Scanning Electron Microscopy (SEM)

The Sf-TCP was affixed on a brass sample holder using conductive carbon paste (DOTITE XC-12, JEOL Ltd., Tokyo, Japan), coated with osmium oxide (FUJIFILM Wako Pure Chemical Corp.) by an osmium coater (Neoc-AN; Meiwafosis, Tokyo, Japan) at 10 mA for 20 s, and then observed under a field emission-scanning electron microscope (FE-SEM, JSM-7600F; JEOL Ltd., Tokyo, Japan) at an accelerating voltage of 2.00 kV.

2.4. Measurement of Specific Surface Area

The specific surface area of Sf-TCP was measured in triplicate with a high-precision multisample gas adsorption amount measuring device (Autosorb®-iQ; Quantachrome Instruments, Kanagawa, Japan) using the nitrogen adsorption method. Before measuring, vacuum degassing was performed at 120 °C for 3 h. Specific surface area was calculated from the range of relative pressures of adsorption-desorption isotherms by 0.05 to 0.35 according to the BET theory formula, as follows:

Monomolecular adsorption:

$$v_m = v\left(1 - \frac{p}{p_0}\right) \quad (1)$$

v : Adsorption amount
$\frac{p}{p_0}$: Relative pressure
Surface area:

$$A_s = \left(\frac{v_m N a_m}{m}\right) \times 10^{-18} \quad (2)$$

N : Avogadro constant
a_m : Molecular occupied cross section
m : Molecular weight of adsorbate

2.5. Measurement of Porosity and the Most Frequent Pore Diameter

We then used MicrotracBEL (Osaka, Japan) to measure porosity and the most frequent pore diameter of Sf-TCP by the mercury intrusion method using an automatic mercury porosimeter (Pascal Model 140; Thermo Scientific, Tokyo, Japan). Density was measured by the helium gas replacement method using a true density measuring apparatus (BELPycno, MicrotracBEL, Osaka, Japan). Before measuring, vacuum degassing was performed at room temperature for 15 min. Each item was measured once. The measured samples were used for subsequent experiments.

2.6. Cell Growth Tests

Mouse calvaria-derived osteoblast-like cells, MC3T3-E1 (RIKEN Cell Bank, Tsukuba, Japan) were used as cultured cells. MC3T3-E1 cells were cultured in an incubator at 37 °C in an atmosphere containing 5% CO_2. αMEM medium supplemented with 10% fetal bovine serum and 1% antibiotic-antimycotic mixed solution was used for cell culture. For passaging of cells, cells were seeded into 10 cm dishes at a density of 4.0×10^4 cells/mL and subcultured twice a week.

For cell proliferation tests, Alamar Blue assay (Alamar Blue Cell Viability Reagent, Remel, Lenexa, KS, USA) was used. MC3T3-E1 cells were seeded into 96-well plates at 6.0×10^3 cells/well. After culturing for 24 h, medium was exchanged and used for experiments.

Sf-TCP and CM were added at 10 mg/well. The control wells contained cells only. As a sample blank, evaluation targets were added to medium and wells without cells were used. Those groups consisted of eight wells, each with two sample blanks.

After addition of evaluation targets, the cells were cultured for 24 h. For the Alamar Blue assay, Alamar Blue Cell Viability Reagent was added to 10% of the total volume and reacted for 1 h. The Alamar Blue assay was performed by reading the fluorescence intensity at an excitation wavelength of 535 nm and an emission wavelength of 590 nm using a Plate Reader (AF2200, Eppendorf, Hamburg, Germany).

2.7. Cell Observation on the Sf-TCP Surface

The Sf-TCP cultured with MC3T3-E1 cells for 24 h was removed with tweezers, fixed by freeze-drying in 2.5% glutaraldehyde (used after dilution of 70% glutaraldehyde, TAAB Laboratories Equipment, Berks, UK), 1% osmium solution (aqueous solution of osmic acid, Nisshin EM, Tokyo, Japan), and t-butyl alcohol (FUJIFILM Wako Pure Chemical Corp.) and observed under an FE-SEM at an accelerating voltage of 15.0 kV.

2.8. Implantation in a Rat Calvaria Defect Model

Experimental animals were male Wistar rats (8 weeks old, weighing 150–200 g, SLC, Hamamatsu, Japan). According to animal experiment guidelines, rats were housed at five rats per cage in a breeding room with controlled room temperature (25 °C ± 2 °C) and humidity (50% ± 10%). Food and water were available ad libitum.

In accordance with the methods described by Tanaka et al. [14], a rat calvaria defect model was established. Briefly, after induction of anesthesia by isoflurane (Forane, ABBOTT JAPAN, Tokyo, Japan) inhalation, pentobarbital (Somnopentyl, Kyoritsu Seiyaku, Tokyo, Japan) was subcutaneously injected at 40 mg/kg, and the operation was performed. Bone defects of 5 mm in diameter were made in the rat calvaria using a trephine bar. No implant was placed in the sham (i.e., control) group, whereas the experimental groups were implanted with 10 mg Sf-TCP or CM. Rats were then housed for 4 or 8 weeks. There were eight rats in each group. Rats that died during the 4–8 week period were not used for evaluation.

2.9. Histological Examination

After euthanasia under anesthesia, the heads of rats were dissected and fixed for 1 week with 10% formalin (FUJIFILM Wako Pure Chemical Corp.). Rat skulls were then decalcified for 3 days in quick dehydrating liquid (K-CX, Pharma, Tokyo, Japan). Samples after demineralization were processed to an appropriate size and embedded in paraffin. After embedding in paraffin, samples were sliced into sections (4 μm thick) using a microtome. Sections were then subjected to hematoxylin (Muto Pure Chemicals, Tokyo, Japan) and eosin (FUJIFILM Wako Pure Chemical Corp.) staining (HE staining) and Masson's Trichrome (MT) staining (Muto Pure Chemicals) and observed with an optical microscope (BX50, Olympus, Tokyo, Japan).

MT-stained specimens were observed with a multispectral automatic tissue section quantitative analysis system (Vectra3, PerkinElmer, Waltham, MA, USA) and photographed as multispectral images. Upon quantification, dye used for MT staining was incorporated and analyzed with inForm (PerkinElmer). For each specimen, we selected and photographed the inForm analysis part manually such that the area occupied by CM and Sf-TCP was maximized within the photograph range of 250 μm × 334 μm defined by Vectra3. The threshold for quantitative analysis was set automatically. Analysis results quantified by pixel number were averaged for each group and calculated for the area of fibrous tissue and cells in the bone defect area.

2.10. Statistical Analysis

For statistical analysis, Student's t-tests were used, and for multiple comparisons, Bonferroni corrections were performed. Statistical results were expressed as means ± standard errors (SEs). The significance level was set at $p < 0.05$.

3. Results

3.1. Size of CM and Sf-TCP

Figure 1 shows photographs of CM, as shown in Figure 1a, and Sf-TCP, as shown in Figure 1b, after sieving according to the catalog size of CM. The catalog CM size was 150–500 μm, but some particles of less than 150 μm were present.

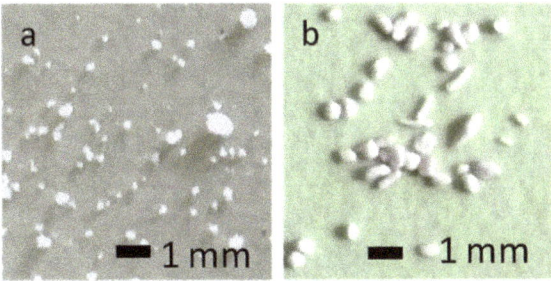

Figure 1. Photographs of the analyzed materials. (**a**) Cerasorb M (CM). (**b**) starfish-derived β-tricalcium phosphate (Sf-TCP) after sieving.

3.2. Surface Structure

Figure 2 shows ×1000 SEM images of CM, as shown in Figure 2a, and Sf-TCP, as shown in Figure 2b. In CM, pores of various shapes were observed on the surface, as shown by the arrows in Figure 2a. In Sf-TCP, pores were found all over the surface, and the structures were connected with other pores inside.

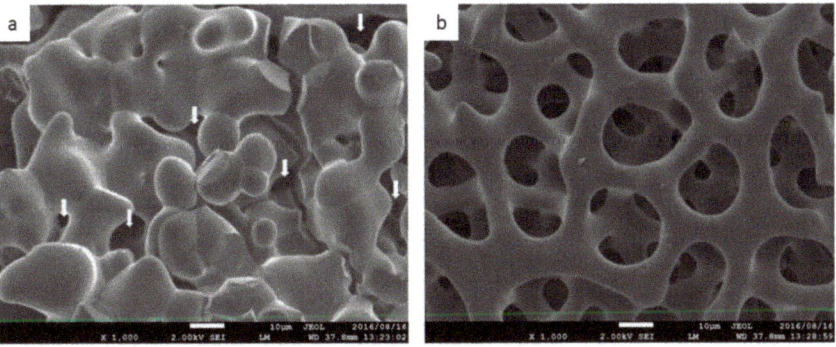

Figure 2. Surface images obtained by scanning electron microscopy. Magnification: ×1000. (**a**) CM. Arrows indicate pores. (**b**) Sf-TCP. Scale bar: 10 μm.

3.3. Density, Most Frequent Pore Diameter, and Porosity

Table 1 shows measurement results of density and porosity. The density of the material was approximately 3 g/cm^3. The total pore volume and total pore surface area for Sf-TCP were 445.42 mm^3/g and 0.130 m²/g, respectively, whereas those for CM were 415.31 mm³/g and 0.062 m²/g, respectively. The Sf-TCP pore surface area was approximately twice that of CM. CM had a median pore diameter of about 57 μm and a most frequent pore diameter of about 100 μm, representing a 2-fold difference. For Sf-TCP, the median diameter and most frequent pore diameter were about 10–12 μm, which showed almost no difference. Additionally, pores that excluded mercury were observed in Sf-TCP in porosity measurements by mercury intrusion.

Table 1. Density, most frequent pore diameter, and porosity.

Sample Name	CM	Sf-TCP
Sample weight (g)	0.1960	0.2061
Density (g/cm³)	3.0665	2.9158
Total pore volume (mm³/g)	415.31	445.42
Total pore surface area (m²/g)	0.062	0.130
Median pore diameter (μm)	57.86	12.37
Most frequent pore diameter (μm)	99.78	10.22
Porosity (mercury penetration was possible) (%)	56.42	52.44
Porosity (mercury penetration was not possible) (%)	−0.72	7.18
Total porosity (%)	55.70	59.62

3.4. Specific Surface Area

Figure 3 shows the adsorption-desorption isotherm resulting from measurement of CM, as shown in Figure 3a, and Sf-TCP, as shown in Figure 3b, using the nitrogen adsorption method.

The calculated specific surface area of CM was 3.693 m²/g, and that of Sf-TCP was 9.676 m²/g; thus, Sf-TCP was about three times larger than CM.

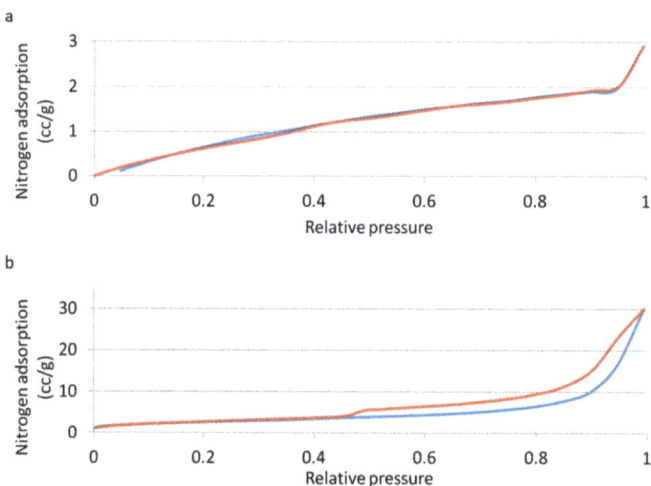

Figure 3. Adsorption-desorption isotherms. The blue line indicates adsorption, and the red line indicates desorption. (**a**) CM; (**b**) Sf-TCP.

From the shape and hysteresis of the materials with a relative pressure range of adsorption-desorption isotherms between 0.45 and 1.00, the pore structure of the surface could be identified by IUPAC classifications [15]. CM was found to be a type II material, having macropores with a diameter of 50 nm or more or no pores, whereas Sf-TCP was a type V material having mesopores with a diameter of 2–50 nm.

3.5. Cell Growth Test

Figure 4 shows the results of the Alamar Blue assay as a ratio of the control when 10 mg Sf-TCP or CM was added. The positive control CM showed no significant differences compared with the control group. In contrast, Sf-TCP-treated cells showed more than a 1.5-fold increase compared with the control and CM groups.

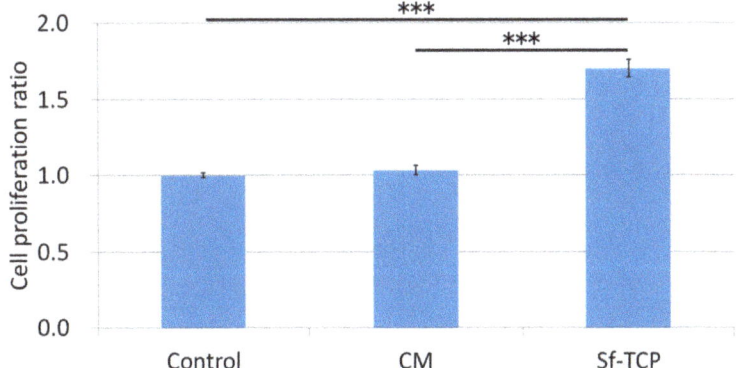

Figure 4. Effects of CM and Sf-TCP on cell proliferation. Values are the means ± standard errors (SEs) ($n = 6$). ***$P < 0.001$.

3.6. Observation of Cells on the Sf-TCP Surface

Figure 5 shows ×3000 SEM images of Sf-TCP after culture with MC3T3-E1 cells. A scaly microstructure was observed on the Sf-TCP surface. MC3T3-E1 cells adhered not only to the Sf-TCP skeleton but also to pores with stretched filopodia.

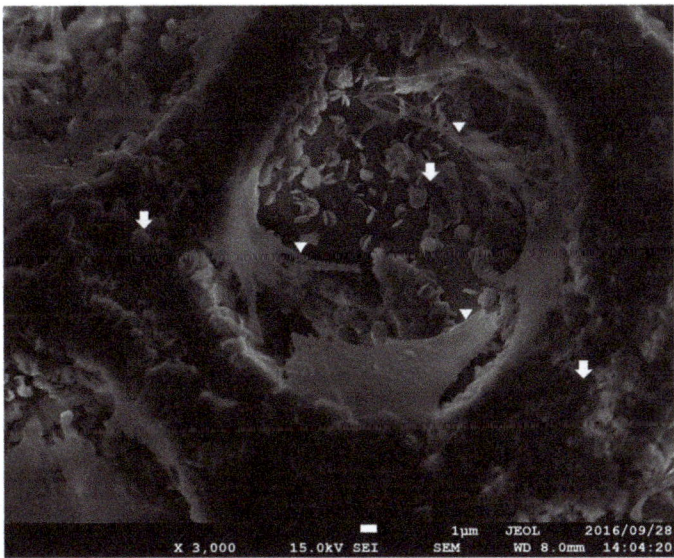

Figure 5. Surface SEM image of Sf-TCP cultured with MC3T3-E1 cells. Magnification: ×3000. The arrows show the scaly structure on the surface, and the triangles show cell filopodia.

3.7. Histological Examination of HE-Stained Specimens

Figure 6 shows photographs of HE-stained specimens. In the sham group, there was a thin tissue layer covering the bone defect. In contrast, in the CM group, the amount of tissue present in CM was nearly unchanged from 4 to 8 weeks. In the Sf-TCP group, tissue was already present in the porous structure of Sf-TCP at 4 weeks.

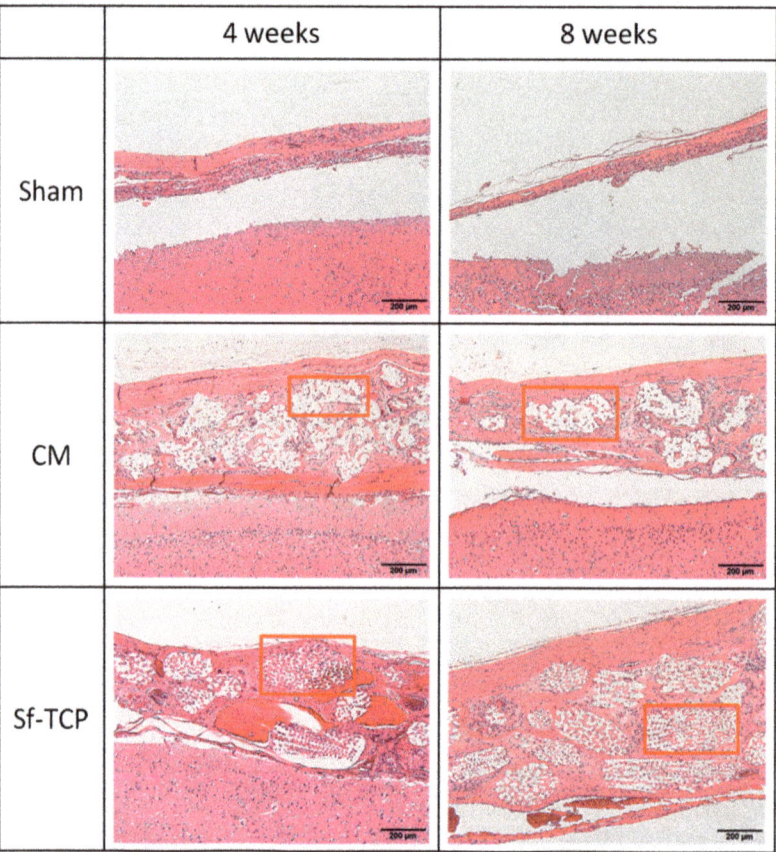

Figure 6. Hematoxylin and eosin (HE) staining. Hematoxylin stains cell nuclei blue, while eosin stains the cytoplasm, connective tissue, and other extracellular substances pink or red. Brain tissues are shown in the lower portion of the images, and the outer side of the skull is shown in the upper portion of the images. Boxed parts are prosthetic materials. Scale bar: 200 μm.

Notably, vascular structures were observed at 8 weeks after implantation in the Sf-TCP group, as shown in Figure 7. Capillary structures were observed only in the Sf-TCP group.

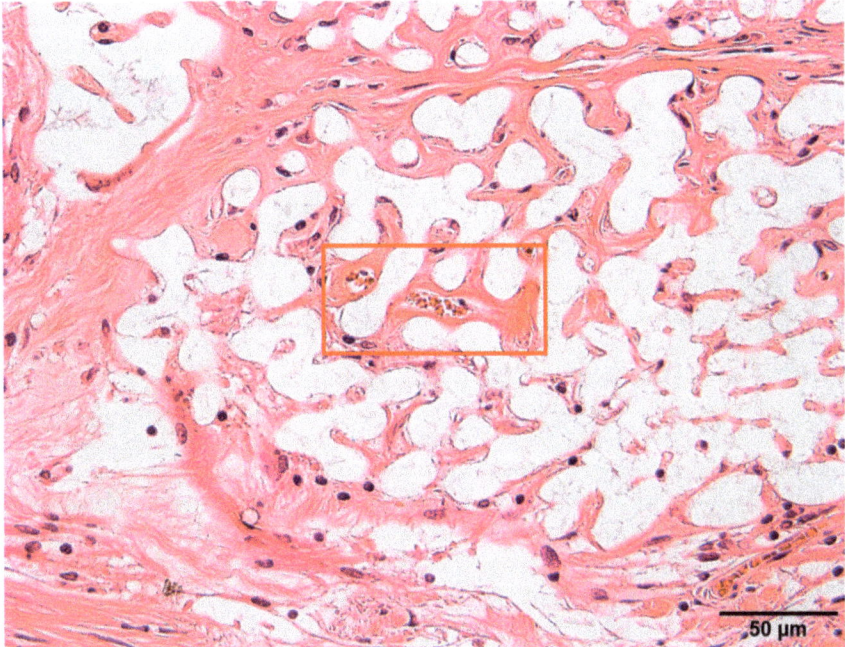

Figure 7. Blood vessel structures (red boxed area) were identified by hematoxylin and eosin (HE) staining.

3.8. Observation and Quantification of MT-Stained Specimens

Figure 8 shows a multispectral image of MT-stained specimens in the CM and Sf-TCP groups. Aniline blue stains fibrous tissue blue, Masson liquid stains cells red, and Orange G stains blood cells orange. For quantitative analysis with inForm, CM and Sf-TCP existed in the bone defects.

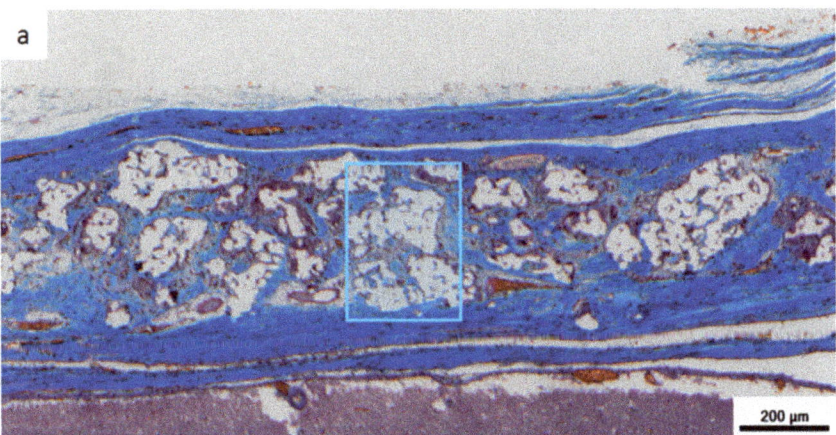

Figure 8. *Cont.*

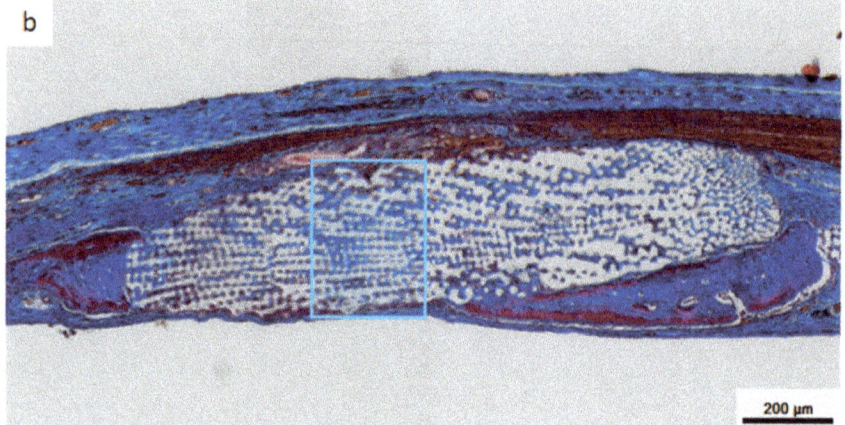

Figure 8. Masson's trichrome (MT)-stained images taken with Vectra3. Aniline blue stains fibrous tissue blue, Masson liquid stains cells red, and Orange G stains blood cells orange. (**a**) CM implant at 4 weeks. (**b**) Sf-TCP implant at 4 weeks. The blue boxes indicate the area quantified in Figure 10.

Figure 9 shows an enlarged image of the region selected in Figure 8 and an image obtained by analyzing the corresponding portion with inForm. Cells and fibrous substances existed mainly in the periphery of CM in the CM group at 4 and 8 weeks, whereas in the Sf-TCP group, cells and fibers entered into the pores of Sf-TCP.

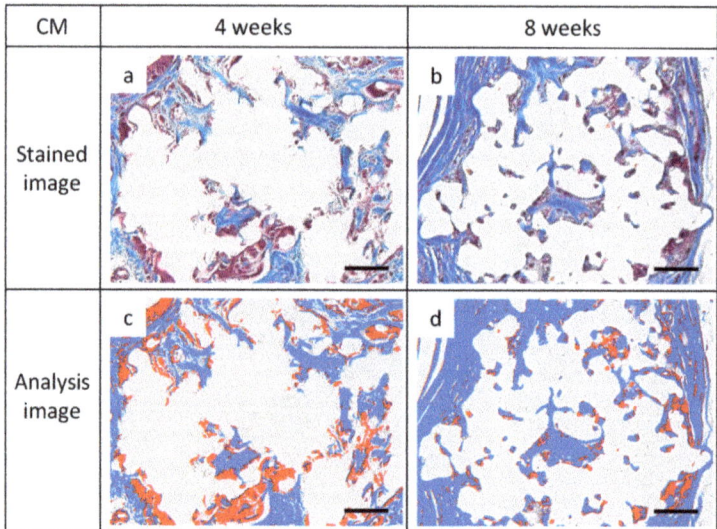

Figure 9. *Cont.*

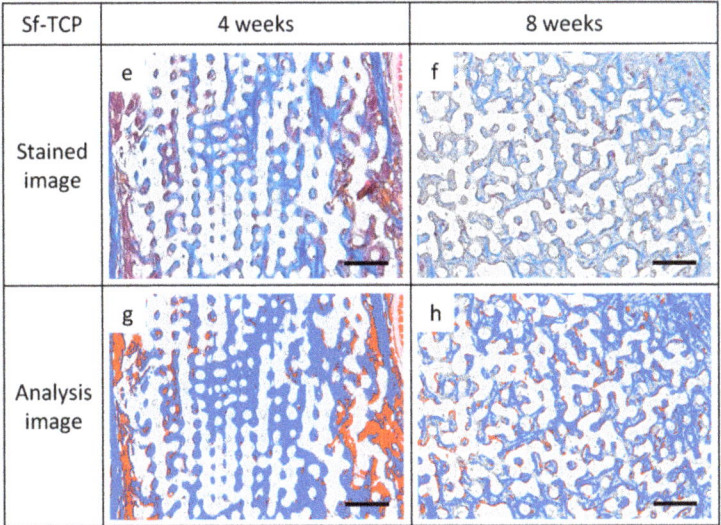

Figure 9. Images of Masson's trichrome (MT)-stained sections and analysis images for quantification. Representative images showing cells and fibrous tissue existing mainly around CM or Sf-TCP pores. Cells appear more uniformly scattered at 8 weeks than at 4 weeks. (**a,b**) CM images with MT staining. (**c,d**) CM images for analysis. (**e,f**) Sf-TCP images with MT staining. (**g,h**) Sf-TCP images for analysis. Aniline blue staining (blue) shows fibrous tissue, and Masson liquid and Orange G staining (red) show cells. Scale bar: 50 µm.

Figure 10 shows the results of quantifying areas of fibrous tissues and cells. Fiber and cell areas in the CM and Sf-TCP groups were compared at 4 and 8 weeks. The results showed that the fiber mass was significantly larger in the Sf-TCP group than in the CM group at both 4 and 8 weeks. However, actual cell amount did not change significantly.

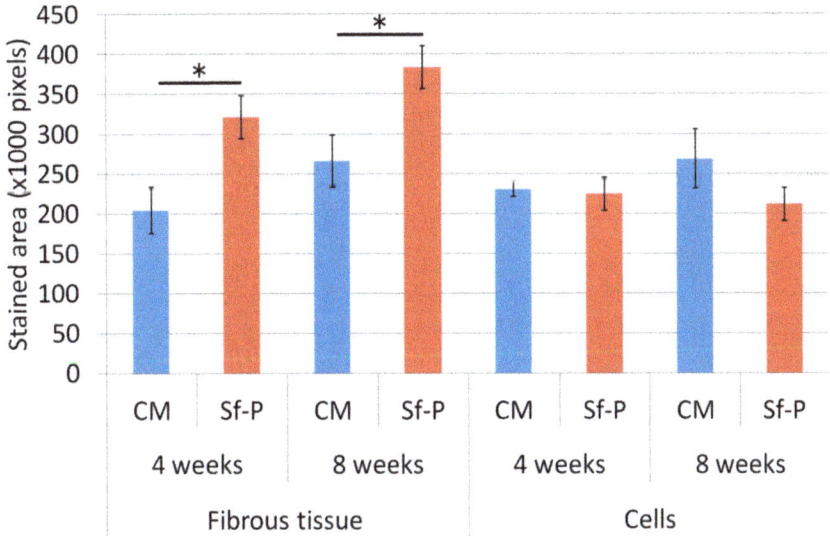

Figure 10. Quantitative analysis using inForm. The numbers of samples used for analysis were five in the Sf-TCP group at 4 weeks and six in other groups. Mean ± SEs. * $P < 0.05$.

Figure 11 shows images of erythrocytes stained with Orange G inside the implant at the time of observation, and Table 2 shows the number of individuals exhibiting erythrocytes in the prosthetic material. These results were observed only in the Sf-TCP group (one of five animals at 4 weeks and three of six animals at 8 weeks).

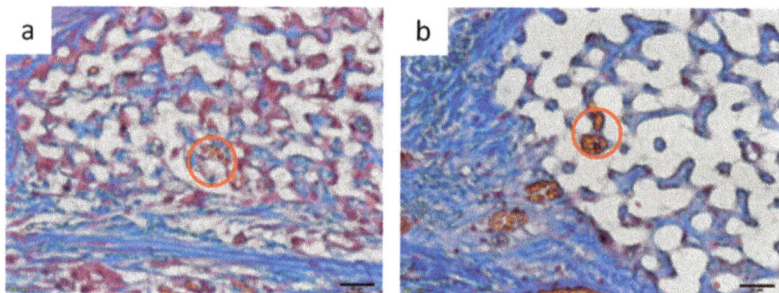

Figure 11. Image of erythrocytes observed inside Sf-TCP. (**a**) 4 weeks. (**b**) 8 weeks. Red circles indicate erythrocytes observed in Sf-TCP. Scale bar: 20 µm.

Table 2. Number of individuals having erythrocytes in prosthetic material (number/total).

Material	4 Weeks	8 Weeks
CM	0/6	0/6
Sf-TCP	1/5	3/6

4. Discussion

In this study, we evaluated the physical properties of Sf-TCP and performed in vitro and in vivo experiments to examine the potential applications of Sf-TCP as a bone substitute material. Our results showed that St-TCP did not induce cytotoxicity, which is important for the development of biomaterials, and facilitated the healing of wounds in vivo, supporting the potential clinical applications of this material.

Our results showed that Sf-TCP induced significant improvement in cell proliferation compared with the control or CM treatments. These findings were thought to be related to the excellent cell adhesion and proliferation properties of Sf-TCP. Indeed, the specific surface area of Sf-TCP was about three times larger than that of CM, consistent with studies demonstrating that cell proliferation increases as the specific surface area increases [16,17]. Because cell proliferation is a main factor indicating biocompatibility in MC3T3-E1 cells [18], our results supported that Sf-TCP was highly biocompatible. Additionally, in morphological observations, MC3T3-E1 cells adhering to Sf-TCP showed well-stretched filopodia, which has been reported to occur during cell migration [19], and the structure of Sf-TCP indicated that the cells could migrate.

In our in vivo experiments, cells and fibrous tissues invaded into the communicating porous structures of Sf-TCP in MT-stained specimens, indicating that cellular tissues could penetrate Sf-TCP, even when the pores were 10–12 µm in size. In addition, increased fibrous tissue area was observed in the Sf-TCP group compared with that in the CM group. The amount of fibrous material has been reported to be related to the abundance of collagen fibers and calcification [20,21]. Thus, the large amount of fibers present in Sf-TCP may indicate progression of bone formation. In porous artificial bone prosthetic materials, new bone invades into the pores and binds to surrounding bone [11]; accordingly, our findings supported the performance of Sf-TCP as a bone substitute material.

In the Sf-TCP group, vascular structures were observed in HE-stained specimens, and erythrocytes were present inside Sf-TCP in MT-stained specimens, suggesting the formation of capillary vessels. The pore size of Sf-TCP (10–12 µm) was smaller than that of previously reported biomaterials (100 µm), which allows invasion by cells and tissues, or that (300 µm) required for angiogenesis, e.g., invasion of capillaries [22]. However, the pore sizes of HAp and β-TCP are typically within the range of 100 to 600 μm (or 150 to 200 μm when coral is used as a biological material) [23–26]. Therefore, it is possible that few studies have evaluated such small pores or that angiogenesis was not observed in CMs, which have pores of 50–100 μm because of differences between artificial and biologically-derived porous structures. Various factors, such as disconnection of communicability by closed pores, variations in pore size, and sharp connecting structures with respect to the size of cells in artificial communication pores, are thought to hinder invasion of cellular tissues. However, in Sf-TCP, there were few of these structures and few elements that inhibited the invasion of cellular tissues; thus, vigorous tissue invasion and formation of capillary vessels would have occurred. In the rat calvaria defect model, capillary vessels are approximately 4–10 μm in diameter in the cerebrum around the defect, and capillaries can invade pores with a diameter of 10–12 μm. This result is important in considering the three-dimensional structure of the potential prosthetic bone filling material. Moreover, artificial preparation of starfish bone-like structures adapted to the capillary diameters of humans may yield artificial bone prosthetic materials capable of invading capillaries and tissues while improving the strength of the material by decreasing the pore size. Additionally, pore sizes increase when β-TCP is dissolved, absorbed, and replaced with autologous bone, and tissues larger than the pore may therefore invade the material.

5. Conclusions

In this study, we evaluated whether Sf-TCP derived from starfish could be used as a bone substitute material. Compared with CM, which is already used clinically, Sf-TCP was found to have comparable porosity but a larger specific surface area. Cell proliferation tests showed that Sf-TCP promoted cell proliferation. Moreover, in animal experiments, Sf-TCP showed high regeneration ability, vigorous invasion of cellular tissue into the pore structure, and introduction of capillary vessels into the prosthetic material, even when the pore size was only approximately 10 μm. However, this study did not examine the pore size required for angiogenesis, so there may be a more appropriate pore size. In the future, we hope to optimize regeneration accompanied by angiogenesis by artificially creating materials with structures similar to that of Sf-TCP but with different pore diameters.

From the above results, we concluded that Sf-TCP obtained by phosphatization of starfish bone may be effective for bone regeneration applications, such as in the treatment of fractures and bone loss. In addition, mimicking the structure of the starfish bone may lead to the development of superior artificial bone substitute materials.

Author Contributions: Conceptualization, H.H. and A.T.; Methodology, H.I., H.H., A.T., K.U., M.S. and M.T.; Formal Analysis, H.I.; Investigation, H.I.; Resources, H.H., A.T. and N.S.; Data Curation, H.I., K.U., M.S., M.T., T.T., A.S. and T.K.; Writing-Original Draft Preparation, H.I.; Writing-Review & Editing, H.H., A.T. and M.S.; Visualization, H.I.; Supervision, H.H., A.T. and N.S.; Project Administration, H.H.

Funding: This research received no external funding.

Acknowledgments: We thank MicrotracBEL Corp. for measurement of the density, most frequent pore diameter, and porosity of Sf-TCP. We thank Ms. Kayo Suzuki (Research Center for Human and Environmental Sciences, Shinshu University) for technical assistance in SEM analysis. We thank Takeuchi laboratory members for their assistance in the preparation of Sf-TCP. We would like to thank Editage (www.editage.jp) for English language editing.

Conflicts of Interest: The authors state that there are no conflicts of interest with regard to the publication of this paper.

References

1. Giannoudis, P.V.; Dinopoulos, H.; Tsiridis, E. Bone substitutes: An update. *Injury* **2005**, *36*, S20–S27. [CrossRef] [PubMed]
2. Bose, S.; Roy, M.; Bandyopadhyay, A. Recent advances in bone tissue engineering scaffolds. *Trends Biotechnol.* **2012**, *30*, 546–554. [CrossRef] [PubMed]
3. Henchel, L.L.; Splinter, R.J.; Allen, W.C.; Greenlee, T.K. Bonding mechanisms at the interface of ceramic prosthetic materials. *J. Biomed. Mater. Res.* **1971**, *5*, 117–141. [CrossRef]

4. Jarcho, M.; Bolen, C.H.; Thomas, M.B.; Bobick, J.; Kay, J.F.; Doremus, R.H. Hydroxylapatite synthesis and characterization in dense polycrystalline form. *J. Mater. Sci.* **1976**, *11*, 2027–2035. [CrossRef]
5. Akao, M.; Aoki, H.; Kato, K. Mechanical properties of sintered hydroxyapatite for prosthetic applications. *J. Mater. Sci.* **1981**, *16*, 809–812. [CrossRef]
6. Sings, S.A.; Bajpai, P.K.; Pantano, C.G.; Driskell, T.D. In vitro dissolution of Synthos ceramics in an acellular physiological environment. *Biomater. Med. Devices. Artif. Organs* **1979**, *7*, 183–190.
7. Urabe, K.; Itoman, M.; Toyama, Y.; Yanase, Y.; Iwamoto, Y.; Ohgushi, H.; Ochi, M.; Takakura, Y.; Hachiya, Y.; Matsuzaki, H.; et al. Current trends in bone grafting and the issue of banked bone allografts based on the fourth nationwide survey of bone grafting status from 2000 to 2004. *J. Orthop. Sci.* **2007**, *12*, 520–525. [CrossRef]
8. Tripathi, G.; Basu, B. A porous hydroxyapatite scaffold for bone tissue engineering: Physico-mechanical and biological evaluations. *Ceram. Int.* **2012**, *38*, 341–349. [CrossRef]
9. Kitamura, M.; Ohtsuki, C.; Ogata, S.; Kamitakahara, M.; Tanihara, M.; Miyazaki, T. Fabrication of α-tricalcium phosphate porous ceramics with controlled pore structure. *J. Soc. Mat. Sci. Japan* **2004**, *53*, 594–598. [CrossRef]
10. Nakahira, A.; Tamai, M.; Sakamoto, K.; Yamaguchi, S. Sintering and microstructure of Porous Hydroxyapatite. *J. Ceram. Soc. Japan* **2000**, *108*, 99–104. [CrossRef]
11. Roy, D.M.; Linnehan, S.K. Hydroxyapatite formed from coral skeletal carbonate by hydrothermal exchange. *Nature* **1974**, *247*, 220–222. [CrossRef] [PubMed]
12. Anavi, Y.; Avishai, G.; Calderon, S.; Allon, D.M. Bone remodeling in onlay beta-tricalcium phosphate and coral grafts to rat calvaria. Microcomputerized tomography analysis. *J. Oral Implantol.* **2011**, *37*, 379–386. [CrossRef] [PubMed]
13. Takeushi, A.; Tsuge, T.; Kikuchi, M. Preparation of porous β-tricalcium phosphate using starfish-derived calcium carbonate as a precursor. *Ceram. Int.* **2016**, *42*, 15376–15382. [CrossRef]
14. Tanaka, M.; Haniu, H.; Kamanaka, T.; Takizawa, T.; Sobajima, A.; Yoshida, K.; Aoki, K.; Okamoto, M.; Kato, H. Physico-chemical, in vitro, and in vivo evaluation of a 3d unidirectional porous hydroxyapatite scaffold for bone regeneration. *Materials* **2017**, *10*, 33. [CrossRef] [PubMed]
15. Sing, K.S.W. Reporting physisorption date for gas/solid systems with special reference to determination of surface area and porosity. *Pure Appl. Chem.* **1985**, *57*, 603–619. [CrossRef]
16. O'Brien, F.J.; Harley, B.A.; Yannas, I.V.; Gibson, L.J. The effect of pore size on cell adhesion in collagen-GAG scaffolds. *Biomaterials* **2005**, *26*, 433–441. [CrossRef] [PubMed]
17. Murphy, C.M.; Haugh, M.G.; O'Brien, F.J. The effect of mean pore size on cell attachment, proliferation and migration in collagen-glycosaminoglycan scaffolds for bone tissue engineering. *Biomaterials* **2010**, *31*, 461–466. [CrossRef]
18. Itakura, Y. In vitro study on biocompatibility of implant materials using mc3t3-e1 osteogenic cell line. *J. Japan Prosthodont. Soc.* **1987**, *31*, 21–29. [CrossRef]
19. Mitchison, T.J.; Cramer, L.P. Actin-based cell motility and cell locomotion. *Cell* **1996**, *84*, 371–379. [CrossRef]
20. Scott, B.L.; Pease, D.C. Electron microscopy of the epiphyseal apparatus. *Anat. Rec.* **1956**, *126*, 465–495. [CrossRef]
21. Robinson, R.A.; Cameron, D.A. Electron microscopy of cartilage and bone matrix at the distal epiphyseal line of the femur in the newborn infant. *J. Biophys. Biochem. Cytol.* **1956**, *2*, 253–263. [CrossRef] [PubMed]
22. Karageorgiou, V.; Kaplan, D. Porosity of 3D biomaterial scaffolds and osteogenesis. *Biomaterials* **2005**, *27*, 5474–5491. [CrossRef] [PubMed]
23. Tsuruga, E.; Takita, H.; Itoh, H.; Wakisaka, Y.; Kuboki, Y. Pore size of porous hydroxyapatite as the cell-substratum controls BMP-induced osteogenesis. *J. Biochem.* **1997**, *121*, 317–324, PMID:9089406. [CrossRef] [PubMed]
24. Kuboki, Y.; Jin, Q.; Takita, H. Geometry of carriers controlling phenotypic expression in BMP-induced osteogenesis and chondrogenesis. *J. Bone Joint Surg. Am.* **2001**, *83*, S101–S115, PMID:11314788. [CrossRef]

25. Zhang, C.; Wang, J.; Feng, H.; Lu, B.; Song, Z.; Zhang, X. Replacement of segmental bone defects using porous bioceramic cylinders: A biomechanical and X-ray diffraction study. *J. Biomed. Mater. Res.* **2001**, *54*, 407–411, PMID:11189048. [CrossRef]
26. Chen, F.; Mao, T.; Tao, K.; Chen, S.; Ding, G.; Gu, X. Bone graft in the shape of human mandibular condyle reconstruction via seeding marrow-derived osteoblasts into porous coral in a nude mice model. *J. Oral Maxillofac. Surg.* **2002**, *60*, 1155–1159. [CrossRef]

© 2019 by the authors. Licensee MDPI, Basel, Switzerland. This article is an open access article distributed under the terms and conditions of the Creative Commons Attribution (CC BY) license (http://creativecommons.org/licenses/by/4.0/).

Article

Mechanobiological Approach to Design and Optimize Bone Tissue Scaffolds 3D Printed with Fused Deposition Modeling: A Feasibility Study

Gianluca Percoco, Antonio Emmanuele Uva, Michele Fiorentino [ID], Michele Gattullo [ID], Vito Modesto Manghisi [ID] and Antonio Boccaccio *

Dipartimento di Meccanica, Matematica e Management, Politecnico di Bari, Via E. Orabona 4, 70126 Bari, Italy; gianluca.percoco@poliba.it (G.P.); antonio.uva@poliba.it (A.E.U.); michele.fiorentino@poliba.it (M.F.); michele.gattullo@poliba.it (M.G.); vitomodesto.manghisi@poliba.it (V.M.M.)
* Correspondence: antonio.boccaccio@poliba.it; Tel.: +39-080-5963393

Received: 21 December 2019; Accepted: 28 January 2020; Published: 1 February 2020

Abstract: In spite of the rather large use of the fused deposition modeling (FDM) technique for the fabrication of scaffolds, no studies are reported in the literature that optimize the geometry of such scaffold types based on mechanobiological criteria. We implemented a mechanobiology-based optimization algorithm to determine the optimal distance between the strands in cylindrical scaffolds subjected to compression. The optimized scaffolds were then 3D printed with the FDM technique and successively measured. We found that the difference between the optimized distances and the average measured ones never exceeded 8.27% of the optimized distance. However, we found that large fabrication errors are made on the filament diameter when the filament diameter to be realized differs significantly with respect to the diameter of the nozzle utilized for the extrusion. This feasibility study demonstrated that the FDM technique is suitable to build accurate scaffold samples only in the cases where the strand diameter is close to the nozzle diameter. Conversely, when a large difference exists, large fabrication errors can be committed on the diameter of the filaments. In general, the scaffolds realized with the FDM technique were predicted to stimulate the formation of amounts of bone smaller than those that can be obtained with other regular beam-based scaffolds.

Keywords: tissue engineering; biomaterials; mechanobiology; scaffold design; geometry optimization

1. Introduction

The development of the recent additive manufacturing techniques and, consequently, the possibility of building constructs with very sophisticated geometries, led many researchers to investigate the scaffold geometries that mostly favor the formation of bone in the shortest time. To this purpose, both regular and irregular scaffold geometries were proposed and investigated. The regular scaffolds include unit cell configurations all with the same shape and dimensions that are regularly replicated in the scaffold volume [1,2]. Analytical solutions were recently developed that put in relationship the equivalent material properties of the entire scaffold with the dimensions and the material properties of the single unit cell [3–6]. The irregular ones include pores differently shaped and dimensioned and present a geometry that can be described with statistical parameters [7–9] but not in a precise form [10]. Although irregular geometry scaffolds are commonly utilized in experiments for bone tissue engineering, regular scaffolds have gained importance in recent decades as they allow precise control of the actual geometry and are, hence, suited to create repeatable physical environments and easier to investigate [5]. Among the possible geometries of regular scaffolds, the geometry realized by means of the fused deposition modeling technique certainly represents an important solution.

The fabrication of scaffolds with additive manufacturing techniques is an issue that recently received recognition and attention from the research community [11,12]. Different rapid prototyping techniques were proposed to fabricate biomaterials scaffolds, based on liquid polymerization [13], material deposition processes [14], powder-based processes [15], sheet lamination [16], binder jetting [17], and material jetting [18]. Among the other rapid prototyping techniques, fused deposition modeling (FDM) is one of the most common 3D printing technologies available on the market, thanks to its low cost, ease of use, and the variety of usable materials. It consists in depositing layers of a polymeric, ceramic, or metallic material; each layer includes cylindrical strands all oriented in a given direction and equally spaced. To guarantee an adequate structural response, the orientation of filaments changes layer by layer. Furthermore, between the cylindrical filaments of adjacent layers, an overlap region exists where the single strand is melted with the adjacent one during the deposition process. Recently, this technique has been successfully utilized to fabricate scaffolds for bone tissue engineering [19–23]. Process parameters of the FDM technique were also optimized to improve the dimensional accuracy of the manufactured components [24]. The typical/traditional approach consists in using the FDM technique by implementing standard process parameters that allow obtaining accurate structures. However, adjusting the strand diameter, and modifying the process parameters is a challenging task and still remains a research topic. In fact, when small modifications on the scaffold geometry must be achieved, acting on process parameters of the traditional FDM technique can be a valid option instead of looking for expensive, more accurate technologies. A recent study proposed a numerical model to simulate the extrusion of a strand of semi-molten material to investigate how the strand cross-section changes for variable process parameters [25]. The porosity and the micro-architecture of parts built with the FDM technique appear very suited to stimulate the colonization and subsequent differentiation of mesenchymal stem cells [26].

Optimization algorithms were implemented to improve the scaffold performance and to minimize the negative effects related to the implantation of the scaffold in the fracture site [27]. Many objective functions were investigated and different optimization strategies were implemented [28–31]. Most of the optimization studies reported in the literature aim to minimize the difference between the equivalent mechanical properties of the scaffold and those of the tissue within which it is implanted. In one word, such studies that consider scaffold stiffness as the design variable aim to minimize the effects of stress shielding at the bone/scaffold interface [32,33]. Optimization strategies based on the compressive modulus expressed as a ratio between third principal stress and the prescribed compressive strain were also utilized to optimize the geometry of scaffolds fabricated with the FDM technique [34]. Singh et al. have proposed a multifactor optimization for the development of biocompatible and biodegradable composite material-based feedstock filament of fused deposition modeling [35]. Only recently, optimization algorithms based on mechanobiological criteria were proposed to design and optimize small volumes of scaffolds with both, regular [36–40] and irregular [41] micro-geometry [42]. In these mechanobiology-based optimization algorithms, the scaffold geometry is perturbed until the micro-architecture that allows maximizing the formation of new bone, is identified. However, to the knowledge of the present authors, no studies are reported in the literature that optimize the geometry of scaffolds fabricated with the FDM technique and that are based on mechanobiological criteria. In this study, we want to bridge this gap.

The objective of this feasibility study is to optimize the geometry of scaffolds fabricated with the FDM technique and to investigate whether this technique can be utilized to fabricate scaffolds designed and optimized to undergo a compression load. The objective function that was utilized is based on the computational mechano-regulation algorithm by Prendergast and Huiskes [43,44]. Such an algorithm models the fracture domain as a biphasic poroelastic material, and hypothesizes the biophysical stimulus that triggers the tissue differentiation process to be a function of the octahedral shear strain and of the fluid velocity. This mechanobiological algorithm was successfully utilized in previous studies investigating the healing process in fractured bones [45,46], in osteochondral defects [47], at the implant/bone interfaces [48] and the regeneration process in scaffolds for bone tissue engineering [49,50].

Other mechano-regulation computational models are reported in the literature, investigating the role of the mechanical environment on the biophysical stimulus that triggers the tissue differentiation process [51–54]. However, the patterns of tissue differentiation predicted by the model of Prendergast and Huiskes were shown to be closest to experimental results compared to other mechanobiological algorithms [55].

We determined, for different values of the filament diameter, the optimal distance between the strands that, for the specific load acting on the scaffold, can maximize the formation of bony tissue. The optimized scaffolds were physically fabricated and successively measured. By utilizing a CMM machine, the distance between the filaments and the filament diameter were measured with high accuracy and compared with the corresponding nominal values.

2. Materials and Methods

2.1. Parametric Finite Element Model

The parametric poroelastic finite element model of a cylindrical scaffold with radius $R = 20$ mm and $h = 5$ mm high was built in Abaqus® (version 6.12, Dassault Systèmes, Vélizy-Villacoublay, France) (Figure 1). Scaffolds with the same dimensions were utilized by Teng et al. [56]. The model consists of layers including aligned cylindrical strands with diameter D equally spaced. The filaments of two adjacent layers form an angle of 90° (Figure 1a). Five different values were considered for the diameter D: 400, 500, 600, 700, and 800 μm while the distance between the filaments d_{fil} was optimized via the optimization algorithm described below (Figure 2). Following Somireddy and Czekanski [57], between two adjacent layers, an overlap region of $0.1 \times D$ was hypothesized. The model of the tissue occupying the scaffold pores (highlighted in red in Figure 1) was built by means of Boolean operations of subtraction, from the entire model volume $V_{TOT} = \pi \times R^2 \times h$, the volume of the scaffold.

According to previous studies [38,58], the volume inside the pores was hypothesized to be occupied by granulation tissue. Exploiting the symmetry of the system, to reduce the computational cost, a one-quarter model was developed. The lower base of the scaffold was clamped while a compression load was applied on the upper surface by means of a rigid plate (highlighted in blue in Figure 1). Different values of the compression load F (Figure 1) were applied to the model, corresponding to the following values of force per unit surface $p = 0.2, 0.5, 1.0,$ and 1.5 MPa. Such values are consistent with those hypothesized in previous studies [36,58]. Symmetry constraints were applied on the lateral surfaces to simulate the continuity of the entire model (Figure 1). Poroelastic four-node tetrahedral finite elements C3D4P available in Abaqus were adopted to mesh the scaffold model (Figure 3). The model of the scaffold and of the granulation tissue included about 5M elements with 1M nodes.

The modeled scaffolds were physically fabricated via the FDM technique by utilizing the polylactic acid (PLA), a biodegradable thermoplastic polyester considered a bioplastic, possessing a Young's modulus of 2300 MPa [59]. This same value of Young's modulus was implemented in the finite element model of the scaffold and utilized in the optimization algorithm described below. The other material properties implemented for scaffold and granulation tissue are the same as those utilized in previous studies [36,40,60] and are listed in Table 1.

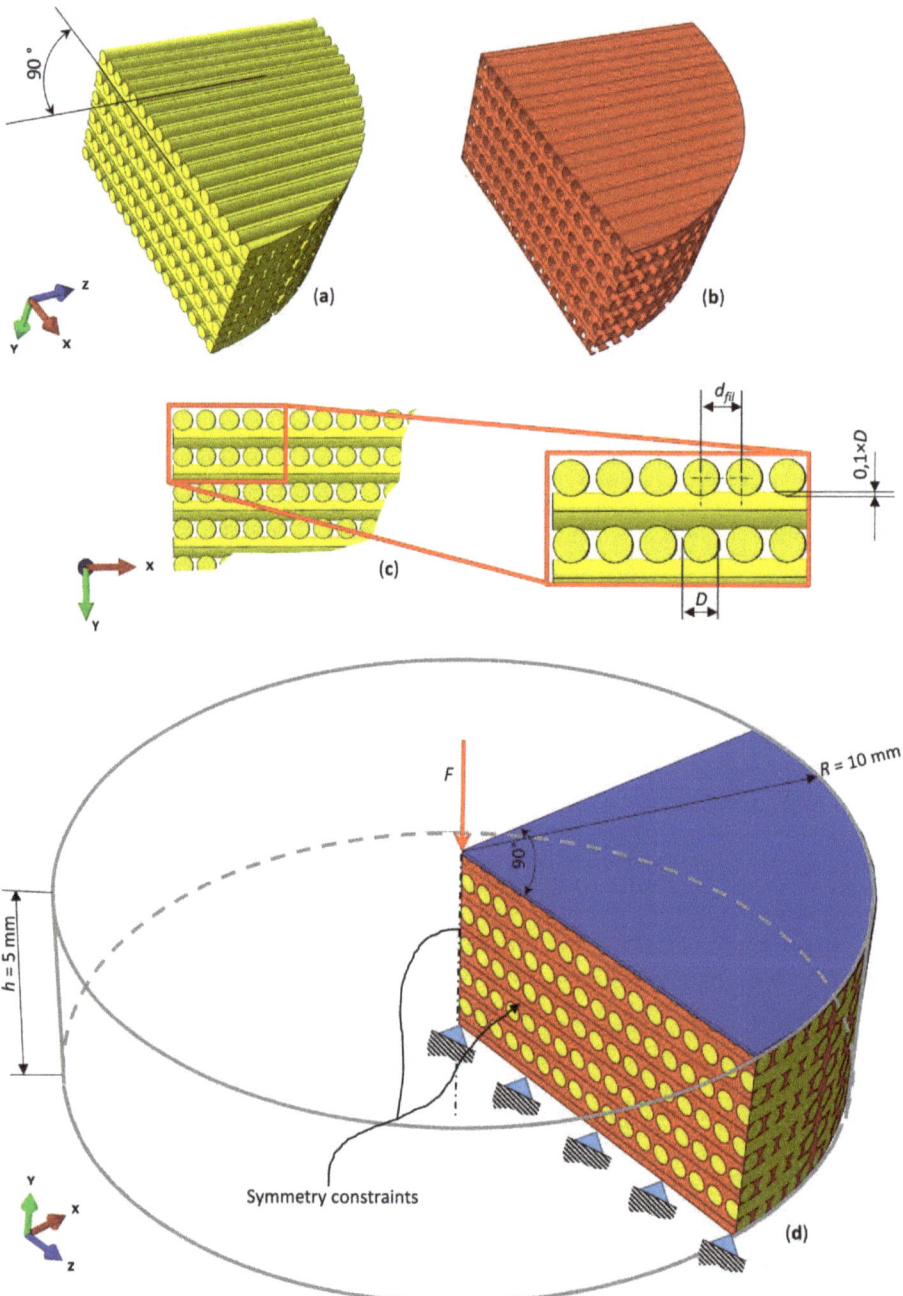

Figure 1. Parametric model of the scaffold (**a**,**c**) and of the granulation tissue (**b**) occupying the scaffold pores. (**d**) Exploiting the symmetry of the system, the one-quarter model was investigated.

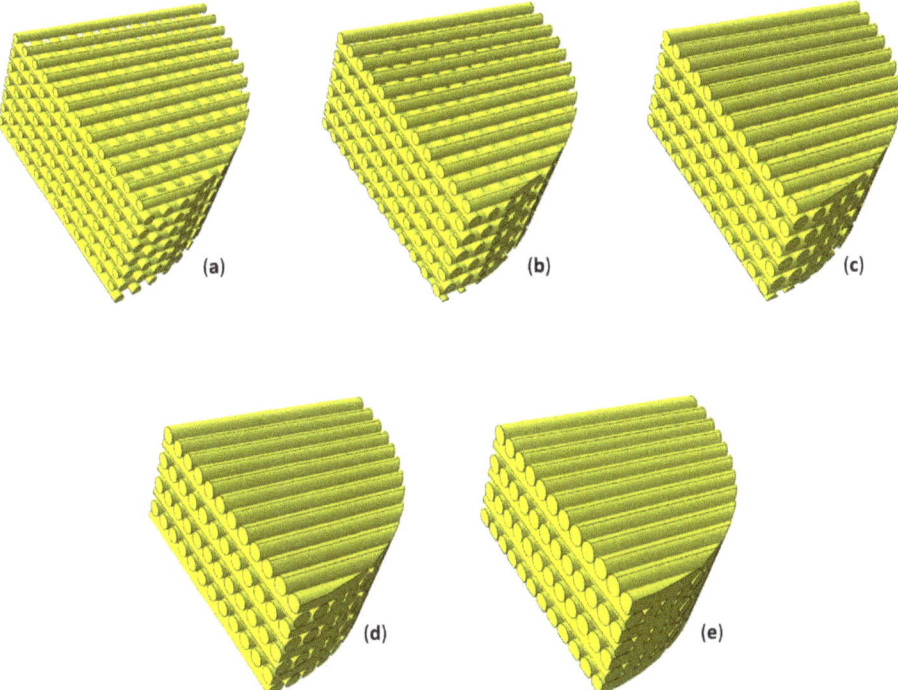

Figure 2. Models of scaffold investigated in the study including strands with diameter $D = 400$ µm (**a**), 500 µm (**b**), 600 µm (**c**), 700 µm (**d**), and 800 µm (**e**). The distance between the filaments was computed by means of the proposed optimization algorithm.

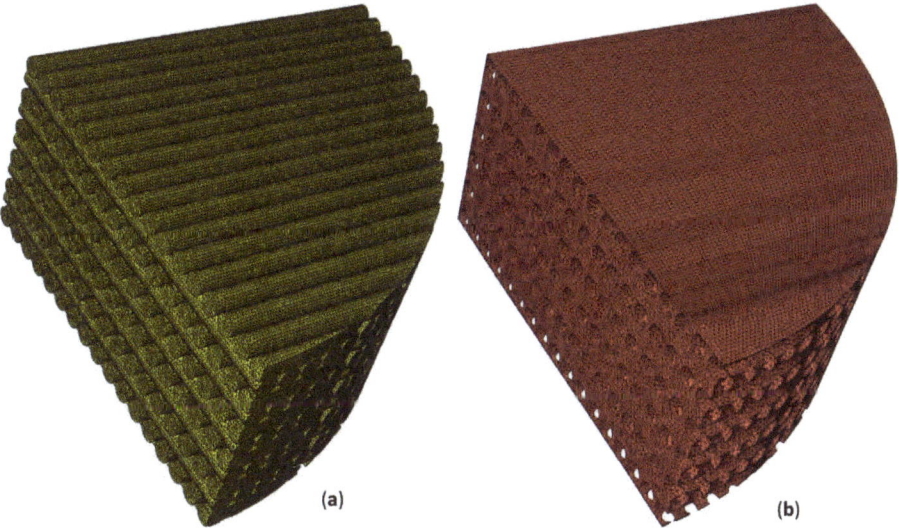

Figure 3. Finite element mesh utilized to model the scaffold (**a**) and the granulation tissue (**b**) including poroelastic four-node tetrahedral elements (C3D4P).

Table 1. Material properties utilized in the model of scaffold and granulation tissue

Material Properties	Scaffold	Granulation Tissue
Young's modulus (MPa)	2300	0.2
Poisson's ratio	0.3	0.167
Permeability (m^4/N/s)	1×10^{-14}	1×10^{-14}
Porosity	0.5	0.8
Bulk modulus grain (MPa)	13920	2300
Bulk modulus fluid (MPa)	2300	2300

2.2. Mechanobiological Model by Prendergast and Huiskes to Describe the Bone Regeneration Process inside the Scaffold

Once the scaffold is implanted in the anatomic region with bone deficiency, the mesenchymal stem cells (MSCs) migrate from the surrounding tissues and invade the scaffold pores. Based on the mechanical stimulus acting on them, MSCs will start to differentiate into different phenotypes. The mechanobiological model by Prendergast and Huiskes hypothesizes that the biophysical stimulus S that triggers the tissue differentiation process is a function of the octahedral shear strain γ and of the interstitial fluid flow v, i.e., the velocity with which the fluid flows through the solid phase, according to the relationship

$$S = \frac{\gamma}{a} + \frac{v}{b} \tag{1}$$

where a = 3.75 % and b = 3 μm/s are two empirical constants determined in a previous study [43]. In particular, the octahedral shear strain γ can be expressed in function of the principal strains ε_I, ε_{II} and ε_{III} as

$$\gamma = \frac{2}{3}\sqrt{(\varepsilon_I - \varepsilon_{II})^2 + (\varepsilon_{II} - \varepsilon_{III})^2 + (\varepsilon_I - \varepsilon_{III})^2} \tag{2}$$

Depending on the value of S, stem cells will be differentiated into the following phenotypes

$$\begin{cases} if\ S > 3 \rightarrow formation\ of\ fibroblasts\ (Fibrous\ tissue) \\ else\ if\ 1 < S < 3 \rightarrow formation\ of\ chondrocytes\ (Cartilage\ tissue) \\ else\ if\ 0.53 < S < 1 \rightarrow formation\ of\ osteoblasts\ (Immature\ bone) \\ else\ if\ 0.01 < S < 0.53 \rightarrow formation\ of\ osteoblasts\ (Mature\ bone) \\ else\ if\ 0 < S < 0.01 \rightarrow bone\ resorption \end{cases} \tag{3}$$

The threshold limits reported in the inequalities (3) are the same as those utilized in a previous study [61].

2.3. Mechanobiology-Based Optimization Algorithm

The task of determining the optimal distance d_{fil} between the filaments was accomplished by implementing an optimization algorithm, a schematic of which is illustrated in Figure 4. The choice of utilizing d_{fil} as a design variable and the strand diameter D as input parameter entered by the user derives from the fact that in the FDM technique, the distance between the strands can be changed with continuity while the strand diameter cannot be controlled with precision as it depends on the nozzle diameter. The algorithm implements the optimization tool available in MATLAB® (Version R2016b, MathWorks, Natick, USA) *fmincon* devoted to finding the minimum of a multivariable scalar function starting at an initial estimate. The objective of the optimization algorithm is to identify the optimal value of the filaments distance d_{fil_optim} that allows maximizing the amounts of mature bone that are predicted to generate within the scaffold pores.

The user has, first, to select the value of the diameter D and of the load per unit area p acting on the scaffold model (Blocks [A] and [B], Figure 4). Then, the user is prompted to specify a tentative initial value of the distance between the filaments d_{fil} (Block [C], Figure 4). At his point, the algorithm writes a python script (Block [D]) and enters into it the tentative value provided by the user (Block [E], Figure 4).

The python script is then given in input to Abaqus (Block [F], Figure 4) that, executing the instructions of the script: (i) builds the CAD model of both, the scaffold and the granulation tissue and applies the boundary and loading conditions above described (Figure 1d) (Block [G], Figure 4); (ii) generates the poroelastic tetrahedral finite element mesh (Block [H], Figure 4); and (iii) runs the finite element (FE) analysis (Block [L], Figure 4). Once the FE analysis has terminated, Abaqus prints, for all the elements inside the scaffold pores (highlighted in red in Figure 1), the volume of the element and the values of strain and interstitial fluid velocity computed in the analysis. At this point, the algorithm reads the document printed by Abaqus and computes, according to the Equation (1) the value of the biophysical stimulus S acting on the single element (Block [M], Figure 4). Then, the algorithm compares all the obtained values of S with the threshold limit reported in the inequalities (3). For the elements where the formation of mature bone is predicted (i.e., the inequality $0.01 < S < 0.53$ is satisfied) the algorithm stores the value of the element volume (Block [N], Figure 4).

Once for all the elements, S was computed, the algorithm determines the total volume of the elements that were predicted to differentiate into mature bone, by summing up all the element volumes previously stored. Then, if V_{BONE} is the total volume of the elements 'mature bone' and $V_{TOT} = \pi \times R^2 \times h$ the total volume of the model, the algorithm computes the percentage $BO_\%$ of the scaffold volume that is predicted to be occupied by mature bone as the ratio between V_{BONE} and V_{TOT} multiplied by 100 (Block [P], Figure 4). As it is clear, the task pursued by the algorithm is to increase as much as possible the percentage $BO_\%$. However, considering that the optimization tool utilized *fmincon* is designed to determine the minimum value of functions, the objective function $\Omega(d_{fil})$ was defined as the opposite value of $BO_\%$ (Block [Q], Figure 4). At this point, the algorithm perturbs the scaffold geometry many times (Block [R], Figure 4), i.e., it proposes different values of d_{fil} as possible candidate solutions (Block [T], Figure 4) and for each of the proposed value, it stores the value of $\Omega(d_{fil})$. Figure 5 shows, for instance, the typical values of $BO_\%$ computed by the algorithm for different values of hypothesized d_{fil}. Once the algorithm has enough points to identify the minimum of $\Omega(d_{fil})$ or, equivalently, the maximum of $BO_\%$, (denoted as $BO_{\%\_max}$ in Figure 5) its stopping criteria are satisfied and hence stops, giving in output, for the above-selected D and p, the optimal value of the strand distance d_{fil_opt} (Block [S], Figure 4).

The domain of variability of d_{fil} was hypothesized to range between the following lower and upper bounds: $d_{fil\_min} = D$ (which corresponds to have the strands in reciprocal contact) and $d_{fil\_max} = 1100$ μm which is approximately the average value of the distance of strands utilized by Neufurth et al. [62] and Bartolo et al. [63].

Each finite element analysis had an average duration of 4 hours on an HP XW6600-Intel®Xeon®DualProcessor E5-5450 3 GHz–32 Gb RAM workstation. Considering that each optimization cycle required about 50 finite element analyses to identify d_{fil_opt} and considering that five values of D (D = 400, 500, 600, 700, and 800 μm) and four values of p (0.2, 0.5, 1.0, and 1.5 MPa) were hypothesized, one can conclude that the total time to carry out all the analyses conducted in this study is: $4 \times 50 \times 5 \times 4 = 4000$ hours. In summary, 5(no. of values of D) × 4(no. of values of p) = 20 values of d_{fil_opt} were computed, i.e., 20 scaffold geometries were optimized.

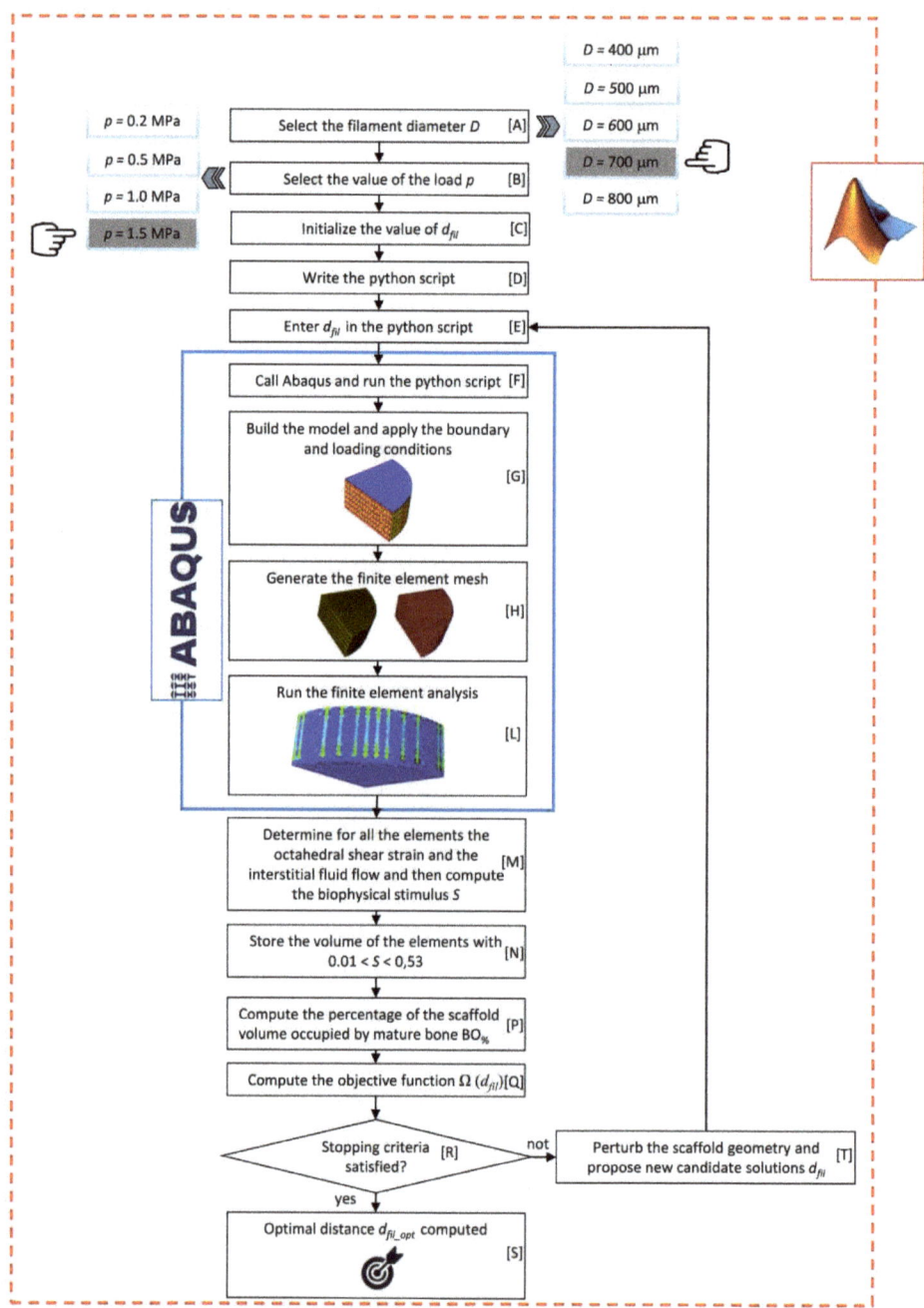

Figure 4. Schematic of the optimization algorithm written in MATLAB environment.

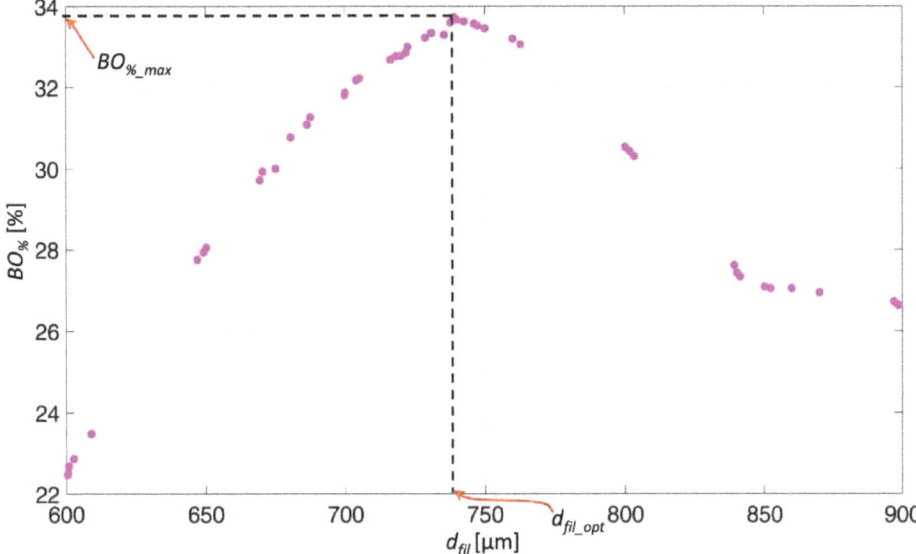

Figure 5. Typical values of $BO_\%$ obtained in an optimization process for different values of d_{fil}. In detail, the diagram refers to the case of D = 600 μm and p = 0.5 MPa.

2.4. Fabrication of the Optimized Scaffolds

For each of the 20 optimized geometries, three samples of scaffold were fabricated via the FDM technique. With the aim of keeping approximately a cylindrical shape of the deposited filament, an Ultimaker 3 was utilized, adjusting the slice height and the flux of extruded filament into the nozzle. If D_n is the nozzle diameter, v_n is the travel speed of the extrusion head, and v_h is the speed of the filament inside the nozzle, slicing software allows modifying the flow rate of extruded strand according to a flow rate coefficient f, usually expressed in terms of percentage

$$D_n^2 v_n = D^2 v_h \tag{4}$$

with

$$v_n = f v_h \tag{5}$$

The deposited filaments are nearly cylindrical if the slice height is equal to D_n, while it is possible to decrease D using f values lower than 100% and consequently reducing the slice height, according to the Comminal's model [25], under the hypothesis of negligible effects on the section circular shape. Considering that, commercial extruders are available for Ultimaker 3D printers with a diameter equal to 400 and 800 μm, intermediate values of the strand diameter have been obtained using nozzles 0.4 and 0.8, lowering f according to Equations (4) and (5), respectively (Table 2).

Table 2. Values of the flow rate f utilized to obtain different strand diameters

Fabricated Strand Diameter D (μm)	Nozzle Diameter D_n (μm)	Flow Rate f (%)
400	400	100
500	800	39
600	800	56
700	800	76
800	800	100

As regards the remaining process parameters default values have been exploited: v_h = 40 mm/s, extrusion temperature 180° and build plate temperature 50° to lower shrinkage.

2.5. Measurement of the Dimensions of the Fabricated Scaffolds

A De Meet 400 Coordinate Measuring Machine CMM, by Schut Geometrical Metrology, Germany, was utilized to measure the diameter of strands and their reciprocal distance d_{fil}. The objective of these measurements was to compare: (i) the distance of filaments actually fabricated and that, denoted as d_{fil_opt}, optimized with the algorithm above described; (ii) the dimension of the strand diameter actually realized with the nominal one. The CMM is equipped with lenses that can be moved and focused by the user. In detail, d_{fil} was measured as the center to center distance between the filaments (Figure 6). For each fabricated sample, 10 measurements were taken of d_{fil} and 10 measurements of D.

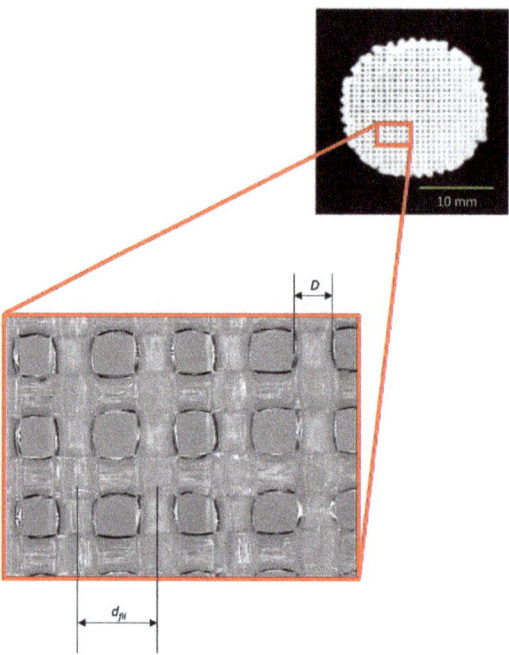

Figure 6. For each scaffold fabricated, the center to center distance between the filaments d_{fil} and the dimension of the diameter D were measured.

In order to evaluate the correctness of the hypothesis that an overlap region of $0.1 \times D$ exists [57] between two adjacent strands of two consecutive layers, three measurements of this region were carried out for each sample fabricated.

3. Results

By implementing the optimization algorithm above described, the optimal geometry of the scaffold for different levels of D and p was computed (Figure 7). As expected, for increasing values of load, decreasing values of the optimal distance d_{fil_opt} were predicted (Figure 8a). In fact, as the load increases, the biophysical stimulus S increases too thus stimulating the formation of 'softer' tissues like the cartilage and the fibrous tissue (see Equations (1) and (3)). To prevent this, the algorithm predicts smaller distances between the strands. This leads to an increase in the global stiffness of the scaffold and hence to a decrease in the value of the biophysical stimulus thus stimulating the formation of 'harder' tissues such as the bone. Interestingly, for an assigned value of load p and for decreasing dimensions of diameter D, the algorithm predicts increasing amounts of bone (Figure 8b).

Figure 7. Optimal scaffold geometries predicted by the optimization algorithm for different values of the filament diameter D and the load per unit area p.

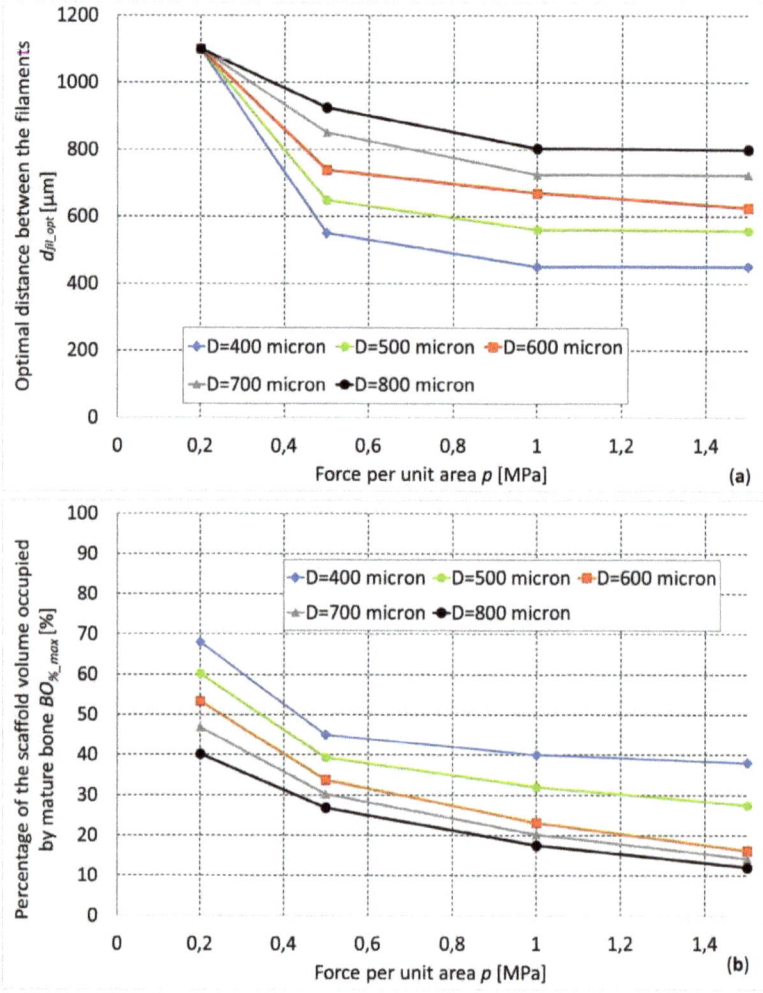

Figure 8. Optimal dimensions of d_{fil} (a) and percentage of the scaffold volume occupied by mature bone (b) predicted for different values of the load.

The measurements carried out on the overlap region of adjacent filaments revealed an average value of this dimension of $(0.12 \pm 0.04) \times D$, which demonstrates the reasonable appropriateness of the hypothesis followed [57]. The values of the distance d_{fil_opt} optimized with the proposed algorithm fell, for almost all the hypothesized values of p, within the range [average ± standard_deviation] of the measured dimensions (Figure 9). In general, it appears that the difference between the optimized distances d_{fil_opt} and the average measured distances never exceeded 8.27% of the optimized distance. Regarding the measured values of the diameter, we noticed that, for $D = 400$ µm and $D \geq 600$ µm, the measured dimensions are very close to the nominal ones (Figure 10a,c–e)), while for $D = 500$ µm, large differences can be seen (Figure 10b). In fact, for $D = 400$ µm and $D \geq 600$ µm, the nominal dimension of the diameter fell, almost in all the values of p investigated, in the range [average ± standard_deviation] of the measured dimensions. In the case of $D = 500$ µm, instead, the nominal value of the diameter is abundantly out of the above-mentioned range, which indicates that significant fabrication errors are committed.

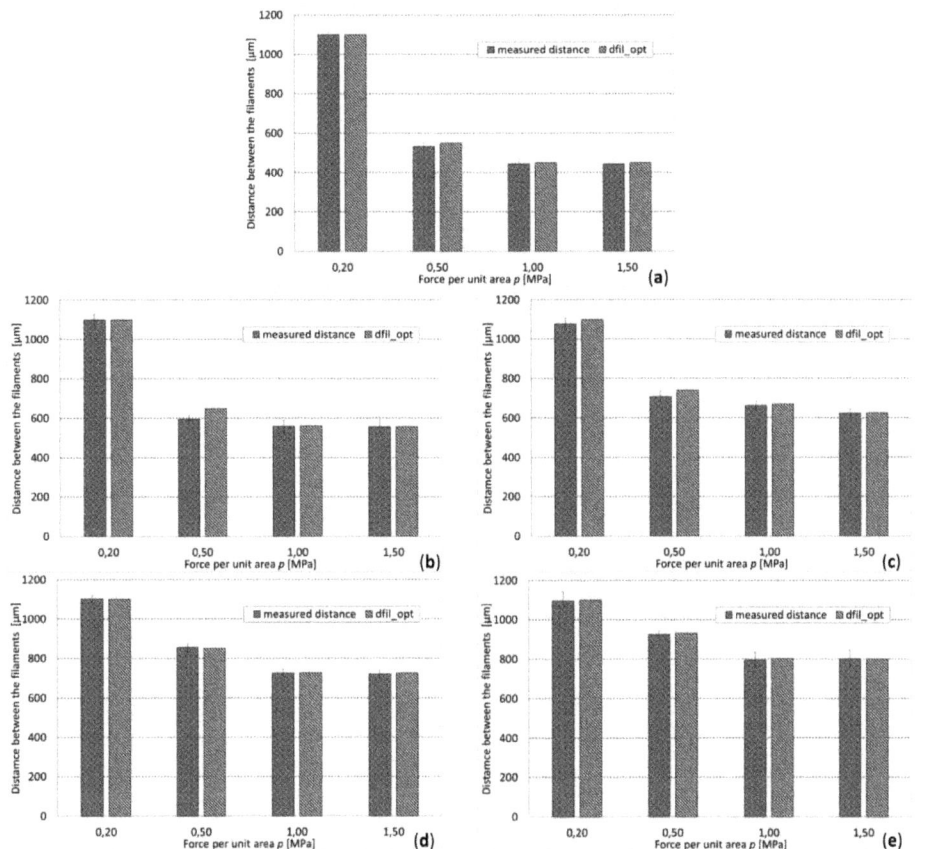

Figure 9. Comparison between measured and optimized distances (between the filaments), for different diameters: (**a**) $D = 400$ μm, (**b**) $D = 500$ μm, (**c**) $D = 600$ μm, (**d**) $D = 700$ μm, (**e**) $D = 800$ μm.

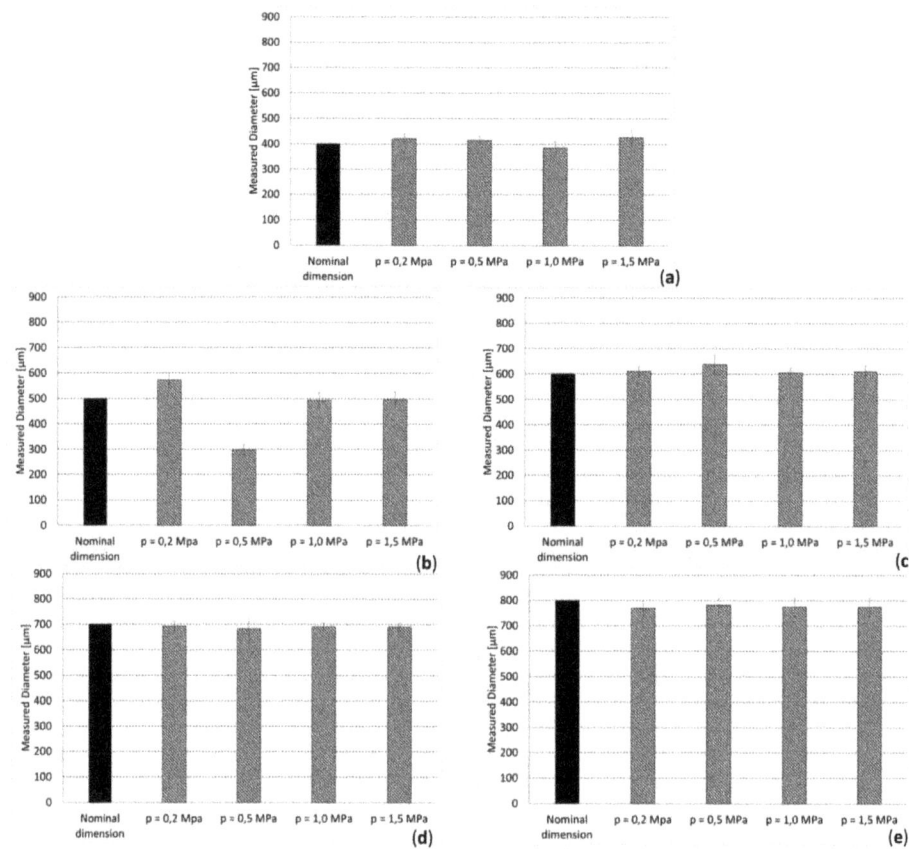

Figure 10. Comparison between measured and nominal dimension of diameter: (**a**) $D = 400$ μm; (**b**) $D = 500$ μm; (**c**) $D = 600$ μm; (**d**) $D = 700$ μm; (**e**) $D = 800$ μm.

4. Discussion

A feasibility study was conducted aimed to investigate whether the FDM technique can be utilized to fabricate scaffolds designed and optimized to undergo a compression load. A mechanobiology-based optimization algorithm was developed and implemented to determine the optimal distance between the filaments of cylindrical scaffolds for bone tissue engineering. The optimal distance was predicted for different hypothesized values of the load acting on the scaffold and diameter of strands. The scaffolds with the optimized dimensions were hence physically fabricated with the FDM technique and successively measured. The precision guaranteed by the FDM technique was finally evaluated by comparing the measured dimensions with the nominal ones.

This study presents some limitations in the model, the computational mechanobiological algorithm, and the experimental measurements. Regarding the model, first, we hypothesized that the strands of a given layer are aligned and form an angle of 90° with those of the adjacent one. Ideally, the model of the scaffold should include the angle formed between the layers as a design parameter that should be optimized via the mechanobiology-based optimization algorithm above described.

Second, we hypothesized that an overlap region of $0.1 \times D$ exists between two adjacent filaments of consecutive layers [57]. In an ideal model, also this dimension should be included as a design parameter to be optimized by means of fmincon. However, including these two additional design variables would make the computational costs tremendously larger than those spent in this study. Third, it would be interesting to investigate how the proposed optimization algorithm works in the case of more complex loading conditions acting on the scaffold model.

However, the hypothesis of more complex loading conditions would lead to losing the symmetry conditions and hence to the impossibility of using the simplified one-quarter model. Regarding the mechanobiological model, we identified the optimal scaffold geometry based on the values of the biophysical stimulus acting on the granulation tissue, i.e., the tissue that was hypothesized to occupy the volume inside the scaffold pores. In reality, this biophysical stimulus changes in time as the granulation tissue is replaced by the other tissues forming during the regeneration process. In other words, in the mechanobiological model, we did not take into account the variable time. However, the inclusion of the time would increase by at least two orders of magnitude the computational cost required to carry out the analyses. Furthermore, at different compression force p, the optimized value dfil_opt is different. In physiological conditions, the compression force acting on a bone may not be constant but variable. A possible strategy that can be adopted to optimize a scaffold subject to a variable compression load consists in designing functionally graded scaffolds, i.e., scaffolds where the geometric parameters change depending on the specific load value acting in the specific scaffold region. For instance, in the regions where the load acting is higher, a functionally graded scaffold may include strands at a shorter distance, in the regions where, instead, the load is smaller the distance between the filaments may be increased. Such a strategy requires including as many design variables as required to provide an adequate structural response to the variable load and is, therefore, different orders of magnitude more expensive than the approach adopted in this study. Increases in computational power will ultimately allow the investigation of the effect of additional geometric parameters on the optimal scaffold geometry and include different design variables and variable time in the optimization analyses. Regarding the experimental measurements, only two dimensions—the strand diameter and the distance between the filaments—were measured and compared with the corresponding quantities obtained via the mechanobiology-based optimization algorithm. The choice of measuring only these two geometrical parameters is due to the fact that by adjusting the diameter and distance between extruded strands, it is possible to design various topologies with variable values of porosity and therefore to have a wide control of the scaffold micro-architecture [34]. All the other geometric parameters involved in the scaffold designing will certainly play a role less relevant than that played by the distance between the strands and the diameter of the filaments. Furthermore, the proposed optimization algorithm was not validated experimentally. The validation requires a large number of experiments as well as an experimental set-up properly studied and organized to make the experimental conditions equivalent to those hypothesized in the numerical model, which goes beyond the scope of this feasibility study. However, should be clarified that the optimization carried out in this study takes into account only mechanobiological aspects and neglects many other chemical and genetic aspects that certainly affect the differentiation process. Therefore, by 'optimized design' we should intend only a design optimized from the mechanobiological point of view. More sophisticated optimizations taking into account the large number of aspects and variables involved in fracture healing should be the object of future studies.

In spite of these limitations, the predictions of the proposed optimization algorithms are consistent with the results of other studies reported in the literature.

(i) Barba et al. [64] implanted cylindrical scaffolds fabricated with FDM into bone defects generated in the limb of adult beagle dogs. In detail, two scaffold types were implanted, one with filaments of 250 µm and the other with filaments of 450 µm. Interestingly, they found that the amounts of bone formed in scaffolds with filaments of 250 µm are larger than those observed in scaffolds with filaments of 450 µm. This result is consistent with the predictions of the proposed optimization algorithm that found increasing amounts of bone in filaments with decreasing values of the diameter D (Figure 8b).

(ii) The typical distribution of the von Mises stresses within the scaffold model displays the presence of stress peaks in the proximity of the point where the generic strand enters in contact with the strand of the adjacent layer (Figure 11). This same mechanical behavior was reported by Uth et al. [34] who observed peaks of stress in alignment with the filaments of the previous layer.

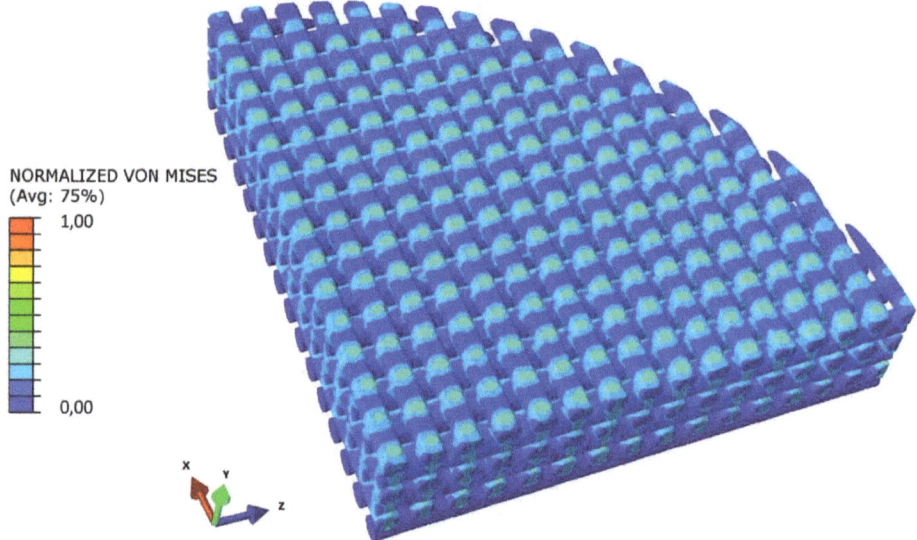

Figure 11. Normalized von Mises stress distribution in a section of the scaffold model. Stress peaks are in alignment with the filaments of the previous layer.

(iii) The amounts of bone predicted with the proposed algorithm are comparable with those predicted with other scaffolds based on different unit cell geometries [36,38,40,41]. However, it appears that scaffolds fabricated with the FDM technique allow the formation of amounts of bone significantly smaller than those obtained with other regular, beam-based scaffolds. In general, in scaffolds fabricated with the FDM technique, the amounts of bone predicted to create are approximately 20% lesser than those generated in other scaffolds [41].

The results obtained are consistent with the physics of the problem. As the load increases, the biophysical stimulus S increases too and with it, the percentage of the scaffold volume occupied by 'soft' tissues such as the cartilage and the fibrous tissue. To counterbalance this tendency, the optimization algorithm tends to decrease the distance between the filaments hence making the scaffold stiffer. The increase in stiffness leads to decreasing levels of S and hence to the formation of harder tissues such as the immature and the mature bone. For very high values of load, the optimal distance d_{fil_opt} tends, asymptotically, towards the dimension of the filament diameter D (Figure 8a). The optimal distance between the filaments d_{fil_opt} was determined for different values of load and for different assigned values of diameter D (Figure 8b). In general, it appears that using smaller diameters, for a fixed value of load p, leads to the formation of larger amounts of bone (Figure 8b). Therefore, if one can choose the nozzle diameter (i.e., if the FDM machine is equipped with nozzles of different diameter), to which the strand diameter is strictly related, should prefer the smaller diameters.

The dimensions measured on the samples were compared with the nominal one, in order to assess the accuracy guaranteed by FDM in the fabrication of scaffolds for bone tissue engineering. For almost all the hypothesized values of load per unit area p, the optimal dimensions dfil_opt fell within the range (average ± standard_deviation) of the measured distances. This leads us to conclude that the FDM is suited to reproduce with high accuracy the designed and optimized distance between the filaments dfil_opt. Furthermore, the values of the standard deviation of the measured distances (between the filaments) never exceeded 6.02% of the optimized dimension, which indicates a reasonably small dispersal of data and hence a rather high reproducibility of the fabrication process, in terms of the distance between the filaments. Regarding the measured dimensions of the filament diameter, we found that the diameters are well reproduced in the case of D = 400 µm and D ≥ 600 µm, but large reproduction errors are made when D = 500 µm. This behavior can be justified with the argument that the strategy of using different flow percentages of PLA to have filaments with different diameters presents a lower limit below which the quality of the filament diameter decreases significantly. Indeed, 500 µm is the diameter that mostly differs with respect to that of the nozzles utilized to extrude the filaments. We can hypothesize that when the material flow is significantly smaller than 100%, the nozzle does not fill correctly and hence the filament cannot form properly. This leads to the deposition of filaments with dimensions significantly different with respect to the nominal ones. However, this limitation can be easily overcome by equipping the FDM machine with a greater number of nozzles. In general, in ideal conditions, to minimize the fabrication errors, nozzles with the same diameter of the strands to be fabricated should be utilized.

5. Conclusions

We conducted a feasibility study which aimed to investigate the potentialities of the FDM technique to be used for the fabrication of scaffolds designed and optimized with mechanobiological algorithms. The present article is the first study ever reported in the literature where the geometry of scaffolds fabricated with the FDM technique is optimized via a mechanobiology-based optimization algorithm. In detail, the optimal distance between the filaments was predicted in function of the filament diameter and of the load acting on the scaffold. The designed and optimized scaffolds have been fabricated and measurements on the dimensions of the samples realized were carried out. We found that the difference between the average dimensions of the fabricated scaffolds and the nominal ones never exceeded 8.27% of the nominal dimension, which demonstrates the rather good accuracy of the FDM technique in reproducing the distance between the filaments. Furthermore, the values of the standard deviation of the acquired distances (between the filaments) never exceeded 6.02% of the optimized dimension, which indicates a reasonably small dispersal of data and hence a rather high reproducibility of the fabrication process in terms of distance between the filaments. However, we found that large reproduction errors are made on the filament diameter when the filament diameter to be realized differs significantly with respect to the nozzle diameter.

In conclusion, we can state that the FDM technique is suitable to build accurate scaffold samples only in the cases where the filament diameter is close to the nozzle diameter. Conversely, when a large difference exists, large fabrication errors can be made on the diameter of the filaments. In general, the scaffolds realized with the FDM technique were predicted to stimulate the formation of amounts of bone smaller than those that can be obtained with other regular beam-based scaffolds.

Author Contributions: Conceptualization, A.B. and G.P.; Methodology, A.B., Experimentation, G.P., Computations, A.B.; Validation, A.B. and G.P.; Investigation, A.B., G.P., and A.E.U.; Data curation, A.B. and G.P.; Writing—original draft preparation, A.B.; Writing—review and editing, G.P., A.E.U., M.F., M.G., and V.M.M.; Visualization, A.B., G.P., A.E.U., M.F., M.G., and V.M.M. All authors have read and agreed to the published version of the manuscript.

Funding: This research received no external funding.

Conflicts of Interest: The authors declare no conflict of interest.

References

1. Hedayati, R.; Sadighi, M.; Mohammadi-Aghdam, M.; Hosseini-Toudeshky, H. Comparison of elastic properties of open-cell metallic biomaterials with different unit cell types. *J. Biomed. Mater. Res. Part B Appl. Biomater.* **2018**, *106*, 386–398. [CrossRef] [PubMed]
2. Ahmadi, S.M.; Yavari, S.A.; Wauthle, R.; Pouran, B.; Schrooten, J.; Weinans, H.; Zadpoor, A.A. Additively Manufactured Open-Cell Porous Biomaterials Made from Six Different Space-Filling Unit Cells: The Mechanical and Morphological Properties. *Materials* **2015**, *8*, 1871–1896. [CrossRef] [PubMed]

3. Ahmadi, S.M.; Campoli, G.; Amin Yavari, S.; Sajadi, B.; Wauthle, R.; Schrooten, J.; Weinans, H.; Zadpoor, A.A. Mechanical behavior of regular open-cell porous biomaterials made of diamond lattice unit cells. *J. Mech. Behav. Biomed. Mater.* **2014**, *34*, 106–115. [CrossRef] [PubMed]
4. Hedayati, R.; Sadighi, M.; Mohammadi-Aghdam, M.; Zadpoor, A.A. Mechanical properties of regular porous biomaterials made from truncated cube repeating unit cells: Analytical solutions and computational models. *Mater. Sci. Eng. C* **2016**, *60*, 163–183. [CrossRef] [PubMed]
5. Hedayati, R.; Sadighi, M.; Mohammadi-Aghdam, M.; Zadpoor, A.A. Mechanics of additively manufactured porous biomaterials based on the rhombicuboctahedron unit cell. *J. Mech. Behav. Biomed. Mater.* **2016**, *53*, 272–294. [CrossRef] [PubMed]
6. Zadpoor, A.A.; Hedayati, R. Analytical relationships for prediction of the mechanical properties of additively manufactured porous biomaterials. *J. Biomed. Mater. Res. Part A* **2016**, *104*, 3164–3174. [CrossRef]
7. Guan, J.; Fujimoto, K.L.; Sacks, M.S.; Wagner, W.R. Preparation and characterization of highly porous, biodegradable polyurethane scaffolds for soft tissue applications. *Biomaterials* **2005**, *26*, 3961–3971. [CrossRef]
8. Doustgani, A.; Vasheghani-Farahani, E.; Soleimani, M. Aligned and random nanofibrous nanocomposite scaffolds for bone tissue engineering Nanofibrous scaffolds for bone tissue engineering. *Nanomed. J.* **2013**, *1*, 20–27.
9. Scaglione, S.; Giannoni, P.; Bianchini, P.; Sandri, M.; Marotta, R.; Firpo, G.; Valbusa, U.; Tampieri, A.; Diaspro, A.; Bianco, P.; et al. Order versus Disorder: In vivo bone formation within osteoconductive scaffolds. *Sci. Rep.* **2012**, *2*, 1–6. [CrossRef]
10. Serra, T.; Navarro, M.; Planell, J. Fabrication and characterization of biodegradable composite scaffolds for Tissue Engineering. In *Innovative Developments in Virtual and Physical Prototyping*; Al, P.J.B., Ed.; CRC Press Taylor & Francis Group: Boca Raton, FL, USA, 2011.
11. Roseti, L.; Parisi, V.; Petretta, M.; Cavallo, C.; Desando, G.; Bartolotti, I.; Grigolo, B. Scaffolds for Bone Tissue Engineering: State of the art and new perspectives. *Mater. Sci. Eng. C* **2017**, *78*, 1246–1262. [CrossRef]
12. Shick, T.M.; Abdul Kadir, A.Z.; Ngadiman, N.H.A.; Ma'aram, A. A review of biomaterials scaffold fabrication in additive manufacturing for tissue engineering. *J. Bioact. Compat. Polym.* **2019**, *34*, 415–435. [CrossRef]
13. Melchels, F.P.W.; Feijen, J.; Grijpma, D.W. A review on stereolithography and its applications in biomedical engineering. *Biomaterials* **2010**, *31*, 6121–6130. [CrossRef] [PubMed]
14. Lužanin, O.; Movrin, D.; Plan, M. Effect of Layer Thickness, Deposition Angle, and Infill on Maximum Flexural Force in Fdm-Built Specimens. *J. Technol. Plast.* **2014**, *39*, 49–58.
15. Ligon, S.C.; Liska, R.; Stampfl, J.; Gurr, M.; Mülhaupt, R. Polymers for 3D Printing and Customized Additive Manufacturing. *Chem. Rev.* **2017**, *117*, 10212–10290. [CrossRef] [PubMed]
16. Zhang, Y.; Wang, J. Fabrication of Functionally Graded Porous Polymer Structures using Thermal Bonding Lamination Techniques. *Procedia Manuf.* **2017**, *10*, 866–875. [CrossRef]
17. Inzana, J.A.; Olvera, D.; Fuller, S.M.; Kelly, J.P.; Graeve, O.A.; Schwarz, E.M.; Kates, S.L.; Awad, H.A. 3D printing of composite calcium phosphate and collagen scaffolds for bone regeneration. *Biomaterials* **2014**, *35*, 4026–4034. [CrossRef]
18. Velasco, M.A.; Lancheros, Y.; Garzón-Alvarado, D.A. Geometric and mechanical properties evaluation of scaffolds for bone tissue applications designing by a reaction-diffusion models and manufactured with a material jetting system. *J. Comput. Des. Eng.* **2016**, *3*, 385–397. [CrossRef]
19. Gregor, A.; Filová, E.; Novák, M.; Kronek, J.; Chlup, H.; Buzgo, M.; Blahnová, V.; Lukášová, V.; Bartoš, M.; Nečas, A.; et al. Designing of PLA scaffolds for bone tissue replacement fabricated by ordinary commercial 3D printer. *J. Biol. Eng.* **2017**, *11*, 1–21. [CrossRef]
20. Turnbull, G.; Clarke, J.; Picard, F.; Riches, P.; Jia, L.; Han, F.; Li, B.; Shu, W. 3D bioactive composite scaffolds for bone tissue engineering. *Bioact. Mater.* **2018**, *3*, 278–314. [CrossRef]
21. Zein, I.; Hutmacher, D.W.; Tan, K.C.; Teoh, S.H. Fused deposition modeling of novel scaffold architectures for tissue engineering applications. *Biomaterials* **2002**, *23*, 1169–1185. [CrossRef]
22. Do, A.V.; Khorsand, B.; Geary, S.M.; Salem, A.K. 3D Printing of Scaffolds for Tissue Regeneration Applications. *Adv. Healthc. Mater.* **2015**, *4*, 1742–1762. [CrossRef]
23. Masood, S.H.; Singh, J.P.; Morsi, Y. The design and manufacturing of porous scaffolds for tissue engineering using rapid prototyping. *Int. J. Adv. Manuf. Technol.* **2005**, *27*, 415–420. [CrossRef]
24. Ceretti, E.; Ginestra, P.; Neto, P.I.; Fiorentino, A.; Da Silva, J.V.L. Multi-layered Scaffolds Production via Fused Deposition Modeling (FDM) Using an Open Source 3D Printer: Process Parameters Optimization for Dimensional Accuracy and Design Reproducibility. *Procedia CIRP* **2017**, *65*, 13–18. [CrossRef]

25. Comminal, R.; Serdeczny, M.P.; Pedersen, D.B.; Spangenberg, J. Numerical modeling of the strand deposition flow in extrusion-based additive manufacturing. *Addit. Manuf.* **2018**, *20*, 68–76. [CrossRef]
26. Sa, M.W.; Nguyen, B.N.B.; Moriarty, R.A.; Kamalitdinov, T.; Fisher, J.P.; Kim, J.Y. Fabrication and evaluation of 3D printed BCP scaffolds reinforced with ZrO2 for bone tissue applications. *Biotechnol. Bioeng.* **2018**, *115*, 989–999. [CrossRef] [PubMed]
27. Boccaccio, A.; Ballini, A.; Pappalettere, C.; Tullo, D.; Cantore, S.; Desiate, A. Finite element method (FEM), mechanobiology and biomimetic scaffolds in bone tissue engineering. *Int. J. Biol. Sci.* **2011**, *7*, 112–132. [CrossRef]
28. Coelho, P.G.; Hollister, S.J.; Flanagan, C.L.; Fernandes, P.R. Bioresorbable scaffolds for bone tissue engineering: Optimal design, fabrication, mechanical testing and scale-size effects analysis. *Med. Eng. Phys.* **2015**, *37*, 287–296. [CrossRef]
29. Dias, M.R.; Guedes, J.M.; Flanagan, C.L.; Hollister, S.J.; Fernandes, P.R. Optimization of scaffold design for bone tissue engineering: A computational and experimental study. *Med. Eng. Phys.* **2014**, *36*, 448–457. [CrossRef]
30. Giannitelli, S.M.; Accoto, D.; Trombetta, M.; Rainer, A. Current trends in the design of scaffolds for computer-aided tissue engineering. *Acta Biomater.* **2014**, *10*, 580–594. [CrossRef]
31. Rainer, A.; Giannitelli, S.M.; Accoto, D.; De Porcellinis, S.; Guglielmelli, E.; Trombetta, M. Load-adaptive scaffold architecturing: A bioinspired approach to the design of porous additively manufactured scaffolds with optimized mechanical properties. *Ann. Biomed. Eng.* **2012**, *40*, 966–975. [CrossRef]
32. Wu, T.; Yu, S.; Chen, D.; Wang, Y. Bionic design, materials and performance of bone tissue scaffolds. *Materials* **2017**, *10*, 1187. [CrossRef]
33. Wieding, J.; Wolf, A.; Bader, R. Numerical optimization of open-porous bone scaffold structures to match the elastic properties of human cortical bone. *J. Mech. Behav. Biomed. Mater.* **2014**, *37*, 56–68. [CrossRef] [PubMed]
34. Uth, N.; Mueller, J.; Smucker, B.; Yousefi, A.M. Validation of scaffold design optimization in bone tissue engineering: Finite element modeling versus designed experiments. *Biofabrication* **2017**, *9*. [CrossRef] [PubMed]
35. Singh, J.; Ranjan, N.; Singh, R.; Ahuja, I.P.S. Multifactor Optimization for Development of Biocompatible and Biodegradable Feed Stock Filament of Fused Deposition Modeling. *J. Inst. Eng. Ser. E* **2019**, *100*, 205–216. [CrossRef]
36. Boccaccio, A.; Fiorentino, M.; Uva, A.E.; Laghetti, L.N.; Monno, G. Rhombicuboctahedron unit cell based scaffolds for bone regeneration: geometry optimization with a mechanobiology – driven algorithm. *Mater. Sci. Eng. C* **2018**, *83*, 51–66. [CrossRef]
37. Boccaccio, A.; Uva, A.E.; Fiorentino, M.; Monno, G.; Ballini, A.; Desiate, A. Optimal load for bone tissue scaffolds with an assigned geometry. *Int. J. Med. Sci.* **2018**, *15*, 16–22. [CrossRef]
38. Boccaccio, A.; Uva, A.E.; Fiorentino, M.; Lamberti, L.; Monno, G. A mechanobiology-based algorithm to optimize the microstructure geometry of bone tissue scaffolds. *Int. J. Biol. Sci.* **2016**, *12*, 1–17. [CrossRef]
39. Boccaccio, A.; Uva, A.E.; Fiorentino, M.; Mori, G.; Monno, G. Geometry design optimization of functionally graded scaffolds for bone tissue engineering: A mechanobiological approach. *PLoS ONE* **2016**, *11*. [CrossRef]
40. Rodríguez-Montaño, Ó.L.; Cortés-Rodríguez, C.J.; Uva, A.E.; Fiorentino, M.; Gattullo, M.; Monno, G.; Boccaccio, A. Comparison of the mechanobiological performance of bone tissue scaffolds based on different unit cell geometries. *J. Mech. Behav. Biomed. Mater.* **2018**, *83*, 28–45. [CrossRef]
41. Rodríguez-Montaño, Ó.L.; Cortés-Rodríguez, C.J.; Naddeo, F.; Uva, A.E.; Fiorentino, M.; Naddeo, A.; Cappetti, N.; Gattullo, M.; Monno, G.; Boccaccio, A. Irregular Load Adapted Scaffold Optimization: A Computational Framework Based on Mechanobiological Criteria. *ACS Biomater. Sci. Eng.* **2019**, *5*, 5392–5411. [CrossRef]
42. Metz, C.; Duda, G.N.; Checa, S. Towards multi-dynamic mechano-biological optimization of 3D-printed scaffolds to foster bone regeneration. *Acta Biomater.* **2019**, *101*, 117–127. [CrossRef] [PubMed]
43. Huiskes, R.; Van Driel, W.D.; Prendergast, P.J.; Søballe, K. A biomechanical regulatory model for periprosthetic fibrous-tissue differentiation. *J. Mater. Sci. Mater. Med.* **1997**, *8*, 785–788. [CrossRef] [PubMed]
44. Prendergast, P.J.; Huiskes, R.; Søballe, K. Biophysical stimuli on cells during tissue differentiation at implant interfaces. *J. Biomech.* **1997**, *30*, 539–548. [CrossRef]

45. Boccaccio, A.; Kelly, D.J.; Pappalettere, C. A mechano-regulation model of fracture repair in vertebral bodies. *J. Orthop. Res.* **2011**, *29*, 433–443. [CrossRef] [PubMed]
46. Boccaccio, A.; Pappalettere, C.; Kelly, D.J. The influence of expansion rates on mandibular distraction osteogenesis: A computational analysis. *Ann. Biomed. Eng.* **2007**, *35*, 1940–1960. [CrossRef] [PubMed]
47. Kelly, D.J.; Prendergast, P.J. Mechano-regulation of stem cell differentiation and tissue regeneration in osteochondral defects. *J. Biomech.* **2005**, *38*, 1413–1422. [CrossRef]
48. Andreykiv, A.; Prendergast, P.J.; Van Keulen, F.; Swieszkowski, W.; Rozing, P.M. Bone ingrowth simulation for a concept glenoid component design. *J. Biomech.* **2005**, *38*, 1023–1033. [CrossRef]
49. Sandino, C.; Checa, S.; Prendergast, P.J.; Lacroix, D. Simulation of angiogenesis and cell differentiation in a CaP scaffold subjected to compressive strains using a lattice modeling approach. *Biomaterials* **2010**, *31*, 2446–2452. [CrossRef]
50. Sandino, C.; Planell, J.A.; Lacroix, D. A finite element study of mechanical stimuli in scaffolds for bone tissue engineering. *J. Biomech.* **2008**, *41*, 1005–1014. [CrossRef]
51. Carter, D.R.; Blenman, P.R.; Beaupré, G.S. Correlations between mechanical stress history and tissue differentiation in initial fracture healing. *J. Orthop. Res.* **1988**, *6*, 736–748. [CrossRef]
52. Bailón-Plaza, A.; Van Der Meulen, M.C.H. A mathematical framework to study the effects of growth factor influences on fracture healing. *J. Theor. Biol.* **2001**, *212*, 191–209. [CrossRef]
53. Claes, L.E.; Heigele, C.A. Magnitudes of local stress and strain along bony surfaces predict the course and type of fracture healing. *J. Biomech.* **1999**, *32*, 255–266. [CrossRef]
54. Gómez-Benito, M.J.; García-Aznar, J.M.; Kuiper, J.H.; Doblaré, M. Influence of fracture gap size on the pattern of long bone healing: A computational study. *J. Theor. Biol.* **2005**, *235*, 105–119. [CrossRef] [PubMed]
55. Isaksson, H.; van Donkellar, C.C.; Huiskes, R.; Ito, K. Corroboration of mechanoregulatory algorithms for tissue differentiation during fracture healing: Comparison with in vivo results. *J. Orthop. Res.* **2006**, *24*, 898–907. [CrossRef] [PubMed]
56. Teng, S.; Liu, C.; Guenther, D.; Omar, M.; Neunaber, C.; Krettek, C.; Jagodzinski, M. Influence of biomechanical and biochemical stimulation on the proliferation and differentiation of bone marrow stromal cells seeded on polyurethane scaffolds. *Exp. Ther. Med.* **2016**, *11*, 2086–2094. [CrossRef]
57. Somireddy, M.; Czekanski, A. Mechanical Characterization of Additively Manufactured Parts by FE Modeling of Mesostructure. *J. Manuf. Mater. Process.* **2017**, *1*, 18.
58. Byrne, D.P.; Lacroix, D.; Planell, J.A.; Kelly, D.J.; Prendergast, P.J. Simulation of tissue differentiation in a scaffold as a function of porosity, Young's modulus and dissolution rate: Application of mechanobiological models in tissue engineering. *Biomaterials* **2007**, *28*, 5544–5554. [CrossRef] [PubMed]
59. Ultimaker Technical Data Sheet PLA 2019. Available online: https://ultimaker.com/download/74970/UM180821%20TDS%20PLA%20RB%20V11.pdf (accessed on 23 September 2019).
60. Castro, A.P.G.; Lacroix, D. Micromechanical study of the load transfer in a polycaprolactone–collagen hybrid scaffold when subjected to unconfined and confined compression. *Biomech. Model. Mechanobiol.* **2018**, *17*, 531–541. [CrossRef] [PubMed]
61. Lacroix, D.; Prendergast, P.J.; Li, G.; Marsh, D. Biomechanical model to simulate tissue differentiation and bone regeneration: Application to fracture healing. *Med. Biol. Eng. Comput.* **2002**, *40*, 14–21. [CrossRef] [PubMed]
62. Neufurth, M.; Wang, X.; Wang, S.; Steffen, R.; Ackermann, M.; Haep, N.D.; Schröder, H.C.; Müller, W.E.G. 3D printing of hybrid biomaterials for bone tissue engineering: Calcium-polyphosphate microparticles encapsulated by polycaprolactone. *Acta Biomater.* **2017**, *64*, 377–388. [CrossRef]
63. Bartolo, P.; Domingos, M.; Gloria, A.; Ciurana, J. BioCell Printing: Integrated automated assembly system for tissue engineering constructs. *CIRP Ann. Manuf. Technol.* **2011**, *60*, 271–274. [CrossRef]
64. Barba, A.; Maazouz, Y.; Diez-Escudero, A.; Rappe, K.; Espanol, M.; Montufar, E.B.; Öhman-Mägi, C.; Persson, C.; Fontecha, P.; Manzanares, M.C.; et al. Osteogenesis by foamed and 3D-printed nanostructured calcium phosphate scaffolds: Effect of pore architecture. *Acta Biomater.* **2018**, *79*, 135–147. [CrossRef] [PubMed]

© 2020 by the authors. Licensee MDPI, Basel, Switzerland. This article is an open access article distributed under the terms and conditions of the Creative Commons Attribution (CC BY) license (http://creativecommons.org/licenses/by/4.0/).

Review

Exploring Macroporosity of Additively Manufactured Titanium Metamaterials for Bone Regeneration with Quality by Design: A Systematic Literature Review

Daniel Martinez-Marquez, Ylva Delmar, Shoujin Sun and Rodney A. Stewart *

School of Engineering and Built Environment, Griffith University, Gold Coast, QLD 4222, Australia; daniel.martinezmarquez@griffithuni.edu.au (D.M.-M.); ylvadelmar@hotmail.com (Y.D.); shoujin.sun@griffith.edu.au (S.S.)
* Correspondence: r.stewart@griffith.edu.au

Received: 23 September 2020; Accepted: 23 October 2020; Published: 27 October 2020

Abstract: Additive manufacturing facilitates the design of porous metal implants with detailed internal architecture. A rationally designed porous structure can provide to biocompatible titanium alloys biomimetic mechanical and biological properties for bone regeneration. However, increased porosity results in decreased material strength. The porosity and pore sizes that are ideal for porous implants are still controversial in the literature, complicating the justification of a design decision. Recently, metallic porous biomaterials have been proposed for load-bearing applications beyond surface coatings. This recent science lacks standards, but the Quality by Design (QbD) system can assist the design process in a systematic way. This study used the QbD system to explore the Quality Target Product Profile and Ideal Quality Attributes of additively manufactured titanium porous scaffolds for bone regeneration with a biomimetic approach. For this purpose, a total of 807 experimental results extracted from 50 different studies were benchmarked against proposed target values based on bone properties, governmental regulations, and scientific research relevant to bone implants. The scaffold properties such as unit cell geometry, pore size, porosity, compressive strength, and fatigue strength were studied. The results of this study may help future research to effectively direct the design process under the QbD system.

Keywords: porous implants; bone implants; metamaterials; titanium; mechanical properties; pore size; unit cell; porosity; elastic modulus; compressive strength; additive manufacturing

1. Introduction

1.1. Current Issues with Traditional Bone Implants and Scaffolds

Many physical conditions necessitate bone tissue replacements and joint implants. Some of these conditions are caused by degenerative diseases, birth defects, and orthopaedic traumas [1]. However, despite the tremendous progress in biomedical engineering, 20% of patients subjected to joint reconstructive surgery experience significant problems [2]. This situation is reflected in the fact that orthopaedic products, such as knee and hip prostheses, are the fifth most recalled medical products; of these recalls, 48% are due to manufacturing issues and 34% to design flaws [3,4]. Some of the main flaws with orthopaedic implants are associated with their longevity, material properties, and mismatch with patient size and shape requirements [5,6]. Stress shielding is one of the main design flaws of load-bearing prostheses. This phenomenon occurs because bone is a self-healing material that requires load application to remodel itself, but a material with a higher modulus of elasticity (*E*) absorbs all the stress generated, leading to bone reabsorption and subsequent loosening of the implant [7].

In the case of bone defects, they can be caused by tumour resection, infections, complex fractures, and non-unions [8]. The most common treatment for bone defects is surgical intervention, where an autograft (bone taken from the patient's body) is used to fill bone defect spaces [9]. However, due to their restricted availability, allografts (bone tissue from a deceased donor) are frequently used to treat critical-size defects [9]. Bone grafting is a common surgical procedure; it has been estimated that 2.2 million grafting procedures are performed worldwide each year [8]. However, late graft rupture has been reported to be as high as 60% 10 years after the grafting procedure [10]. Allograft transplantation has a success rate of approximately 70%. The low success rate of allografts is caused by the prevalence of infection, rejection by the host's immune system, fatigue fractures, delayed union, non-union, and incomplete graft resorption [11,12]. In the case of autografts, the disadvantages are increased post-operative morbidity, lack of available tissue, chronic pain, infection, nerve injury, and weakened bone donor graft sites [12,13].

To solve these grafting problems, several scaffold traditional techniques have been used without much success: solvent-casting particulate-leaching, gas foaming, fibre meshes (fibre bonding), phase separation, melt moulding, emulsion freeze drying, solution casting, and freeze drying [14]. Some of the disadvantages of traditional scaffold fabrication techniques are their lack of control over porosity characteristics, such as pore size, pore distribution, and interconnectivity; the toxic by-products of scaffold degradation; and their lack of consistent mechanical properties [14]. Hence, traditional techniques for bone reconstruction including grafting and prostheses are not sufficiently effective, which represent a medical challenge that comes with several limitations and risks [9]. Moreover, no material yet exists with the ideal properties for bone tissue replacement [15–17]. To overcome these issues, tissue engineering has focused on additive manufacturing technologies to produce the next generation of bone implants and scaffolds.

1.2. Additive Manufacturing

Additive manufacturing (AM) technologies, supported by computer-aided design (CAD) software, progressively build 3D physical objects from a series of cross-sections, which are joined together to create a final shape [18]. With AM, it is possible to create complex interconnected and porous structures with controlled pore size, shape, and distribution and properties resembling bone mechanical properties, such as a modulus of elasticity to induce bone ingrowth [19,20]. This capability permits the fabrication of hierarchical structures at the microscale and the manipulation of material properties to create metamaterials. In terms of implant design, this advance means that products can be designed with a biomimetic approach according to the patient's anatomy and the bone tissue's mechanical properties [21]. The design freedom of AM allows its use in difficult clinical scenarios in which bone diseases, deformities, and trauma usually necessitate the reconstruction of bone defects with complex anatomical shapes, which is extremely difficult even for the most skilled surgeon [22]. The complex reconstruction of bone defects is possible through combining the advantages of AM with CAD and medical imaging technologies, such as computed tomography and magnetic resonance, to fabricate implants according to the patient's specific anatomy, thus achieving an exact adaptation to the region of implantation [23]. In the search of suitable materials for AM, bone regeneration, and implant application tissue engineering has focused on developing a variety of different types of synthetic and natural materials.

1.3. Materials for Bone Regeneration and Implant Applications

Materials appropriate for implantation within the human body require distinct biocompatible properties. Therefore, in the selection of appropriate materials for implant applications, several factors must be considered. First, the intended implant location must be considered to predict host response, which is governed by the biochemical and physical environments in contact with the medical device [24,25]. Second, the material should possess appropriate biological and mechanical properties for its specific purpose to prevent physical damage to the body. Third, from the perspective of tissue engineering, materials should mimic one or multiple characteristics of the natural region of repair. In the case of bone repair, the desired characteristics are osteoconductivity, osteoinductivity, and osseointegration. As a result, for an optimum scaffold and prosthesis design, material science may combine several technologies to create suitable materials that fulfil these needs.

1.3.1. Polymers

Polymers for AM and tissue engineering applications are biocompatible materials that offer several advantages over other materials, including biodegradability, cytocompatibility, easy processability, and flexibility in the tailoring of their properties [26]. Polymers can be classified as natural or synthetic and some of them already have regulatory approval [27].

Natural polymers are made from proteins such as alginate, gelatine, collagen, silk, chitosan, cellulose, and hyaluronic acid [28]. The advantages of natural polymers are their excellent biodegradability, low production costs, and superior chemical versatility, as well as their improved biological performance that allow better interactions with cells than other biomaterials, improving their attachment and differentiation [29]. However, natural polymers can be expensive to produce, due to the difficulty in controlling their mechanical properties, biodegradation rate, and quality consistency [30].

Due to the disadvantages of natural polymers, different synthetic polymers, such as polycaprolactone (PCL), polylactic acid (PLA), and poly Lactic-co-Glycolic Acid (PLGA), have been developed. Their advantages include low immunogenic potential, large scale low production cost, and good quality consistency [31]. Moreover, their mechanical properties, microstructure, and degradation rate can be tuned according to needs [27]. Despite the advantages of natural and synthetic polymers, they are unsuitable for load-bearing applications due to their lower modulus of elasticity compared to bone, unstable mechanical strength, and tendency to creep [32,33]. Hence, in recent years, a variety of polymers have been combined with different materials to such as bioceramics (e.g., bioglasses, tri-calcium phosphates, and carbon nanotubes) and metals to create composite materials with tunable mechanical properties as well as with the capacity to deliver drugs, exosomes, and growth factors, to name a few [34–37].

1.3.2. Bioceramics

Bioceramics are a large group of materials used for bone substitution and regeneration. Calcium phosphate (CaP) ceramics is one of the main groups of bioceramics. Calcium phosphate ceramics are abundant in bone, constituting between 80% and 90% of bone's anorganic matter. This group of bioceramics is widely used as implant coating, bone grafting, and more recently have been fabricated for bone scaffolding applications with AM [38,39]. Hydroxyapatite HAP and β-tricalcium phosphate (β-TCP), are the most-studied CaP bioceramics. The main advantages of calcium phosphate materials are their osteoinductive and osteoconductive properties, as well as their dissolution in body fluids [40]. For load-bearing applications, the major disadvantage of CaPs is their poor mechanical properties. Despite their good compressive strength, CaPs lack plastic deformation, making them brittle and prone to cracking. Consequently, these materials are not yet suitable for load-bearing applications [41]. Nevertheless, the lower wear rate of CaPs makes these materials the preferred choice for surface coating to reduce wear in joint prostheses [42]. They are also commonly used for spinal fusion, maxillofacial and cranio-maxillofacial reconstruction, as well as bone filler and bone cement due to their excellent biocompatibility and osteoconductivity [43].

Discovered in 1969 by Larry Hench, bioglasses are ceramic materials composed of calcium, phosphorus, and silicon dioxide [16]. Bioglasses are bioactive ceramic materials with strong osteointegrative and osteoconductive properties, as well as higher mechanical strength than calcium phosphate ceramics [44]. Hence, bioglasses have been intensely investigated with AM for bone tissue engineering applications [45,46]. The advantage of these materials is that by changing the proportions of their basic components, different forms with different properties can be obtained; for example, non-resorbable bioglasses can be transformed into resorbable bioglasses [44]. Moreover, they can be designed with controlled biodegradability and drug and cell delivery capabilities [47,48]. Their applications also include bioglass scaffolds produced using AM with controlled porosity architecture and improved mechanical properties for bone regeneration [49]. However, bioglasses are limited for use in practical load-bearing applications due to their low resistance to cyclic loading and their brittleness [50].

1.3.3. Metals and Titanium as a Bio-Metamaterial

Metals have been the common choice to replace hard tissue in load-bearing applications due to their mechanical properties, corrosion resistance, and biocompatibility. Most of these materials are alloys, such as 316L stainless steel (316LSS), cobalt chromium (Co–Cr), and titanium (Ti) alloys [5]. Among all metallic materials, the titanium alloy Ti–6Al–4V is the gold standard for orthopaedic applications [51,52] because of its high biocompatibility [53], high corrosion resistance, low modulus of elasticity [5], and high strength-to-weight ratio [19]. Furthermore, Ti is a reactive metal that naturally forms a thin layer of oxide, which blocks metal ions from reaching its surface, increasing its biocompatibility [54]. The biomedical applications of Ti–6Al–4V are quite broad and encompass dental implants; hip, shoulder, knee, spine, elbow, and wrist replacements; bone fixation components; and cardiovascular applications [5].

Nevertheless, the most common problems of metallic materials are wear and the stress-shielding effect caused by their high modulus of elasticity compared with bone [52,55–57]. Moreover, despite the excellent biocompatibility and mechanical properties of Ti and Ti alloys, they usually require long healing periods to create a stable interface with the surrounding bone [58], with insufficient implant osseointegration as a potential outcome [59]. Hence, to further augment the biological, mass transport, and mechanical performance of Ti and Ti alloys different metamaterials have been developed. For example, metallic bone implants with a modulus of elasticity similar to that of bone can drastically reduce wear, shear stress, and bone resorption and consequently prevent implant loosening and revision surgery [60]. This may translate into enhanced quality of life for the patient, reductions in hospital expenses and recovery time, and improvement in joint dynamic performance [61]. With porous Ti and Ti alloy bio-metamaterials, osseointegration is also improved, and superior results have been accomplished in relation to mechanical properties. Nonetheless, pores act as stress concentrators, reducing the material load capacity [23]. As a result, for the design of load-bearing prostheses, it is crucial to balance mechanical properties with biological stimulation. Consequently, there have been several efforts to find the optimal balance between pore size and porosity percentage in different materials. For example, Zaharin et al. [62] investigated the effect of pore variation on the porosity and mechanical properties of several Ti–6A–l4V porous scaffolds. According to their results, scaffolds based on cube and gyroid unit cells with a pore size of 300 µm provided similar properties to bone. Moreover, they concluded that increments in porosity decreased the scaffolds' elastic modulus and yield strength. In an earlier study Bobyn at al. [63] investigated the effects of pore size variation of cobalt-base alloy implants on the rate of bone growth. For this purpose, casted cobalt-base alloy implants were coated with powder particles and implanted into canine femurs for several weeks. The results indicated that pore sizes between 50 and 400 µm provided the maximum bone ingrowth and fixation strength.

Despite the excellent biocompatibility and mechanical properties of Ti and Ti alloys, they usually require long healing periods to create a stable interface with the surrounding bone, frequently resulting in insufficient osseointegration [64]. Hence, to further augment Ti's bioactivity, corrosion resistance, and mechanical properties different mechanical, chemical, and physical surface modification methods have been developed [65–67]. Depending on the surface treatment used to modify Ti substrate, different topographic features can be achieved at the macroscale, microscale, and nanoscale. There is a large amount of evidence that rough Ti surfaces with topographic microfeatures better protein adsorption and provide higher osteoblasts attachment growth proliferation and activity than surface smooth surfaces [68]. Nonetheless, it has been demonstrated that nanoscale topography outperforms macro and micro-scale surface features towards augmenting cellular functions [69]. More recently, at has been proposed that a combination of different topographic features at the macro, micro, and nanoscale with local drug delivery functions can further enhance the biological, chemical, tribological, and mechanical performance of Ti bone implants [70–73].

1.4. Purpose and Objectives

The purpose of this research is to provide researchers and industry with an in-depth adaptation of the Quality by Design (QbD) system for the fabrication of additively manufactured porous Ti implants considering the QbD guidelines for 3D printed bone implants and scaffolds [74]. The QbD system is composed by eight main steps that need to be systematically followed to acquire a complete comprehension of the product and its manufacturing process, including the identification and control of all variables to achieve the desired quality. Specifically, the scope of this present study was limited to the first step of the QbD framework (Figure 1). Thus, the objectives of this study are:

1. Define the ideal mechanical, geometrical and dimensional characteristics of the internal architecture of Ti bone scaffolds from a biomimetic perspective.
2. Compare the results of different studies on fully porous Ti structures in relation to the ideal quality attributes of bone scaffolds.
3. Identify the studies on fully porous Ti structures that satisfies the critical quality attributes of Ti porous bone implants and scaffolds.

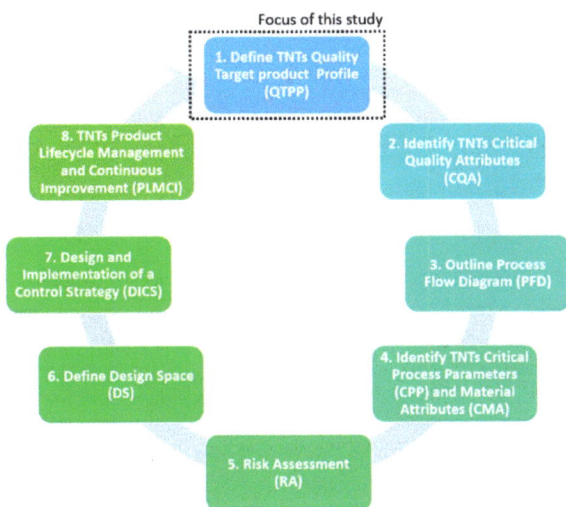

Figure 1. Schematic showing the focus of this study within the QbD system.

2. Materials and Methods

This systematic research study is part of the implementation of the QbD approach for porous metal implants. Therefore, a constructive research approach was used to further extend the QbD system for patient-specific bone implants and scaffolds produced by AM [75,76]. For an in-depth interpretation and synthesis the researchers immersed themselves in the contextual literature [77]. This was an exploratory qualitative study which requires the collection of secondary data from various datasets of peer-reviewed publications following the Preferred Reporting Items for Systematic Reviews and Meta-Analyses (PRISMA) statement [78].

2.1. Data Collection

A systematic search was conducted on 20 January 2020 in the Science Direct and Google Scholar databases according to objectives and the PRISMA statement. Relevant keywords were connected with the Boolean operators "OR" and "AND". Terms relevant to this research included the following: Titanium, Ti, additive manufacturing, 3D printing, rapid prototyping, bone tissue engineering, bone implant, implant, scaffold, prostheses, porous, porosity, and mechanical properties. To specify the search further, the terms were connected with Boolean operators (AND, OR): Implant(s); scaffold(s); prosthes(is, es); titanium; Ti; additive manufacturing; additively manufactured; 3D printing; and 3D printed.

The full phrase used was: (Implant* OR scaffold* OR prosthes?s) AND (titanium OR Ti) AND ("additive* manufactur*" OR "3D print*") AND "mechanical properties" AND (porous OR porosity) AND "pore size" AND "elastic modulus" AND "fatigue".

2.2. Study Selection

Selected studies in the systematic literature search were limited to the following inclusion criteria: (1) peer-reviewed papers with full-text published within the last 20 years (2000 ± 2020); (2) empirical studies reporting the mechanical properties of Ti and Ti alloy porous scaffolds produced by AM for bone repair; (3) published in the English language; (4) the first 10 pages of the search results were assessed; and (5) the search results were sorted by relevance. From the systematic literature search in both the Science Direct and Google Scholar databases, a total of 5941 results were generated from which 83 articles were fully assessed, as presented in Table 1 and Figure 2.

Table 1. Search strategy, studies between 2000 and 2019.

Database	Records Identified	Total
Google Scholar	3020	5941
Science Direct	2921	
Duplicates	80	5861

2.3. Data Extraction and Analysis

The systematic literature search conducted in this study aimed to gather results of different studies regarding the characteristics of natural bone tissue, as well as the mechanical, geometrical, and dimensional properties of additively manufactured Ti porous implants with controlled porosity and/or pore size. The classification topics used in this study were pore size, pore shape, porosity, interconnectivity, multi-scaled, elastic modulus, compressive yield strength (σ_y), ultimate compressive strength (σ_{tu}), and fatigue strength. The references from the collected articles were systematically reviewed to identify further articles relevant to the subject. A full-text screening was performed by Y.D and D.M to avoid potential bias. A consensus meeting resolved any discrepancies between the reviewers.

Once all applicable literature had been identified, the extracted data were used to further extend the first step of the QbD system for fully porous Ti bone implants. Moreover, the data were also used to compare the different characteristics of Ti porous bone implants with natural bone tissue, and also to identify the most relevant characteristics that need to be imitated in the development of fully porous Ti bone implants.

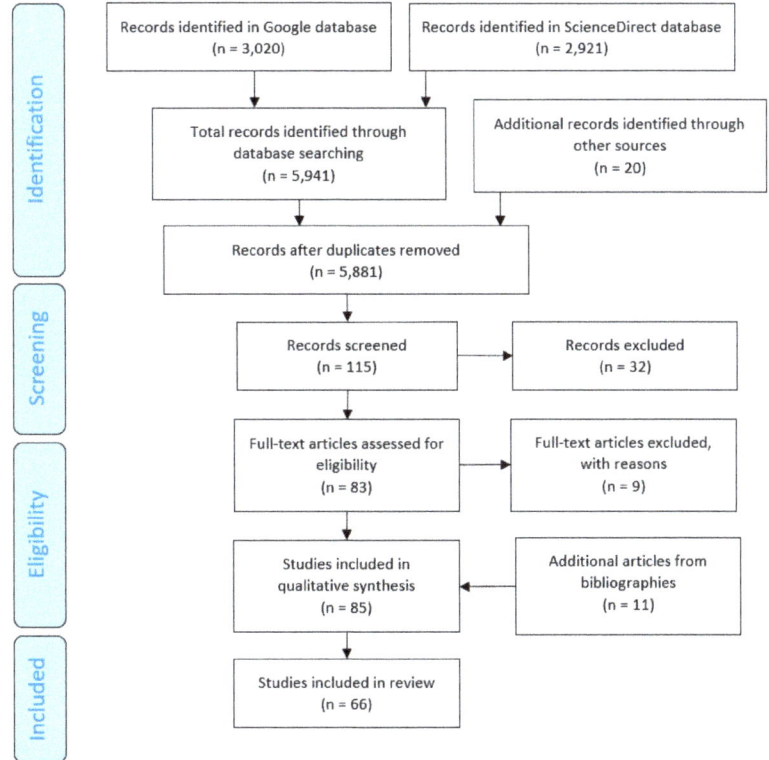

Figure 2. Search strategy and selection of studies in the Google Scholar and ScienceDirect databases.

3. Results

The systematic search identified a total of 64 different studies with data relevant to additively manufactured (AMd) porous Ti implants and scaffolds, as presented in Tables 2 and 3. From the 64 studies identified, 14 studies were used to extract information related to bone structure and mechanical properties (Table 2). The remaining 50 studies provided detailed information in relation to different characteristics of porous Ti scaffolds fabricated by AM for bone implant purposes. A total of 807 experimental data from these studies was extracted, analysed, and categorised in the following categories: pore size, pore shape, porosity, multi-scaled porosity, elastic modulus, interconnectivity, yield strength, ultimate compressive strength, and fatigue strength, as shown in Table 3. However, from these nine categories multi-scaled porosity was excluded from the analysis due to insufficient data for further analysis. Therefore, a total of eight different implant features were selected due to availability of recorded data in scientific research articles. The selected eight features are: unit cell, porosity, pore size, interconnectivity, elastic modulus, compressive yield strength, ultimate compressive strength, and fatigue strength.

Table 2. Studies selected in the systematic search according to information related to bone properties.

Ref	Cod.	Pore Characteristics				Mechanical Properties		
		Pore Size	Pore Shape	Porosity	Interconnectivity	Young's Modulus	Yield Strength	Compressive Strength
[79]	S51	✔	✔	✔	✔	✔	✔	
[80]	S52			✔		✔	✔	✔
[81]	S53	✔	✔	✔	✔	✔	✔	✔
[82]	S54			✔	✔	✔	✔	✔
[83]	S55					✔		
[84]	S56							✔
[85]	S57			✔	✔	✔		
[86]	S58			✔	✔			
[87]	S59	✔						
[88]	S60	✔	✔	✔	✔			
[89]	S8	✔		✔	✔	✔		
[90]	S62			✔		✔		
[91]	S63					✔	✔	✔
[84]	S64	✔		✔		✔		
[92]	S65	✔						
[93]	S66					✔	✔	✔

Table 3. Studies selected in the systematic search according to information related to additively manufactured (AMd) porous Ti structures' mechanical, geometrical, and dimensional properties.

Ref	Cod.	Pore Characteristics					Mechanical Properties			
		Size	Unit Cell Geometry	Porosity%	Connectivity	Multi-scaled	Young's Modulus	Compressive Yield Strength	Ultimate Compressive Strength	Fatigue
[19]	S1	✔	✔	✔	✔		✔		✔	
[94]	S2	✔	✔	✔			✔	✔		
[82]	S3	✔	✔	✔	✔		✔	✔	✔	
[95]	S4	✔	✔			✔	✔	✔		✔
[62]	S5	✔	✔	✔	✔		✔	✔		
[96]	S6	✔	✔	✔			✔			
[97]	S7		✔	✔			✔		✔	
[98]	S8	✔	✔	✔	✔		✔		✔	
[99]	S9	✔	✔	✔	✔					
[100]	S10	✔	✔	✔	✔	✔	✔	✔	✔	
[101]	S11	✔	✔	✔			✔	✔		
[102]	S12	✔	✔	✔	✔	✔	✔	✔		
[103]	S13	✔	✔	✔	✔		✔	✔	✔	✔
[104]	S14		✔	✔	✔		✔	✔	✔	✔
[105]	S15	✔	✔	✔			✔			
[106]	S16	✔	✔	✔	✔			✔	✔	✔
[107]	S17	✔	✔	✔	✔			✔	✔	✔
[108]	S18	✔	✔	✔	✔		✔	✔	✔	
[109]	S19	✔	✔	✔	✔	✔				
[110]	S20	✔	✔	✔		✔	✔		✔	
[111]	S21		✔	✔	✔		✔	✔	✔	
[112]	S22	✔	✔	✔			✔	✔		✔
[113]	S23	✔	✔	✔	✔		✔	✔		
[114]	S24	✔	✔	✔	✔				✔	
[115]	S25	✔	✔	✔			✔	✔		
[116]	S26	✔	✔	✔	✔		✔	✔		✔
[117]	S27	✔	✔	✔			✔	✔		✔
[118]	S28	✔	✔	✔			✔	✔		✔
[119]	S29	✔	✔	✔		✔	✔	✔		
[120]	S30	✔	✔	✔			✔	✔	✔	
[121]	S31	✔	✔	✔	✔	✔	✔	✔	✔	
[122]	S32	✔	✔	✔	✔		✔	✔		
[123]	S33	✔	✔	✔	✔	✔	✔	✔		
[124]	S34	✔	✔	✔	✔		✔		✔	
[125]	S35	✔	✔	✔			✔	✔	✔	✔
[126]	S36	✔	✔	✔			✔		✔	
[127]	S37	✔	✔	✔			✔	✔		
[128]	S38	✔	✔	✔	✔		✔	✔	✔	
[79]	S39	✔	✔	✔			✔	✔		✔
[129]	S40	✔		✔			✔		✔	
[130]	S41	✔	✔	✔	✔		✔			
[131]	S42	✔	✔	✔			✔	✔		
[132]	S43	✔	✔	✔	✔			✔		
[133]	S44	✔	✔	✔	✔		✔	✔		
[134]	S45	✔	✔	✔			✔		✔	
[135]	S46		✔	✔						✔
[136]	S47	✔	✔	✔	✔	✔	✔	✔		
[137]	S48		✔	✔			✔	✔		✔
[138]	S49	✔	✔	✔			✔	✔		
[139]	S50	✔	✔	✔			✔	✔		

Through the systematic research performed in this study and by reviewing the medical device regulations from the Food and Drug Administration (FDA), a Quality Target Product Profile has been established including proposed values for the Ideal Quality Attributes for mechanical and dimensional properties of porous bone implants. These target properties are aimed for porous metal implant structures designed for load bearing implant applications. Following this, the results of the selected studies were compared and discussed, from a biomimetic point of view, with the characteristics of natural human bone to identify implants with properties similar or superior to human bone and current medical standards.

3.1. QbD Step 1: Ideal Quality Target Product Profile

The Quality Target Product Profile (QTPP) is critical for formulating the ideal features of a product considering both performance and safety. To direct the product development process, it is vital to understand user needs. Using QTPP, design failures can be identified early in the product development process to reduce costs and time. According to Martinez-Marquez et al. [36], the quality of bone implants should be defined from three perspectives: product-based, manufacturing-based, and user-based. In the context of fully porous Ti microstructures, the dimensions of quality considered most relevant are performance, features, reliability, conformance, durability, and perceived quality. The requirements relating to bone implants corresponding to these quality dimensions have been identified through studies of existing scientific research results.

3.2. Bone Implant Requirements

An additively manufactured bone implant must act as a stable scaffold that is biocompatible without causing inflammation or leaching material toxins into surrounding tissue. It requires a suitable surface that promotes cell adhesion and differentiation as well as provides a constant flow of cell nutrients and metabolic waste. This allows for bone tissue formation [81]. The scaffold material must have mechanical properties matching those of the surrounding tissues to avoid stress shielding and mechanical failure [81,82]. The bone pore size, geometry, interconnectivity, and porosity are microscopic features that make for the foundation of bone regeneration, cell growth, osteoconduction, and cell proliferation [74]. Designing implants with adequate pore dimensions allows for a constant flow of cell nutrients and waste. It also allows for sufficient connections to establish between the local bone area and the scaffold [74]. If the right attributes are chosen for the implant microstructure, it can mimic human bone's natural characteristics, which is the end goal of biomimetic implant design [79,80]. Vasireddi and Basu [140] completed a list of general requirements for 3D printed implants:

- "A 3D, highly porous structure to support cell attachment, proliferation and extracellular matrix production;
- An interconnected pore network to promote oxygen, nutrient and waste exchange;
- A biocompatible and bioresorbable substrate with suitable degradation rates;
- An appropriate surface chemistry for cell attachment, proliferation and differentiation;
- Mechanical properties to support, or match, those of the tissues at the site of implantation; an architecture which promotes formation of the native anisotropic tissue structure; and
- An adapted geometry of clinically relevant size and shape."

In similar research, Jabir et al. [141] described the main fundamental requirements for implants as biocompatibility, good manufacturability, geometric precision, appropriate design, biomechanical stability, resistance to implant wear, corrosion and aseptic loosening, bioactivity, and osteoconduction. Since the environment in the human body is highly corrosive and biomaterials are usually bioactive, the implant will interact with its environment after implantation [140]. The implant must therefore be designed with both useful functions and biological safety, providing utmost biocompatibility. The implant design must have a high degree of reproducibility, which will ensure faster and cheaper manufacturing as well as predictability and reliability.

Furthermore, the implant must be durable and of initial strength for safe handling during sterilisation, transport, and surgery, and to survive physical forces in vivo after implantation [142]. The implant will be subjected to constant load in the body, from walking and further strenuous movements [127]. Its mechanical strength is vital for its viability, where the implant must last the entirety of its expected lifetime without defects or failure [127].

Taking all the above into consideration in conjunction with the ideal eight quality dimensions of AMd bone implants proposed by Martinez-Marquez at al. [74], we proposed seven ideal quality dimensions of porous internal architecture of Ti bone implants. These quality dimensions are based on three quality perspectives, namely product, manufacturing, and user-based, as presented in Table 4.

Table 4. The ideal quality dimensions of porous internal architecture of Ti bone implants.

Quality Approach	Dimension	Description
Product-based approach	Performance	The porous microstructure should provide an environment ideal for bone ingrowth and endow the implant with a stiffness similar to natural human bone while maintaining sufficient strength.
	Features	Tailored internal architecture with specific properties, including but not limited to pore size, unit cell, porosity, elastic modulus, interconnectivity, compressive yield, and ultimate strength, as well as fatigue strength.
	Reliability	Optimised fabrication of porous Ti structures with high mechanical strength as well as a high degree of reproducibility, minimal defects, and zero failure rates (within their life expectancy).
Manufacturing-based approach	Manufacturability	The scaffold's micro-geometry should be designed in such a way that it is easy to manufacture with high accuracy and definition.
	Conformance	The mechanical, geometrical, and dimensional characteristics should comply with medical regulations and quality standards.
	Durability	Porous Ti structures should withstand mechanical forces experienced during handling, implantation surgery, and operation thereafter in a traumatised bone microenvironment constantly under load.
User-based approach	Perceived quality	Clinicians should have access to relevant characteristics through medical reports and statistical data where implant performance can be seen.

3.3 QbD Step 1.1: Ideal Quality Attributes

The Ideal Quality Attributes (IQA) are the tissue or biological construct characteristics that must be mimicked to imitate the desired tissue biological architecture and functions. The IQA can be dimensional, physicochemical, mechanical, biological, or functional. However, if any technological and regulatory limitations exist it is important to consider that these IQA serve just as 'ideal models' to imitate even if it is not possible to achieve them. Therefore, the IQA can provide an ideal goal for the product development process in any tissue engineering project.

By studying the properties of natural human bone, we can find the different IQAs for bone implants [79,80]. As mentioned in Section 3.1, the implant must be of adequate strength while the elastic modulus must be similar to that of surrounding tissue to avoid damage [81]. The structure must be porous, using unit cells that allow for fluid movement and bone ingrowth, and the right porosity will reduce the implant stiffness. Interconnectivity completes the flow within the structure and facilitates bone ingrowth [82]. A combination of cancellous and cortical bone properties applied on bone implants can allow the implant to function like a natural bone, yet stronger without impairing surrounding native tissue. Another specific property of bone is its ability to heal itself when fractured [90,143]. The process involves cell migration, differentiation, and cellular proliferation [125]. The ability of natural bone to selfheal must be considered when designing a Ti implant, since the structure will not be able to do the same.

3.3.1. Structure and Composition of Bone

To fully understand the microstructure requirements of fully porous Ti implants, one must first understand the properties of human bone. Bone is essentially an open-cell composite material of fibrous protein, collagen, and calcium phosphate crystals, with an intricate vascular system forming various structures and systems in a five-level hierarchically organised structure [84,144]. According to Rho, et al. [145], these five hierarchical levels are dimensional scales ranging from the macro to the sub-nano levels (Figure 3).

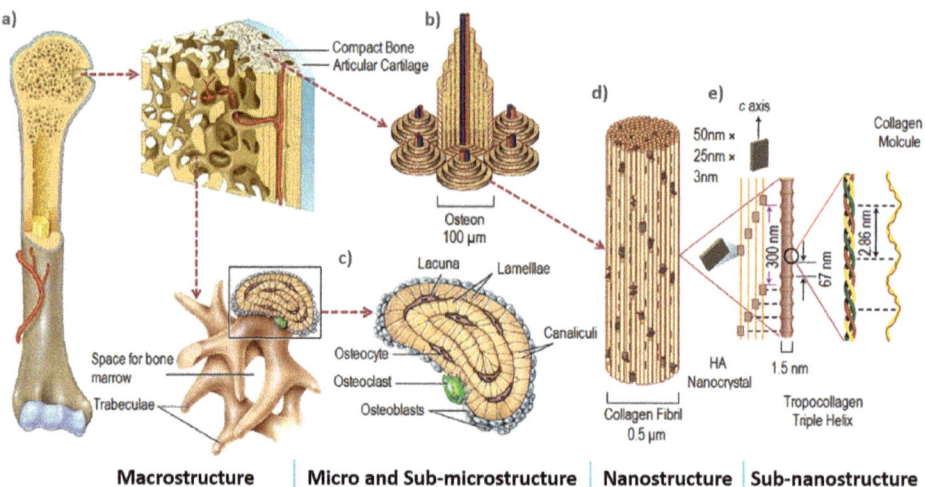

Figure 3. Hierarchical structural organization of bone: (**a**) cortical and cancellous bone; (**b**) osteons with Haversian systems; (**c**) lamellae; (**d**) collagen fiber assemblies of collagen fibrils; (**e**) bone mineral crystals, collagen molecules, and non-collagenous proteins. Image reproduced from Ref [81].

Bone's macrostructure is composed of cancellous and cortical bone, which are two different regions of bone with different density. Cortical bone forms the outside layer of the bone providing a strong, compact structure, leaving only 3–10% of the volume for its biological elements, such as blood vessels, osteocytes, erosion cavities, and canaliculi [81]. Cancellous bone forms the inside of the bone and is spacious and highly porous [81]. The pores are filled with bone marrow and the spacious architecture allows space for metabolic waste and nutrients to flow. Cancellous bone has an active metabolism and regenerates quicker compared with cortical bone [146]. By changing its density, cancellous bone can reorganise its structure depending on the stress direction [107]. These features cause cancellous bone mechanical properties to vary from bone to bone and change longitudinally [144].

The microstructure of bone ranges from 10 to 500 μm and it contains three major cavities. These are Haversian canals, osteocyte, lacunae, and canaliculi [88]. Cortical bone microstructure is composed of cylindrical structures called osteons with diameters ranging between 70 to 140 μm [81,87,92]. Along osteons' central axis are pores called haversian systems, with diameters ranging between 20 to 50 μm, containing blood vessels and nerves [92]. Cancellous bone micro-architecture is composed of irregular units called trabeculae which create its porous structure. The pore geometry and pore size of cancellous bone porous structure is critical for cell distribution and cell migration [81]. Cancellous bone is naturally stochastic, with random pore distribution of pores of different size.

The pores in cancellous bone are ellipsoidal in the natural direction of loading and are usually 300–600 μm wide [147]. The interconnectivity between the pores of cancellous bone is essential for nutrient and waste diffusion.

At the micro level are also located the three types of differentiated bone cells osteoclasts, osteoblasts, and osteocytes, as shown in Figure 4. Osteoclasts and osteoblasts are vital for the functions of developing and healing bone tissue [86]. Osteoclasts (Figure 4a) are the main cells responsible for resorption of old bone tissue. Osteoblasts (Figure 4b) are bone cells responsible for bone formation, remodelling, fracture healing (for which they are critical), and bone development [148]. Osteocytes (Figure 4c) are osteoblasts cells present inside mature bone and serve as mechanosensory cells to control the activity of osteoclasts and osteoblasts [149].

Bone's sub-microstructure, from 1 to 10 μm, is composed of lamellae which in cortical bone compromise the concentric layers of osteons and in cancellous bone lamellae forms the trabeculae volume [150,151]. The sub-nanostructure of bone, below a few hundred nanometres, is composed molecular constituent elements, such as collagen, non-collagenous organic proteins, and mineral. From a few hundred nanometres to 1 μm is bone's nanostructure comprising fibrillar collagen with embedded hydroxyapatite nanocrystals [152].

3.3.2. Bone Mechanical Properties

Most bones in the body are load bearing and require high mechanical strength. Bone tissue is anisotropic and stronger in compression than in tension. The mechanical properties must be measured in two orthogonal directions: longitudinal, which is the natural loading direction, and transverse [86]. The mechanical strength of bone is complex to measure as it varies with age, health, activity, and position in the body [86]. Bone becomes stiffer and less ductile with age and its ability to heal decreases [155]. It is also likely to weaken with immobilisation, such as for a person with movement disabilities or limited physical activity.

The two different types of bone, cortical and cancellous, have completely different mechanical properties; therefore to specify the properties of bone as one material, both bone types must be considered. Cortical bone is highly dense an act like a shell that provides the greatest stiffness and resistance to bending. In contrast, the mechanical properties of cancellous bone are determined by its apparent density and trabecular architecture. The trabecular structure of cancellous bone is arranged accordingly to the stress distribution of load, as shown in Figure 5. As a result, the least material is used in the most strategic locations to carry the greatest loads with the least strain [156].

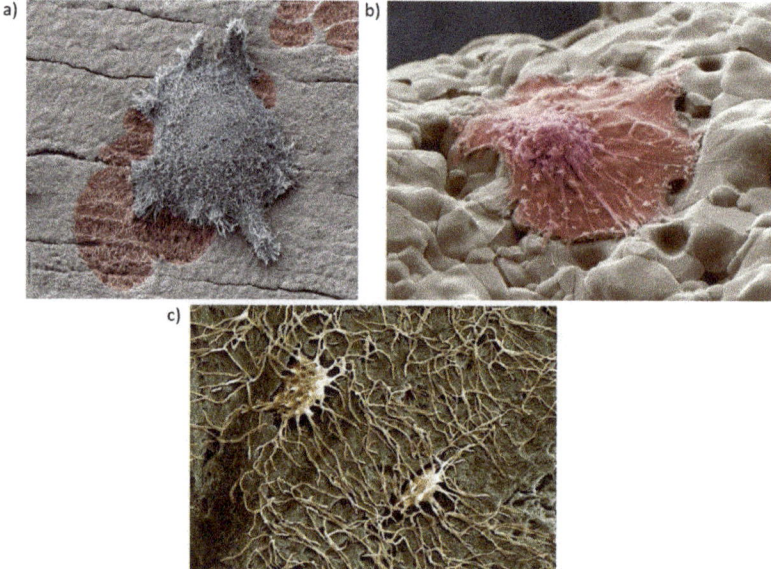

Figure 4. Coloured scanning electron micrographs of bone cells: (**a**) Activated osteoclast and resorption pit by kind permission of Timothy Arnett Ref [153]; (**b**) Osteoblast growing on a bone scaffold made of calcium oxide and silicon dioxide with added strontium and zinc by kind permission of Guocheng Wang from [154]; (**c**) Osteocytes embedded in the bone matrix with long cytoplasmic extensions reaching into the bone tissue, by kind permission of Kevin Mackenzie. Here, the minerals in the bone have been removed by embedding in resin and etching with perchloric acid. This reveals the spaces in the bone and the shape of the osteocyte cells.

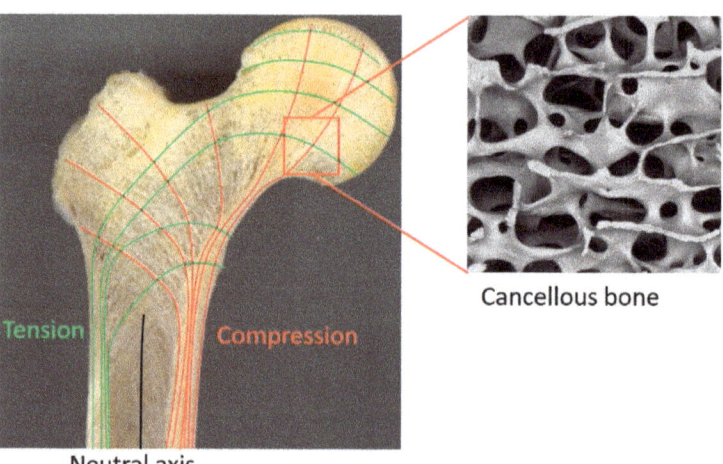

Figure 5. Cortical and cancellous bone; image adapted from [157,158] by kind permission of Alan Boyde.

The mechanical properties of cortical and cancellous bone are difficult to measure, and tend to vary depending on bone orientation, test methods, mathematical formulas, and assumptions [146,151,159]. According to Chen and Thouas [57], the elastic modulus of cortical bone is approximately 11–21 GPa in the longitudinal direction and 5–13 GPa in the transverse direction. Another study suggested

18–22 GPa [147], whereas Wang et al. [81] suggested a range of 3–30 GPa. Cortical bone has a porosity of less than 10% [85]. The elastic modulus of cancellous bone is estimated to be 0.02–6 GPA and it has a high porosity of 50–90% [85,86].

To achieve adequate strength in bone implant design, one must understand the strength requirement of bone. Especially cortical bone must be considered, because the strength of such tissue is the minimum strength required by the implant. The compressive yield point of bone represents the threshold from where the structure accumulates irreversible deformation. Unlike metals such as steel, the yield point cannot be clearly distinguished, and it is rather associated with a continuous transition zone [160]. Furthermore, it has been proven that the compressive yield strength of bone varies depending on the anatomic site [161]. The challenges in determining the yield strength of cortical bone results in varying values in the literature. Researchers have estimated cortical bone to have a compressive yield strength of 133.6 ± 34.1 MPa [162]. The same characteristic was estimated to be 108–117 MPa by Yeni and Fyhrie [163]. Further studies have tested compressive yield strain using uniaxial compression and achieved 141.0 ± 5.0 [164], 111.0 ± 18.6 [165], 112.5 ± 9.5 [166], and 115.1 ± 16.4 MPa [167].

If the loading surpasses the yield point for bone, it will eventually reach the ultimate point. This point represents the ultimate compressive strength the bone can withstand until irreversible strains and damage occur. Past this point, macrocracks are formed and fracture occurs [160]. Unlike yield strength, the ultimate compressive strength of cortical bone can be exactly determined, using a stress–strain experiment [160]. However, due to bone properties differentiating, values vary in the literature. For example, Wang et al. [84] suggested that cortical bone has an ultimate compressive strength of 180–200 MPa, whereas Calori et al. [8] suggested a wider range of 130–290 MPa and Henkel et al. suggested 100–230 MPa [168]. Tables 5 and 6 summarise the mechanical and dimensional properties of natural bone considered in this study.

Table 5. Summary of dimensional properties of natural human bone.

Material	Pore Size	Pore Shape	Porosity	Interconnectivity	Ref
Cancellous bone	300–600 μm	Spongy, ellipsoidal pores	50–90%	55–70%	[82,84,89,91, 169]
Cortical bone	10–50 μm	Cylindrical canals	3–10%	-	

Table 6. Summary of mechanical properties of natural human bone.

Material	Young's Modulus	Compressive Yield Strength	Ultimate Compressive Strength	Compression Fatigue Strength	Ref
Cancellous bone	0.02–6 GPa	7.2–23.2 MPa	17–33 MPa	72.6–124 MPa at 10^6 cycles	[8,82,84,89,91, 93,162,165,168–171]
Cortical bone	3–30 GPa	92.4–167.7 MPa	100–290 MPa	137 MPa at 10^6 cycles	

3.4. Comparison of Properties of Porous Ti Scaffolds Fabricated by AM and Ideal Quality Attributes

3.4.1. Unit Cell Geometry

Metamaterials can be rationally designed by changing their geometry at the microscale of the constituting unit cells of the porous structure. In this systematic search, a total of 169 porous scaffolds were identified as a rationally designed and fabricated for bone implant applications. It was found that there are three preferred strategies for fabricating bio-metamaterials: beam-based, sheet-based, and including irregular porous structures [172].

According to Figure 6, the preferred design approach was beam-based, which represent 74.6% of the total scaffolds produced in the selected studies. The beam-based bio-metamaterials' micro-architecture is composed of a lattice structure created using unit cells based on platonic solids, Archimedean solids, prisms and anti-prisms, and Archimedean duals [173–177] to mimic bone porous macro structure and mechanical properties such as modulus of elasticity. Nevertheless, the biological performance of bio-metamaterials created with beam-based geometries is limited by their inaccurate description of complex natural shapes due to their straight edges and sharp turns [109].

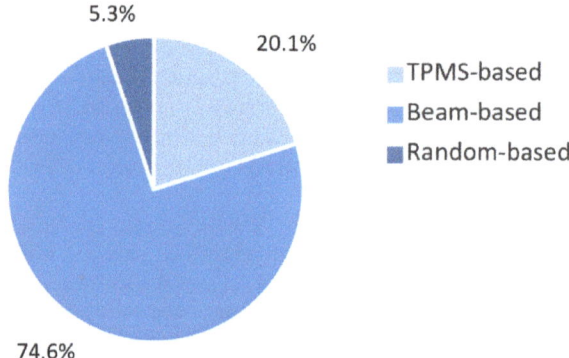

Figure 6. Distribution of experimental studies on beam-based, triply periodic minimal surfaces (TPMS), and random based geometries for bone implant applications.

The unit cells of sheet-based geometries, on the other hand, are based on triply periodic minimal surfaces (TPMS), which are present in different organisms and cellular structures [178]. Therefore, it is no surprising that the second most used design strategy identified in this systematic search was sheet-based representing 20.1% of the total scaffolds produced, from which all used a TPMS as unit cell. Bio-metamaterials based on TPMS can mimic the various properties of bone to an unprecedented level of multi-physics detail in terms of mechanical properties and transport properties [79,172]. Moreover, the bone-mimicking mean surface curvature of zero of TPMS eliminates the effect of stress concentrators at nodal points [120].

In the case of irregular porous structures, are created in a random way to generate irregular porous structures that mimic trabecular bone geometry and mechanical properties [179]. These irregular structures are generated using the Voronoi and Delaunay tessellation methods. Irregular structures have been found to further enhance scaffold's permeability and bone ingrowth compared with porous structures designed with regular unit cells [180]. According to our results only 5.3% of the studies used a randomised design approach to create irregular porous structures to mimic trabecular bone. This result was surprising considering that irregular porous structures were the first additively manufactured coatings used for orthopaedic implants in the medical industry. However, unlike random porous scaffolds, the great advantage of using regular repeating arrays of unit cells made of beam or sheet-based geometries is that they allow the creation of metamaterials with properties that can accurately be predicted [181]. This explain why these design strategies are preferred in research.

From all the different possible unit cells that can be used to produce metamaterials, a total of 17 types of unit cell were used by the selected studies, as presented in Figure 7. According to Figure 7, the beam-based diamond unit cell (59 studies) was the most used, followed by the cubic (18 studies), and the gyroid TPMS (17 studies). These results correlate with the opinion of different experts who have stated that the diamond unit cell is the most studied for the development of metamaterials due to its biomimetic mechanical properties [121,182]. The high mechanical properties and self-supporting properties of the diamond unit cell are due to its unique geometrical arrangement, where one node is tetrahedrally surrounded by four other nodes coming from the crystal structure of the diamond crystal [113], as shown in Figure 8a. Moreover, this arrangement gives 48 symmetry elements to the diamond structure, making this unit cell invariant to different symmetry operations such as translations, reflections, rotations, and inversion [183]. Similarly, extensive research for bone regeneration has been performed to study the cubic unit cell (Figure 8b). The research interest in the cubic unit cell is because it is based on one of the simplest and easy–to–manufacture platonic solids thanks to its struts at an angle of 90° [62]. In the case of porous metamaterials based on the gyroid TPMS (Figure 8c), they have been found to exhibit similar topology to human trabecular bone, and also superior mechanical properties compared with metamaterials based on other types of TPMS [120]. For example, according to Yang et al. [184] metamaterials based on the gyroid TPMS have a more homogeneous stress distribution, which can provide equal mechanical stimulation to bone cells.

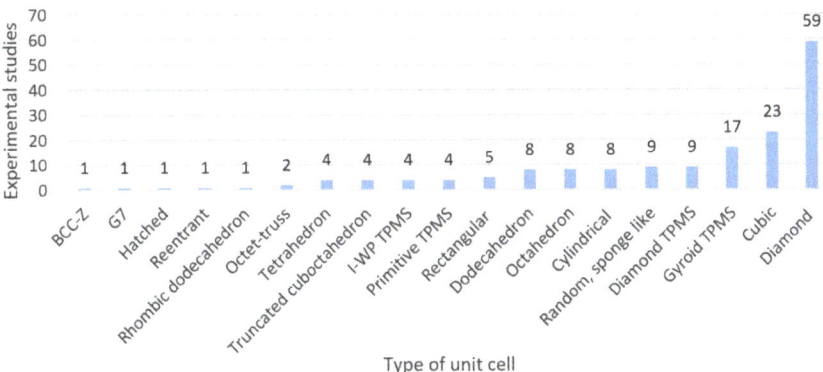

Figure 7. Comparison of the number of experimental studies of each unit used for bone regeneration identified in the reviewed articles.

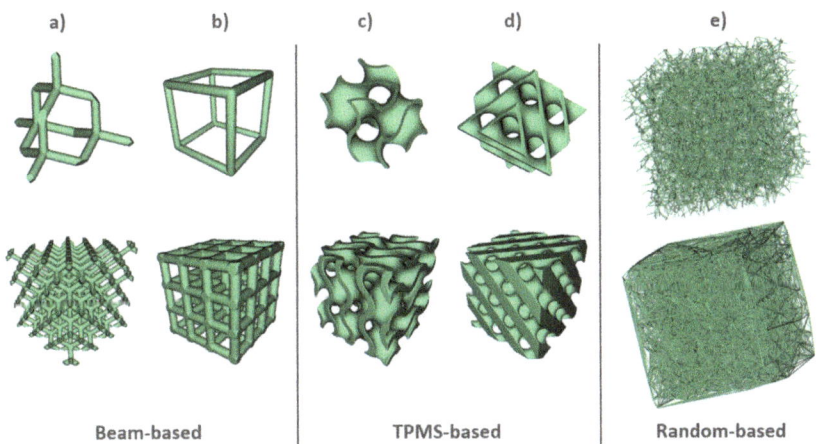

Figure 8. Some examples of unit cells with their corresponding meta-biomaterial scaffold below: (**a**) diamond beam-based unit cell; (**b**) cubic beam-based unit cell; (**c**) Gyroid triply periodic minimal surface based (TPMS-based) unit cell; (**d**) Diamond TPMS-based unit cell; and (**e**) Voronoi (top) and Delaunay (bottom) irregular porous structures.

In the case of random or stochastic structures, Kou et al. [185] suggested that scaffolds based on this structures are more realistic; that is, they look more like natural bone with random non-uniform pore distribution and pore size [185]. Such structures are believed to provide benefits such as improved mechanical properties, including strength, fluid dynamics, surface area, and surface-to-weight ratio.

They combine advantages of small and large pores without necessarily decreasing the mechanical strength or reducing the bone in-growth to levels that are inappropriate in application [186]. Figure 8 presents the five most representative unit cells identified in this study.

3.4.2. Porosity

It is known that the degree of micro-porosity in bone implants directly affects their biological and mechanical properties. The porosity of natural bone is crucial for vascularisation, diffusion of cell nutrients and metabolic waste, and cell migration [187], and in a similar way it is important for metal bone implants. Moreover, several studies have considered porosity as the main parameter affecting stiffness and strength of porous biomaterials. Increased porosity reduces the strength of the implant [187,188]. As a result, porous metallic biomaterials are used as coatings in many medical applications, but more recently porous biomaterials have been proposed for load-bearing applications beyond surface coatings [138,142]. Ti and Ti alloys are commonly used for load-bearing implant applications due to their relatively low elastic modulus fatigue resistance, high strength to weight ratio, and corrosion resistance [82,94]. However, bulk Ti and Ti alloys do not completely match all the mechanical properties of natural bone such as modulus of elasticity. Therefore, it is a need of the hour to accomplish specific mechanical properties for Ti or Ti-based alloys by controlling the porosity and pore characteristics for customised implants [131]. However, the ideal porosity for medical implants seems controversial in the literature [181].

In this systematic review a total of 49 articles out of 50 recorded porosity of various degrees, as shown in Figure 9. For example, Stamp et al. [186] recommend using a porosity above 65% in medical implants whereas Ghanaati et al. [189] found that vascularisation increased in vivo when reducing the porosity from 80 to 40%. Sarhadi et al. [190] and Schiefer et al. [191] have recommended using a porosity of approximately 50%. According to Will et al. [192], the porosity that best promotes vascularization in porous scaffolds is 40–60%. Pattanayak et al. [193] manufactured porous Ti implants and reported an increase in compressive strength from 35 MPa to 120 MPa when reducing the porosity from 75 to 55%. Murr et al. reduced the porosity from 88 to 59% with an increase in stiffness from 0.58 GPa to 1.03 GPa [131]. As mentioned, natural cancellous bone has a porosity of 50–90% [85,86].

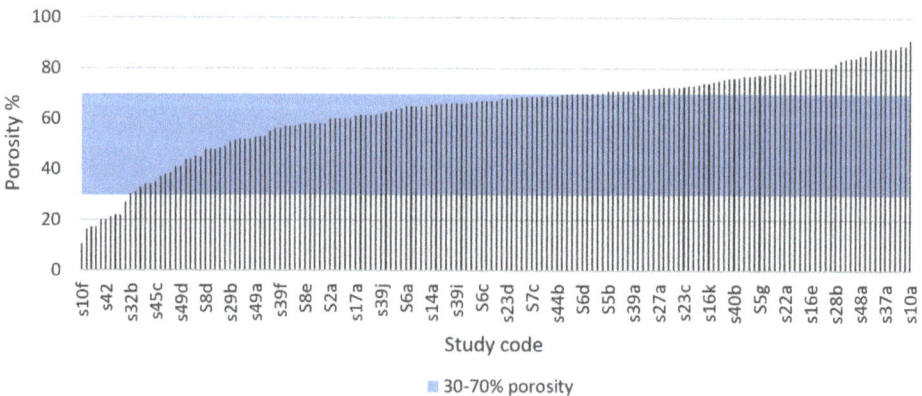

Figure 9. Porosity of implant specimens in the reviewed articles, compared with the Ideal Quality Attribute (IQA) target zone.

Zou et al. [102] designed three implants with similar pore size and shape. By reducing the porosity from 72 to 53% they achieved a compressive strength of 200 MPa and Young's modulus of 4.3 GPa, instead of a compressive strength of 60 MPa and Young's modulus of 2 GPa. Even though both implants achieved a stiffness close to that of human bone, only the implant with 53% porosity achieved a compressive strength greater than human bone. Hence, the implant with 72% porosity would not qualify as a load-bearing bone implant. These results, along with previous research by Pattanyak et al. and Murr et al., confirm the influence of porosity on mechanical properties [193]. They also show that the implant porosity may be adjusted within limits to increase strength and adjust bone stiffness, which varies for each patient's characteristics.

From a medical regulatory perspective, the U.S Food and Drug Administration (FDA) only approves implant porosities of 30–70% for porous coatings on solid Ti implants [194]. The range is relatively large but can be used to one's advantage since both the elastic modulus and strength of the implant can be adjusted by adjusting the porosity. Implants with porosity outside of this range do not comply with FDA regulations and cannot enter the market. It is known that medical regulations, especially for implantable medical devices, are based on strong scientific evidence. Therefore, it is vital to adhere to these regulations when designing bone implants.

Taking into consideration the medical regulations for porous implants, we selected a porosity range of 30–70% as the IQA for porous metal implants to identify studies in the systematic search that fabricated Ti scaffolds with porosity values within this range. From the 49 articles that recorded porosity of various degrees, a total of 167 results were extracted and compared with the selected IQA porosity range. The results of this comparison are presented in Figure 9. According to our results, 56.6% of the porous scaffolds studied in the 49 selected articles had porosity values within the acceptable porosity range (30–70%) required to satisfy medical regulations such as the FDA. By contrast, a total of 26 studies explored the properties of porous scaffolds with porosity values above the acceptable porosity range representing 37.3% of the total results extracted in this systematic literature review. There were several reasons for these studies to explore porosity levels higher than 70%. For example, Zhang et al. [100] and Amin Yavari et al. [106] fabricated different porous structures with various porosities to explore their mechanical properties and deformation mechanisms. Moreover, porous Ti scaffolds with high levels of porosity can serve as storage for mesenchymal stem cells to facilitate bone tissue regrowth, and also to improve cell oxygenation and nutrition [100]. On the other hand, Ti porous scaffolds with porosity levels lower than 30% can provide similar mechanical properties to cortical bone [98,100].

3.4.3. Macropore Size

Since macro pore size is directly related to the strength, porosity, and stiffness of the implant, it is an important property for implant design [195]. Pore size has a profound effect on the behaviour of osteogenic cells even in an organ culture system [196]. The implant's macro porosity determines whether bone cells can successfully penetrate and grow within the structure, and many studies have discussed the influence of pore size on the biological properties of implants [197]. Furthermore, several studies have shown that a minimum pore size of 100 μm is required for vascularization and bone ingrowth, but pores larger than 100 μm increase bone in-growth by allowing improved vascularization and oxygenation [86,91,193]. A minimum macropore size limit of 100 μm is supported by further research as vascular penetration has been found to be restricted in smaller pore interconnections [85,168,188,192].

Studies have found that pores greater than 300 μm are required for vascularisation and bone ingrowth [86,168]. Tang et al. [188] found that 200–350 μm is the optimal macropore size, and various studies have found that bone ingrowth is less likely to occur beyond 400 μm [90,195,198]. However, research that used pore sizes of 300, 600, and 900 μm in porous Ti scaffolds found that those with macropores sizes of 600 and 900 μm had much higher bone ingrowth compared with the scaffolds with 300 μm pores [101]. Bose et al. [85] suggested that all macropore sizes between 100 and 600 μm are osteoconductive. Fukuda et al. [199] experienced greater results in 500 and 600 μm pores compared with 900 and 1200 μm pores. Xue et al.'s [98] results showed that macropore sizes in the range of 100–600 μm possess the optimum ability for cell growth into the pore structure of porous titanium.

According to the FDA regulations, macropore sizes of 100–1000 um are approved for coatings for Ti implants [194]. Large macropores have a smaller surface area than do small pores, decreasing the cell attachment on the implant [86]. However, large macropores increase scaffold vascularisation, which is vital for supplying oxygen and nutrients to the tissue as well as osteoblast proliferation and migration [80], but they decrease the mechanical strength of the material. The limit of how much the macropore size can be increased while maintaining sufficient mechanical strength depends on both the material and the processing conditions. Therefore, regulatory guidelines for surface coating may be misleading for fully macroporous implants, since the strength-to-weight ratio differs between a porous and solid structure. Since the porosity decreases the strength of the implant, and large pore sizes decrease the strength of the internal architecture, large pore sizes must be avoided to increase the structure's strength. A more defined pore size range is therefore sought [86].

In this study, the macroporosity used in different studies was explored. It was found that no consensus currently exists on what upper limit to macropore size that is ideal, but somewhat of a consensus on the lower limit exists (100 µm). According to FDA regulations, porous implants should have macropore sizes between 100 and 1000 µm [194]. It has further been found that macropores start to lose their osteogenic functionality when larger than 500–600 µm [8,16,85,98,199]. Considering that a fully porous structure is weaker than a solid structure, and that high strength is vital for implants, it can be assumed that there is no need to design a structure with pores larger than what is needed to cater for all functions within the implant.

These findings made us choose a macropore size range of 100–600 µm as the IQA for porous metal implants to identify the studies in the systematic search that fabricated Ti scaffolds with pore size values within this range. From the 42 articles that recorded pore size of various degrees, a total of 144 results were extracted and compared with the selected IQA pore size range, as seen in Figure 10. According to our results, 51.4% of the results of all studies had a macropore size within 100–600 µm. It was further noted that 86.8% of the experimental results of all studies had a macropore size within the FDA recommended range of 100–1000 µm. From these results, we could infer that most of the research studies identified through the systematic search somewhat considered the macroporosity range required to satisfy medical regulations.

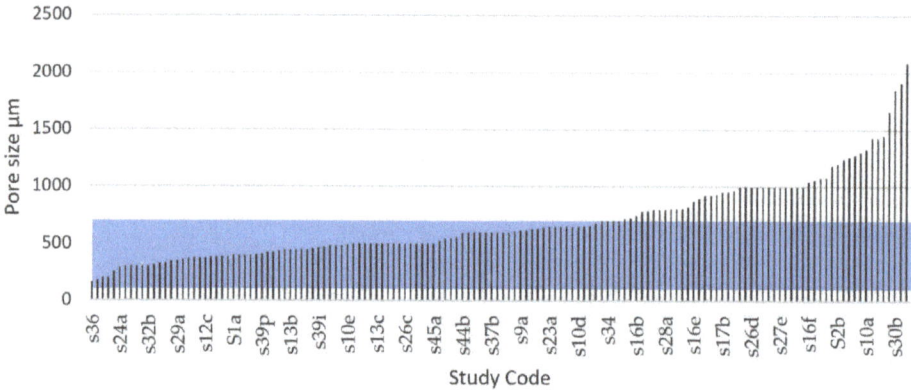

Figure 10. The pore sizes recorded in the reviewed articles, compared with the Ideal Quality Attribute IQA target zone.

By contrast, a total of 68 results out 144 showed pore size values above the acceptable macropore size range representing 48.6% of the total results extracted in this systematic literature review. There were several reasons for these studies to explore pore sizes above 600 µm. For example, the FDA approves macropore sizes between 100 and 1000 µm [194]. Hara et al. [96] tested four porous structures with different macropore size to explore their mechanical properties. Taniguchi tested 300, 600, and 900 µm pore sizes and found that the structures with 600 and 900 µm pore size exhibited higher bone ingrowth. Large macropores increase scaffold vascularisation, which is vital for supplying oxygen and nutrients to the tissue as well as osteoblast proliferation and migration [80]; however, larger pores also have a smaller surface area compared with small pores, decreasing the cell attachment on the implant [86]. Larger macropores also result in higher porosity, and this reduces the strength of the implant [187,188].

The ideal macropore size for bone implants is controversial and undefined, and according to Otsuki et al., a reason for the varied pore size data may be that the interconnectivity of pores was not considered [198]. Macropore size regulations have been developed for the first generation of porous implants, which use a single-scaled porous network with repeated, equally sized, and shaped pores. However, bone grows in a naturally random structure with pores of different sizes, shapes, and directions similar to the structure of a sponge [130,155]. The use of a multiscale porous scaffold that combines smaller and larger pore sizes within the same structure is a recent strategy to optimise the internal architecture of implants [142,188]. This method combines advantages of both small and large macropores without decreasing the strength or reducing the bone in-growth to levels that are inappropriate in application [188]. According to our results, a total of 8 out of 50 studies used some sort of multiscale pore approach, but it was observed that the design method varied, and that the researchers failed to provide the percentage of the total structure that used each pore size. It was further observed that a multiscale porous structure occurred in some implants where the manufacturing of a single-scaled structure resulted in varying pore sizes due to unprecise manufacturing tolerances.

3.4.4. Pore Inter Connectivity

According to our results, 46% of the collected studies registered pore interconnectivity. Interestingly, all of these studies designed their porous scaffolds with an interconnectivity of 100%. The pores in a porous bone implant must be interconnected to ensure movement and the supply of necessary nutrients through ingrowth of tissue and bone [200]. Interconnected pores tend to facilitate the flow of fluids and biological cells through the structure which is essential for bone tissue formation [185]. According to Nyberg et al. [201] the integration of artificial material tissue with native tissue can be improved by interconnected pores. Tang et al. [188] suggested that an increased pore interconnectivity increases the number and size of blood vessels formed in scaffolds. The interconnectivity is also a critical factor for ensuring that all cells within the structure are within a 200 μm range from a blood supply to provide transfer of nutrients and oxygen [202]. According to the FDA's recommendations for porous metal coatings, pores in such structures must be interconnected [194]. Although this requirement is for surface coatings, it also indicates the importance of an interconnected porosity for fully porous implants. In the systematic review, it appears as though a vast majority of studies had indicated the importance of an interconnected porosity. Therefore, to guarantee all processes and fluid movements necessary for tissue and bone ingrowth, the selected IQA for pore interconnectivity would ideally be 100%.

3.4.5. Elastic Modulus

It was observed in this systematic review that an elastic modulus is a property commonly reported in AM porous scaffolds studies (by 89% of all studies). A controlled modulus of elasticity has proved to be critical in prostheses and scaffolds to avoid stress shielding [81,82]. Stress shielding occurs when there is a stiffness mismatch between the implant and surrounding bone, and it can cause inflammation and the need for revision surgery [197]. Ti and common implant Ti alloys have an elastic modulus of roughly 100–120 GPa [81,84,138,197]. A reduced modulus is necessary to avoid stress shielding and can be achieved by designing implants with a porous structure [90].

Defining an ideal specific modulus of elasticity for porous bone implants is not practical because Since the mechanical properties of human bone, especially the elastic modulus, change drastically with factors, such as age, physical activity, and health. For example, femoral bone specimens from patients aged 3, 5, and 35 years had an elastic modulus of 7, 12.8, and 16.7 GPa, respectively, indicating a dramatic change with age [90]. As previously shown in Table 6, the elastic modulus of human bone varies in the literature. Chen and Thouas [57] estimated the elastic modulus of cortical bone to be approximately 11–21 GPa in the longitudinal direction, whereas Lee et al. [147] suggested 18–22 GPa.

Wang et al. [81] suggested a wider range of 3–30 GPa. These findings indicate that the stiffness of an implant may need to be adjusted specifically for the person it is intended for, and that the target value for the elastic modulus may be specific to each patient. Therefore, it is more practical to think that for the design of porous scaffolds, an ideal target area exists for the modulus of elasticity. Based on this, the IQA for elastic modulus is proposed to be 3–30 GPa for fully porous Ti implants.

Figures 11 and 12 show the elastic modulus that was reported in the reviewed articles and these values were compared with the proposed IQA. From the extracted of elastic modulus results, 55.5% reached the target area of 3–30 GPa. These implants achieved an elastic modulus within the range of natural bone and would therefore eliminate risk of stress shielding. By contrast, 40% of the results exhibited an elastic modulus below 3 GPa, and only 3.6% of the results reported an elastic modulus higher than 30 GPa. These results clearly demonstrated that most studies are aiming towards a modulus of elasticity closer to the bone modulus.

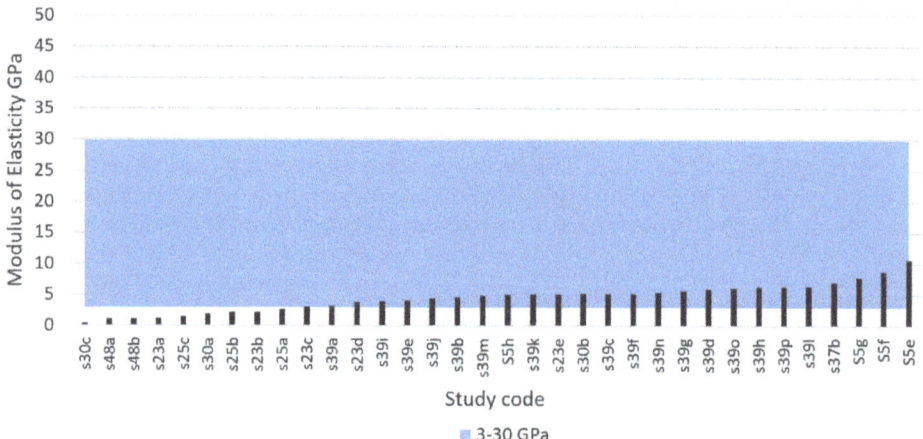

Figure 11. Elastic modulus of porous metamaterials based on triply periodic minimal surfaces (TPMS) compared with the Ideal Quality Attribute (IQA) target zone.

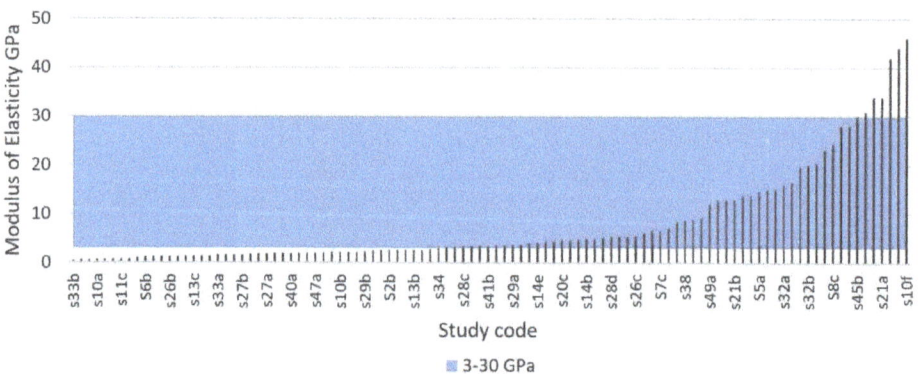

Figure 12. Elastic modulus of porous beam-based metamaterials compared with the Ideal Quality Attribute (IQA) target zone.

The elastic modulus of metals such as titanium and its alloys naturally have a much higher elastic modulus compared with bone [81]. However, research shows that the elastic modulus of metals can be readily adjusted by modifying their porosity. Porous metals with a low modulus of elasticity correspond to high levels of porosity. For example, Wang et al. [84] explored five types of porous structures using the same material (TiNbZr) and pore size (550 µm) but ranging porosity. His results revealed that four of the implants with porosities ranging from 42% to 69% all achieved an elastic modulus within the approved range of 3–30 GPa; however, the implant with the highest porosity (74%) achieved the lowest elastic modulus of 1.6 GPa. Similarly, Li et al. [203] used a porosity of 91% resulting in a low elastic modulus of 0.8 GPa. Furthermore, Chen et al. [204] received an elastic modulus of 44.4 GPa for a porous titanium structure using 30% porosity but by increasing the porosity to 40% the elastic modulus was reduced to 24.7 GPa.

Using the data obtained through the systematic literature search we calculated two multiple linear regressions to predict the modulus of elasticity of beam and TPMS-based AMd Ti scaffolds. The regression model used the independent variables of pore size, relative density (porosity), and the interaction of pore size–porosity. According to our results, a regression equation was found for beam-based AMd Ti scaffolds ($F(3,75) = 54.139$, $p < 0.0001$), with an R^2 adj of 0.671, as shown in Figure 13a. The residuals of the multiple linear regression are randomly scattered around the centre line of zero with no obvious pattern. The predicted compressive yield strength of beam-based scaffolds is equal to $27.738 - 0.078$ (pore size) $- 27.417$ (porosity) $+ 0.0689$ (pore size*porosity), where pore size is coded or measured in µm, and relative density expressed as porosity as a percentage. The beam-based scaffolds' modulus of elasticity decreased 0.078 MPa for each µm, 27.417 MPa per 1% of porosity increment, and increased 0.0689 MPa for the interaction pore size*porosity. Both pore size ($p < 0.0005$) and porosity ($p < 0.0001$) were significant predictors of beam-based scaffolds' modulus of elasticity, including the interaction between pore size and porosity ($p < 0.0001$).

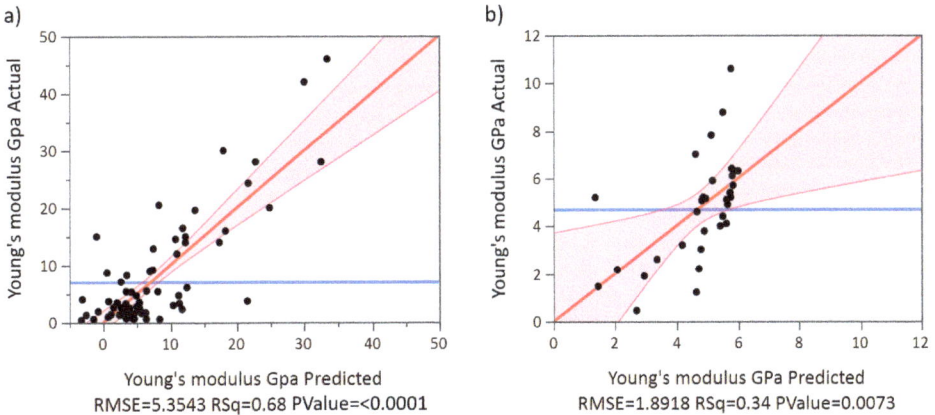

Figure 13. Overall predicted model of elastic modulus actual versus an elastic modulus predicted (a) beam-based scaffolds, and (b) TPMS-based scaffolds.

In the case of the multiple linear regression of TPMS based AMd Ti scaffolds a significant regression equation was also found ($F(3,28) = 4.897$, $p < 0.0073$), with an R^2 adj of 0.273, as shown in Figure 13b. The residuals of the multiple linear regression are randomly scattered around the centre line of zero with no obvious pattern. The scaffolds' predicted modulus of elasticity is equal to $0.008 - 0.002$ (pore size)-2.342 (porosity) $+ 0.002$ (pore size*porosity). The TPMS based scaffolds' modulus of elasticity decreased 0.002 MPa for each µm, 2.342 MPa per 1% of porosity increment, and 0.002 MPa for the interaction pore size*porosity. Pore size was a significant predictor of TPMS-based scaffolds' modulus of elasticity with a p-values < 0.0563. However, porosity was not a significant predictor with p-values < 0.553 and 0.843, respectively. Moreover, no interaction between pore size and porosity was found regarding to modulus of the elasticity.

3.4.6. Compressive Yield Strength

For an adequate functioning of any load-bearing implant, it is vital that its design withstand the required forces and loading cycles. Mechanical strength is one of the implant's most crucial features for avoiding implant failure. To withstand the loads of daily activities, load-bearing implants must have at least the same yield strength as the bone that they replace [81]. The yield point of bone represents the threshold from where the structure accumulates irreversible deformation. Unlike bulk metals such as steel, the yield point of bones cannot be clearly distinguished; it is rather associated with a continuous transition zone [160]. Strain beyond the yield point will deform the structure beyond its point of resilience causing material damage, usually occurring as micro-cracks [205]. Bone tissue has evolved to mainly support compressive stress [206,207]. Bone is 30% weaker under tensile stress, and 65% weaker under shear stress [208]. Therefore, load-bearing implant scaffolds require a high compressive strength to prevent fractures and improve functional stability [209]. The compressive yield strength of cortical bone varies in the literature. As previously shown in Table 6, the compressive yield strength of cortical bone varies approximately between 90 and 170 MPa. To replace like with like, using a biomimetic approach for comparison purposes, a minimum and a maximum compressive yield strength of 90 MPa and 170 MPa were selected as the IQA for fully porous Ti implants.

The systematic search identified that 37 out of 50 studies recorded compressive yield strength, from which a total of 133 experimental results were extracted and compared, as shown in Figures 14 and 15. Figure 14 presents the results of studies using TPMS structures and Figure 15 presents the results of studies using porous beam-based metamaterials. Both comparisons show high numbers of studies resulting in a compressive yield strength below 90 MPa. A total of 55.7% of all results had a compressive yield strength below the defined IQA target and 25% of the studies achieved a compressive yield strength within the bone region. On the other hand, only 19% of the extracted experimental results had a strength above the bone region. Such implants would have strengths similar to or higher than cortical bone and are expected to not experience permanent deformation caused by the expected bone compressive loading conditions in the human body.

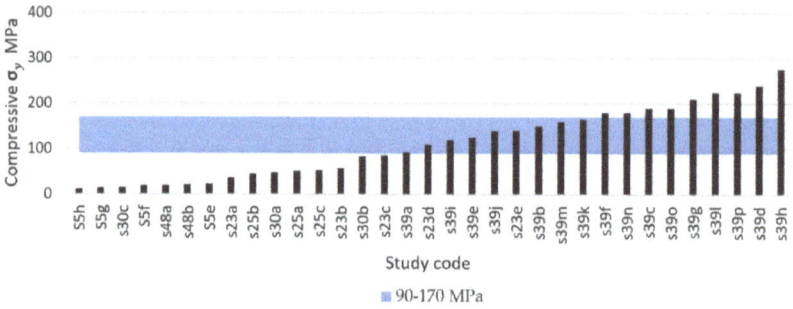

Figure 14. Compressive yield strength of porous metamaterials based on triply periodic minimal surfaces (TPMS) compared with the Ideal Quality Attribute (IQA) target zone.

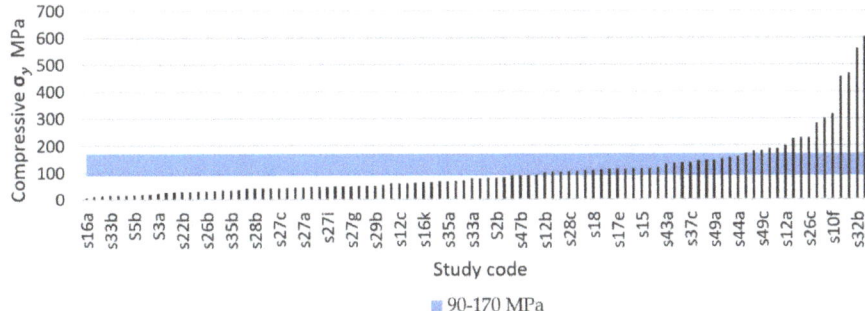

Figure 15. Compressive yield strength of porous beam-based metamaterials compared with the Ideal Quality Attribute (IQA) target zone.

Decreased strength of a porous implant can result from high porosity and large pore sizes, [81,188]. For example, Zhang et al. [113] fabricated porous scaffolds based on the TPMS diamond unit cell with a wide range of compressive yield strengths from 36 MPa to 140 MPa just by varying the scaffolds' porosity and maintaining the pore size constant. The type of unit cell used to design porous scaffolds can also drastically change it mechanical properties. For example, Zhao et al. [116] fabricated four porous scaffolds with the same pore size and similar porosities using two different unit cells (tetrahedron and octahedron). However, the scaffolds based on the octahedron unit cell registered almost double the compressive strength compared with those based on the tetrahedron unit cell [116]. The compressive yield strength of porous metals can also be enhanced by gradually changing the porosity level along the radial direction of the scaffold. This was demonstrated by Zhang et al. [100], who reported functionally graded porous scaffolds based on the diamond unit cell with superior comprehensive mechanical properties to the biomaterials with uniform porous structures.

Using the data obtained through the systematic literature search, we calculated two multiple linear regressions to predict compressive yield strength based on pore size, porosity, and the interaction of size–porosity for beam-based and TPMS-based AMd Ti scaffolds, respectively.

According to our results, a regression equation was found for beam-based AMd Ti scaffolds ($F_{(3,75)}$ = 31.452, $p < 0.0001$), with an R^2 adj of 0.539, as shown in Figure 16a. The residuals of the multiple linear regression are randomly scattered around the centre line of zero, with no obvious pattern. The scaffolds' predicted compressive yield strength is equal to 380.557 − 0.075(pore size)−350.828 (porosity) + 0.557 (pore size*porosity), where pore size is coded or measured in μm, and porosity is measured as a percentage. The beam-based scaffolds' compressive yield strength decreased 0.075 MPa for each μm, 350.828 MPa per 1% of porosity increment, and increased 0.557 MPa for the interaction of pore size*porosity. Both pore size ($p < 0.0378$), porosity ($p < 0.001$), and the interaction between pore size and porosity ($p < 0.0048$) were significant predictors of beam-based scaffolds' compressive yield strength. The interaction between pore size and porosity was found to be significant with a p-value < 0.0048.

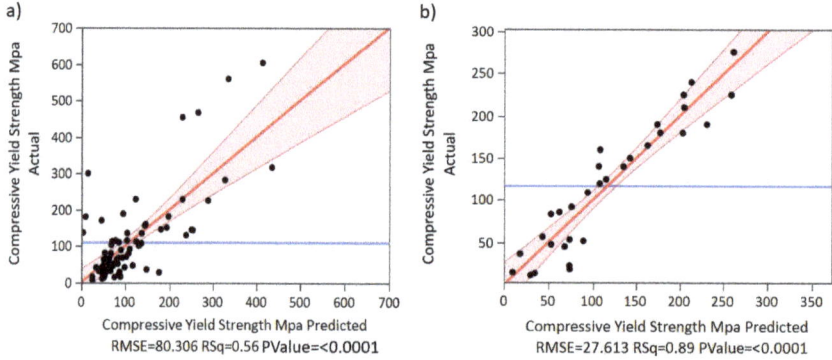

Figure 16. Overall predicted model of compressive yield strength Actual versus compressive yield strength Predicted of (**a**) beam-based scaffolds, and (**b**) TPMS-based scaffolds.

Regarding the multiple linear regression of TPMS-based AMd Ti scaffolds, a significant regression equation was also found (F(3,27) = 65.547, $p < 0.0001$), with an R^2 adj of 0.872, as shown in Figure 16b. The residuals of the multiple linear regression are randomly scattered around the centre line of zero, with no obvious pattern. The scaffolds' predicted compressive yield strength is equal to 524.780 − 0.008(pore size)−625.266 (porosity) + 0.183 (pore size*porosity). The TPMS-based scaffolds' compressive yield strength decreased 0.008 MPa for each μm, 625.266 MPa per each 1% of porosity increment, and increased 0.183 MPa for the interaction of pore size*porosity. Porosity was a significant predictor of TPMS based scaffolds' compressive yield strength with p-values < 0.0001. However, pore size and the interaction between pore size and porosity were non-significant, with p-values < 0.7018 and 0.3260, respectively.

3.4.7. Ultimate Compressive Strength

If loading surpasses the yield point of bone, it will eventually reach the ultimate point. This point represents the maximum compressive strength that a material can withstand without irreversible strains and damage occurring. Past this point, macrocracks are formed and fracture occurs [160]. Bone implants in load-bearing applications must withstand high stress within the body, to a degree where no permanent deformation occurs during the load that the implant is expected to be exposed to. Hence, controlled ultimate compressive strength is a crucial property to study in bone implant research. Natural bone is estimated to have an ultimate compressive strength of 180–200 MPa [84]. However, results vary in the literature. For example, Calori et al. [8] suggested a more widespread range of 130–290 MPa, whereas Henkel et al. suggested 100–230 MPa [168]. To replace like with like using a biometric approach, bone implants should have an ultimate compressive strength similar to that of bone [8]. Taking into consideration the compressive yield strength suggested previously and the three results presented in Table 5, the proposed IQA region for the ultimate compressive strength is between 180 MPa and 290 MPa.

In this systematic search a total of 60 experimental results of ultimate compressive strength from 19 different studies were extracted. Figures 17 and 18 show the ultimate compressive strength of the different studies compared with the defined IQA target of between 180 MPa and 290MPa. Figure 17 corresponds to experimental results of porous scaffolds composed of TPMS unit cells compared with the IQA target. According to Figure 17, only one study with three different experimental results measured the ultimate strength of porous scaffolds based on TPMS unit cells. In this study by Yanez et al. [120], three ultimate compressive strengths of 17, 47.5, and 83.5 MPa were achieved using the gyroid unit cell. Dramatic improvement in the ultimate compressive strength of the scaffold was achieved. This improvement in mechanical properties was possible by slightly changing the gyroid unit cell into an elongated gyroid. However, none of the experimental results obtained by Yanez et al. [120] were able to reach the minimum IQA ultimate compressive strength proposed in this study. The low strength of Yanez at al.'s [120] samples can be attributed to their high porosity values which ranged between 75% and 90%. This increased the stress concentration, reducing the ultimate compressive strength.

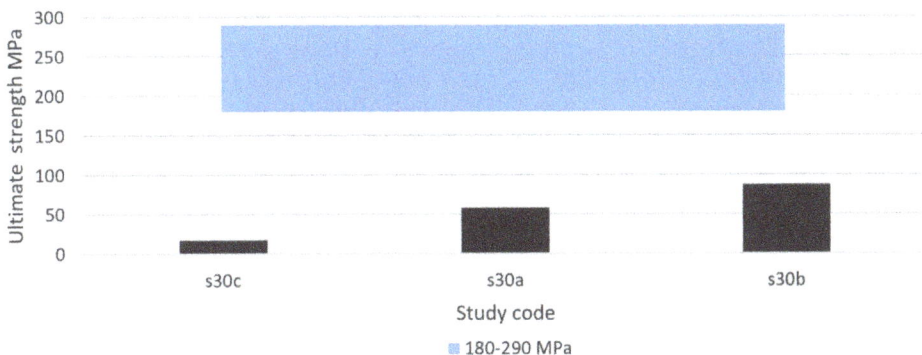

Figure 17. Ultimate compressive strength of porous metamaterials based on triply periodic minimal surfaces (TPMS) compared with the Ideal Quality Attribute (IQA) target zone.

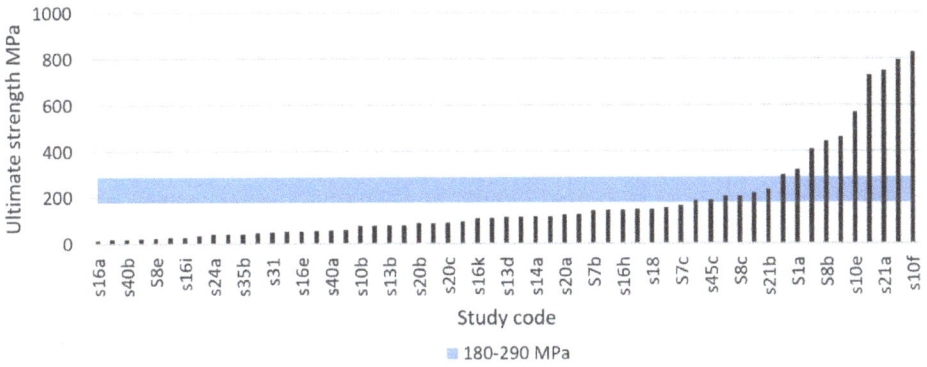

Figure 18. Ultimate compressive strength of porous beam-based metamaterials compared with the Ideal Quality Attribute (IQA) target zone.

Figure 18 is composed of 19 different studies on beam-based unit cells where 57 experimental results are compared with the IQA target. According to Figure 18, 10 experimental results reached higher ultimate compressive strengths than bone. The highest ultimate compressive strength (830 MPa) recorded was achieved with a scaffold based on the diamond unit cell by Zhang et al. [100]. From all results recorded, a total of 16.7% achieved an ultimate compressive strength above the proposed IQA ultimate compressive strength; 10% had similar ultimate compressive strengths to bone; and 73.3% had lower ultimate compressive strengths than bone. The scaffolds that did not fulfil the required IQA ultimate compressive strength would risk fracturing due to macrocracks occurring during high loads. Porosity within a structure has been proven to decrease the strength of a structure [81,188], which may explain the high number of scaffolds with low ultimate strength.

The goal of designing a porous structure with enough porosity and pore size without diminishing strength is a difficult task, and as researchers aim to create highly porous structures with low elastic modulus, many structures experienced low ultimate strength. Attar et al. [111] manufactured a porous titanium structure by SLM with rectangular pores and 17% porosity and achieved an ultimate compressive strength of 747 MPa. Using the same material, manufacturing method, and unit cell, a different structure with 37% porosity achieved an ultimate strength of 235 MPa. In a similar manner, Chen et al. [204] designed three structures of the same material and manufacturing method. Using porosities of 30%, 40%, and 50%, the ultimate compressive strengths recorded were 524, 301.7, and 120.3 MPa, respectively, indicating that increased porosity reduces the strength of the structure.

3.4.8. Fatigue Strength

During normal daily activities, load-bearing implants experience just a fraction of the material's ultimate stress [209–211]. However, after years of use, the high cyclic loading to which load-bearing implants are subjected eventually leads to the accumulation of small stresses, causing progressive and localised material damage that results in implant failure [212]. For instance, one of the most critical mechanical properties for load-bearing implants is fatigue strength. However, fatigue strength is the most difficult mechanical property to determine [213].

The required fatigue resistance of a load bearing implant and its components mainly depends on their cyclic loading conditions and the required life span. For example, it is estimated that lower limb prostheses are subjected to up to 2 million gait cycles per year [214], and in the case of orthodontic prostheses these can reach up to 300,000 loading cycles per year [213]. Therefore, a large variety of medical standards exist for testing fatigue strength. Some of these fatigue tests differ depending on the type of load applied such as tension–tension, compression–compression, and tension–compression. In the case of load-bearing bones, their loading conditions in real-life activities are complex [135]. However, bone is mainly loaded in compression [206,207]. Therefore, to test the fatigue life of metamaterials for bone implant applications, compression fatigue tests are preferred due to the simplicity of the test setups [215].

Regarding the number of cycles that load bearing implants and their components need to have tested for fatigue strength, all the different medical standards agree that such products need to have a fatigue life within the high-cycle fatigue region ($N > 10^4$ cycles). For example, the ASTM standard F2777 – 16 recommends testing tibial inserts' endurance and deformation under high flexion with a minimum number of cycles of 2.2×10^5, and in the case of dental implants they are typically tested up to 5 million cycles [213,216]. Nevertheless, for a component of a load-bearing implant to have at least 25 years of life span [217], the highest number of cycles that must be tested is 10^7 cycles [213,216]. Taking into consideration current medical standards for load-bearing implants, the high-cycle fatigue region between 10^4 cycles and 10^7 cycles was selected as the IQA fatigue life for porous titanium metamaterials for bone regeneration.

Using the selected high-cycle fatigue region, this systematic literature search identified a total of 13 different studies on fatigue resistance, among which 11 studies performed compression–compression fatigue tests. Then, for comparison purposes, a total of 51 experimental results were extracted and compared. Moreover, to facilitate the comparison of the results of the studies, they were classified according to the type of unit cell used to produce porous structures as beam and TPMS-based as resented in Figures 19–21. According to our results, the TPMS porous structures that withstood the highest stresses at the high-cycle fatigue region were achieved by Bobbert et al. [79]. The primitive TPMS structure presented the highest stress within the high-cycle fatigue region, with 232 MPa at 3×10^4 cycles, as shown in Figure 19. The TPMS porous structures that were able to withstand the second and third highest stresses within the high-cycle fatigue region were the I-WP and diamond structures, with 227 MPa at 3×10^5 cycles and 204 MPa at 3×10^6 cycles as shown in Figures 19 and 20. Remarkably, the primitive TPMS structure was the only one to pass the 10^7 threshold with 80 MPa at 3×10^7 cycles, as presented in Figure 20.

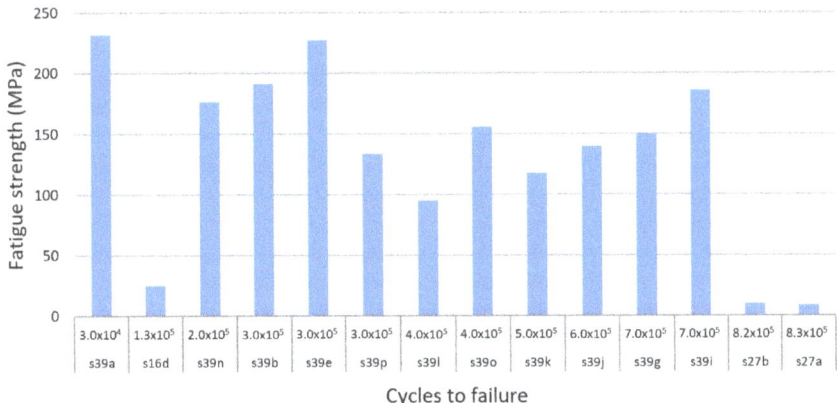

Figure 19. Comparison of the results of different studies on the fatigue strength of porous metamaterials based on triply periodic minimal surfaces (TPMS) within the high-cycle fatigue region between 10^4 and 10^5 cycles.

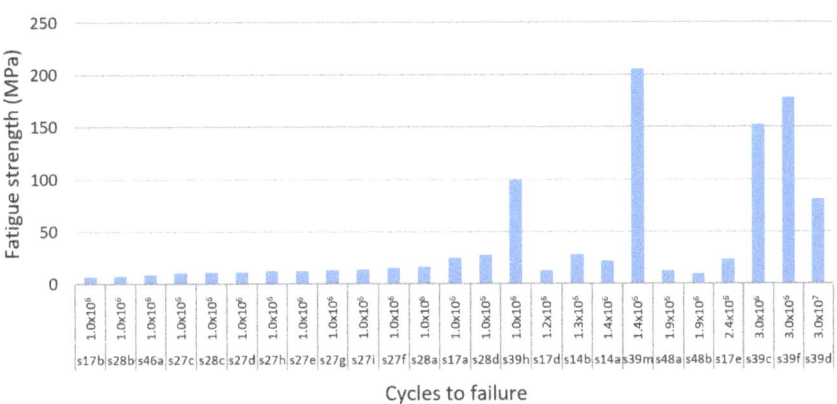

Figure 20. Comparison of the results of different studies on the fatigue strength of porous metamaterials based on triply periodic minimal surfaces (TPMS) within the high-cycle fatigue region between 10^6 and 10^7 cycles.

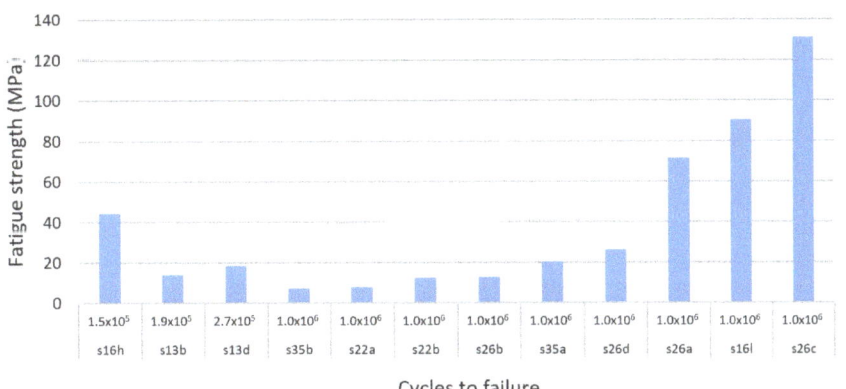

Figure 21. Comparison of the results of different studies on the fatigue strength of porous beam-based metamaterials within the high-cycle fatigue region between 10^6 and 10^7 cycles.

In the case of beam-based metamaterials the study performed by Zhao et al. [116] achieved the highest fatigue strength with 130 MPa at 10^6 cycles using a lattice structure based on the octahedron unit cell, as shown in Figure 21. In second and third place with the highest fatigue strength are the tetrahedron and the cubic porous lattice structures with 90 MPa at 1×10^6 cycles and 90 MPa at 1×10^6 cycles fabricated by Amin Yavari et al. [106] and Zhao et al. [116], as presented in Figure 21.

According to the results of this systematic search it could be seen that TPMS structures provide superior fatigue strength to porous bio-metamaterials compared with beam-based unit cells. Moreover, in terms of fatigue resistance, it was identified that the primitive, I-WP, and diamond TPMS provided the best performance, whereas the octahedron, cubic, and tetrahedron are the best-performing lattice unit cells. However, it is crucial to note that several factors can affect the fatigue life of additively manufactured Ti metamaterials. Some of these factors are residual stresses and stress concentrators caused by high surface roughness and manufacturing defects [218]. Moreover, the fatigue strength of bulk materials significantly degrades when they are in porous form or when voids and pores are developed during fabrication [219]. In the case of additively manufactured components, it has been proven that their fatigue strength is extremely sensitive to localised and nonuniform heat and uncontrolled cooling cycles during fabrication [220,221].

3.5. Discussion and Summary of the Findings

Table 7 presents a summary of proposed IQA target values for porous Ti and Ti alloy bone implants aimed at load-bearing applications. These values are based on scientifically supported values found in human bone research, federal regulations (FDA), and research articles on porous Ti implants manufactured using AM between 2000 and 2020. These properties are part of the first step of the QbD framework of "Define the Quality Target Product Profile." The IQAs are necessary for a systematic and qualitative design approach. These properties will provide benchmark guidance to facilitate future research on porous bone implant design.

Table 7. Summary of Ideal Quality Attributes proposed in this study.

Ideal Quality Attributes	
Porosity	30–70%
Pore size	100–600 μm
Elastic modulus	3–30 GPa
Compressive yield strength	90–170 MPa
Ultimate compressive strength	180–290 MPa
Fatigue resistance	72.6–137 MPa at 10^6
Interconnectivity	100%

An IQA target zone for the porosity of porous Ti implants has been proposed as 30–70% based on results found in research articles as well as in current FDA regulations. We found that 56.6% of all studies in this review achieved a porosity within this range. AM was found to produce porous structures with highly controlled porosity. Numerous studies as well as FDA regulations have discussed the importance of using porous structures for tissue ingrowth, which have numerous advantages to non-porous implants. Whereas porous coatings are used in some instances, it was observed in this study that many researchers believe in using a fully porous structure to achieve a biomimetic structure mimicking natural bone. All data used extracted in this systematic literature review came from studies that used scaffolds with constant porosity. However whether a repeated lattice structure throughout the entire structure is sufficient or a biomimetic "sponge-like" structure with irregular, elongated pores should be used has not yet been confirmed in research; nevertheless implants using a biomimetic design approach replicating natural bone received attention in recent research [86]. Since bone has a random and stochastic structure with pores of different sizes, a multiscale porous structure with pores of different sizes deserves further research. Multiscale porous scaffold structures have been proven to perform better than one-dimensional structures [142]. Nonetheless, to properly compare implants with a multiscale structure, pore distribution and size must be recorded.

In the case of scaffolds' pore size, the IQA was proposed to be 100–600 µm based on numerous performed research trials. Whereas the FDA requirement is a pore size between 100–1000 µm, a more defined pore size range of 100–600 µm is supported by research due to numerous research trials having experienced reduced bone ingrowth in pores larger than 600 µm. Strut thickness is the thickness of the pore walls within a porous structure. It may have a substantial impact on implant mechanical properties [90] and can serve as a unit cell's characteristic to compare different mechanical properties in relation to it. However, it was found that struct thickness is not commonly reported in the literature.

Research has proven that the elastic modulus of bone changes drastically with age and that patient customisation is a necessity in specific cases. As a result, in this study, an IQA for elastic modulus was proposed to be 3–30 GPa based on values from research studies. According to data analysed in this study a porous structure has a significant influence on mechanical properties in Ti-based implants. Therefore, a porous structure can be altered to provide an elastic modulus comparable to that of natural bone and be adjusted to modify the elastic modulus according to the patient's age and health condition. This will further reduce the risk of stress shielding between the implant and surrounding tissue.

As reported by most studies it was found in the regression analysis that the elastic modulus of Ti and its alloys is directly influenced by the implant porosity, and that by increasing the porosity the elastic modulus is increased and vice versa. A total of 55.5% of the studies in this review recorded an elastic modulus within this range. Regarding the effect of unit cells on the modulus of elasticity of porous structures, it was found that the studies that used beam-based unit cells covered a wider range of modulus of elasticity values than did those than used TPMS-based unit cells (Figure 22a). Moreover, the studies that used beam-based unit cells also obtained modulus of elasticity values within the whole range of human bone Young's modulus, as shown in Figure 22a.

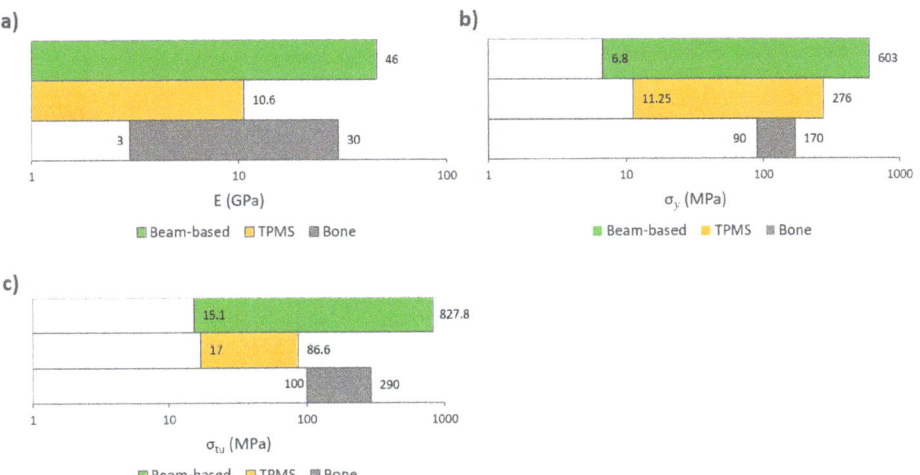

Figure 22. Comparison in logarithmic scale between mechanical properties of bone, beam-based, and triply periodic minimal surfaces (TPMS): (a) modulus of elasticity; (b) compressive yield strength; and (c) ultimate compressive strength.

Compressive yield and ultimate strength were studied. The compressive strength values for cortical bone were found to vary in the literature, which is why a range of values was selected to represent this bone characteristic. The IQA for compressive yield strength was proposed to be a minimum of 90 MPa and the IQA target for ultimate compressive strength a minimum of 180 MPa. A total of 44.1% of the studies achieved a compressive yield strength above 90 MPa, whereas only

26.7% achieved an ultimate compressive strength above 180 MPa. Both TPMS-based and beam-based scaffolds provided a wide range of compressive yield strengths covering the whole range of human bone compressive yield strength, as presented in Figure 22b. However, according to Figure 22c only beam-based scaffolds provided an adequate range of ultimate compressive yield strength values that include bone properties. The narrow range of ultimate compressive yield strength values of TPMS-based scaffolds can be explained by the lower number of studies that have addressed this mechanical with this type of unit cell. This indicates that a need exists for TPMS-based scaffold research to report the ultimate compressive yield strength.

Just as with the elastic modulus, the strength of implants was found to be directly affected by scaffold porosity. Increased porosity was found to rather dramatically decrease the strength of the implant as demonstrated by the regression model performed with the results of several different studies. Moreover, it was found that large pores are directly related to lowered strength and elastic modulus, and that several studies have struggled to produce highly porous structures with strength higher than or close to natural bone. Therefore, to ensure the adequate strength of implants, pores larger than required (>600 μm) and porosity values higher than necessary (>70%) are recommended to be avoided as these have been found to likely result in low-strength structures.

4. Conclusions

This systematic literature review presented an overview of the reported properties in research studies of fully porous Ti bone implants manufactured with AM received in the last two decades. The study focused on implants made of Ti and Ti alloy since they are considered ideal biomaterials for load-bearing applications. This study followed a QbD approach and includes the first step of the QbD system that defines the QTPP for properties relating to the porous internal architecture of fully porous Ti implants designed for load bearing applications. IQA, part of the QbD system, have been proposed supported by properties of natural human bone, governmental regulations, and scientific research relevant to bone implants. Unit cell geometry, porosity, elastic modulus, compressive yield strength, ultimate compressive yield strength, and compressive fatigue strength were systematically reviewed and benchmarked against the proposed IQA.

This study found that many implant geometrical, mechanical, and dimensional characteristics were directly related to each other. Scaffolds' pore size influences the porosity of the structure, and the porosity alters the elastic modulus as well as the strength of the implant. The unit cell geometry was also found to directly affect the Young's modulus and strength of porous scaffolds, and naturally would also impact the structures' interconnectivity. Moreover, by using ranges rather than set values where possible, such as for elastic modulus, porosity, and pore size, there is a flexibility to the design that allows the implant to be adjusted to its purpose and patient. Such design flexibility is necessary as bone properties vary with patients' age and anatomic site.

Despite the variety of scaffold characteristics reviewed in this study, future systematic literature searches should also focus on other properties, such as fluid dynamics, surface finish (topography), creep, and hardness, as well as surface coatings. Moreover, it should be considered that bone ingrowth may modify implants' mechanical properties; hence, they may be dependent on the level of bone ingrowth. Therefore, it is important to measure the changes in strength and stiffness in metallic bone scaffolds after bone ingrowth has occurred. Since implant location may determine the importance of each implant property, future studies would ideally find IQA for different implant locations within the human body. For example, mechanical features are important to study in load bearing implants, whereas fluid dynamics and vascularisation might be more important features for facial implants that require good aesthetic results and that are placed close to fragile tissue and nerves.

This study was possible due to the abundance of data available in research articles. However, to further develop an effective QbD engineering strategy for the development of specific bone implants it is important to identify the degree of importance of each bone scaffold characteristic according to the implant's future location within the human body. This can be identified by following the second step in the QbD system, namely Critical Quality Attributes (CQA), as detailed described in previous work [222]. CQA are product characteristics that must fall within specific limits to comply with the quality standards defined in the QTPP. They can be identified through prior knowledge and experimental data from systematic research based on scientific and risk management rationale that considers regulatory and business requirements [223]. Each step of the QbD system must be further developed to facilitate the strategic, qualitative development of bone implants and increase the rate of successful research studies.

Overall, Ti and Ti alloy porous bone implants are well underway to achieving the ideal properties that will fully allow them to replace natural bone. With the help of the QbD system, consistent and qualitative design of medical implant devices can be achieved.

Author Contributions: Conceptualisation: D.M.-M., Y.D., and R.A.S.; Formal analysis: D.M.-M. and Y.D.; Funding acquisition: R.A.S.; Investigation: D.M.-M. and Y.D.; Methodology: D.M.-M., Y.D., S.S., and R.A.S.; Project administration: D.M.-M. and R.A.S.; Resources: D.M.-M. and Y.D.; Supervision: D.M.-M., S.S., and R.A.S.; Visualisation: D.M.-M. and Y.D.; Writing—original draft: D.M.-M. and Y.D.; Writing—review and editing: D.M.-M., S.S., and R.A.S. All authors have read and agreed to the published version of the manuscript.

Funding: This research received no external funding.

Acknowledgments: This work was supported by Griffith School of Engineering and Built Environment.

Conflicts of Interest: The authors declare no conflict of interest.

References

1. Cox, S.C.; Thornby, J.A.; Gibbons, G.J.; Williams, M.A.; Mallick, K.K. 3D printing of porous hydroxyapatite scaffolds intended for use in bone tissue engineering applications. *Mater. Sci. Eng. C* **2015**, *47*, 237–247. [CrossRef] [PubMed]
2. Reimann, P.; Brucker, M.; Arbab, D.; Lüring, C. Patient satisfaction - A comparison between patient-specific implants and conventional total knee arthroplasty. *J. Orthop.* **2019**, *16*, 273–277. [CrossRef]
3. Wafa, H.; Grimer, R.J. Surgical options and outcomes in bone sarcoma. *Expert Rev. Anticancer Ther.* **2006**, *6*, 239–248. [CrossRef] [PubMed]
4. US Food and Drug Administration. *Understanding Barriers to Medical Device Quality*; US Food and Drug Administration: Silver Spring, MD, USA, 2011.
5. Geetha, M.; Singh, A.; Asokamani, R.; Gogia, A. Ti based biomaterials, the ultimate choice for orthopaedic implants—A review. *Prog. Mater. Sci.* **2009**, *54*, 397–425. [CrossRef]
6. Maji, P.K.; Banerjee, A.J.; Banerjee, P.S.; Karmakar, S. Additive manufacturing in prosthesis development–a case study. *Rapid Prototyp. J.* **2014**, *20*, 480–489. [CrossRef]
7. Katti, K.S. Biomaterials in total joint replacement. *Colloids Surf. B Biointerfaces* **2004**, *39*, 133–142. [CrossRef]
8. Calori, G.; Mazza, E.; Colombo, M.; Ripamonti, C. The use of bone-graft substitutes in large bone defects: Any specific needs? *Injury* **2011**, *42*, S56–S63. [CrossRef]
9. Griffin, K.S.; Davis, K.M.; McKinley, T.O.; Anglen, J.O.; Chu, T.-M.G.; Boerckel, J.D.; Kacena, M.A. Evolution of bone grafting: Bone grafts and tissue engineering strategies for vascularized bone regeneration. *Clin. Rev. Bone Miner. Metab.* **2015**, *13*, 232–244. [CrossRef]
10. Cancedda, R.; Giannoni, P.; Mastrogiacomo, M. A tissue engineering approach to bone repair in large animal models and in clinical practice. *Biomaterials* **2007**, *28*, 4240–4250. [CrossRef]
11. Hornicek, F.J.; Gebhardt, M.C.; Tomford, W.W.; Sorger, J.I.; Zavatta, M.; Menzner, J.P.; Mankin, H.J. Factors affecting nonunion of the allograft-host junction. *Clin. Orthop. Relat. Res.* **2001**, *382*, 87–98. [CrossRef] [PubMed]
12. Burchardt, H. The biology of bone graft repair. *Clin. Orthop. Relat. Res.* **1983**, *174*, 28–34. [CrossRef]
13. Seiler, J., 3rd; Johnson, J. Iliac crest autogenous bone grafting: Donor site complications. *J. South. Orthop. Assoc.* **1999**, *9*, 91–97.
14. Sachlos, E.; Czernuszka, J. Making tissue engineering scaffolds work. Review: The application of solid freeform fabrication technology to the production of tissue engineering scaffolds. *Eur. Cells Mater.* **2003**, *5*, 39–40. [CrossRef] [PubMed]

15. Saijo, H.; Igawa, K.; Kanno, Y.; Mori, Y.; Kondo, K.; Shimizu, K.; Suzuki, S.; Chikazu, D.; Iino, M.; Anzai, M. Maxillofacial reconstruction using custom-made artificial bones fabricated by inkjet printing technology. *J. Artif. Organs* **2009**, *12*, 200–205. [CrossRef] [PubMed]
16. Lichte, P.; Pape, H.; Pufe, T.; Kobbe, P.; Fischer, H. Scaffolds for bone healing: Concepts, materials and evidence. *Injury* **2011**, *42*, 569–573. [CrossRef]
17. Parthasarathy, J. 3D modeling, custom implants and its future perspectives in craniofacial surgery. *Ann. Maxillofac. Surg.* **2014**, *4*, 9–18. [CrossRef] [PubMed]
18. Prince, J.D. 3D Printing: An industrial revolution. *J. Electron. Resour. Med. Libr.* **2014**, *11*, 39–45. [CrossRef]
19. Ryan, G.E.; Pandit, A.S.; Apatsidis, D.P. Porous titanium scaffolds fabricated using a rapid prototyping and powder metallurgy technique. *Biomaterials* **2008**, *29*, 3625–3635. [CrossRef]
20. Barbas, A.; Bonnet, A.-S.; Lipinski, P.; Pesci, R.; Dubois, G. Development and mechanical characterization of porous titanium bone substitutes. *J. Mech. Behav. Biomed. Mater.* **2012**, *9*, 34–44. [CrossRef]
21. Wong, K.-C.; Scheinemann, P. Additive manufactured metallic implants for orthopaedic applications. *Sci. China Mater.* **2018**, *61*, 440–454. [CrossRef]
22. Martinez-Marquez, D.; Mirnajafizadeh, A.; Carty, C.P.; Stewart, R.A. Facilitating industry translation of custom 3d printed bone prostheses and scaffolds through Quality by Design. *Procedia Manuf.* **2019**, *30*, 284–291. [CrossRef]
23. Martinez-Marquez, D.; Jokymaityte, M.; Mirnajafizadeh, A.; Carty, C.P.; Lloyd, D.; Stewart, R.A. Development of 18 quality control gates for additive manufacturing of error free patient-specific implants. *Materials* **2019**, *12*, 3110. [CrossRef] [PubMed]
24. Kulkarni, M.; Mazare, A.; Schmuki, P.; Iglič, A. Biomaterial surface modification of titanium and titanium alloys for medical applications. In *Nanomedicine*; Seifalian, A., de Mel, A., Kalaskar, D.M., Eds.; One Central Press (OCP): Altrincham, UK, 2014; p. 111.
25. Heness, G.; Ben-Nissan, B. Innovative bioceramics. *Mater. Forum* **2004**, *27*, 104–114.
26. Singh, M.; Jonnalagadda, S. Advances in bioprinting using additive manufacturing. *Eur. J. Pharm. Sci.* **2020**, *143*, 105167. [CrossRef]
27. Donnaloja, F.; Jacchetti, E.; Soncini, M.; Raimondi, M.T.J.P. Natural and Synthetic Polymers for Bone Scaffolds Optimization. *Polymers* **2020**, *12*, 905. [CrossRef]
28. Asghari, F.; Samiei, M.; Adibkia, K.; Akbarzadeh, A.; Davaran, S. Biodegradable and biocompatible polymers for tissue engineering application: A review. *Artif. Cells Nanomed. Biotechnol.* **2017**, *45*, 185–192. [CrossRef]
29. Rodríguez, G.R.; Patrício, T.; López, J.D. Natural polymers for bone repair. In *Bone Repair Biomaterials*; Elsevier BV: Amsterdam, The Netherlands, 2019; pp. 199–232.
30. Haugen, H.J.; Lyngstadaas, S.P.; Rossi, F.; Perale, G. Bone grafts: Which is the ideal biomaterial? *J. Clin. Periodontol.* **2019**, *46*, 92–102. [CrossRef]
31. Dwivedi, R.; Kumar, S.; Pandey, R.; Mahajan, A.; Nandana, D.; Katti, D.S.; Mehrotra, D. Polycaprolactone as biomaterial for bone scaffolds: Review of literature. *JOBCR* **2020**, *10*, 381–388. [CrossRef]
32. Liu, Y.; Lim, J.; Teoh, S.-H. Review: Development of clinically relevant scaffolds for vascularised bone tissue engineering. *Biotechnol. Adv.* **2013**, *31*, 688–705. [CrossRef]
33. Bose, S.; Roy, M.; Bandyopadhyay, A. Recent advances in bone tissue engineering scaffolds. *Trends Biotechnol.* **2012**, *30*, 546–554. [CrossRef]
34. Huang, B.; Vyas, C.; Roberts, I.; Poutrel, Q.-A.; Chiang, W.-H.; Blaker, J.J.; Huang, Z.; Bártolo, P. Fabrication and characterisation of 3D printed MWCNT composite porous scaffolds for bone regeneration. *Mater. Sci. Eng. C* **2019**, *98*, 266–278. [CrossRef] [PubMed]
35. De Moura, N.K.; Martins, E.F.; Oliveira, R.L.M.S.; de Brito Siqueira, I.A.W.; Machado, J.P.B.; Esposito, E.; Amaral, S.S.; de Vasconcellos, L.M.R.; Passador, F.R.; de Sousa Trichês, E. Synergistic effect of adding bioglass and carbon nanotubes on poly (lactic acid) porous membranes for guided bone regeneration. *Mater. Sci. Eng. C* **2020**, *117*, 111327. [CrossRef] [PubMed]
36. De Witte, T.M.; Wagner, A.M.; Fratila-Apachitei, L.E.; Zadpoor, A.A.; Peppas, N.A. Immobilization of nanocarriers within a porous chitosan scaffold for the sustained delivery of growth factors in bone tissue engineering applications. *J. Biomed. Mater. Res. Part A* **2020**, *108*, 1122–1135. [CrossRef] [PubMed]
37. Swanson, W.B.; Zhang, Z.; Xiu, K.; Gong, T.; Eberle, M.; Wang, Z.; Ma, P.X. Scaffolds with Controlled Release of Pro-Mineralization Exosomes to Promote Craniofacial Bone Healing without Cell Transplantation. *Acta Biomater.* **2020**. [CrossRef]

38. Ghayor, C.; Weber, F.E. Osteoconductive Microarchitecture of Bone Substitutes for Bone Regeneration Revisited. *Front. Physiol.* **2018**, *9*, 960. [CrossRef]
39. Kim, J.-W.; Yang, B.-E.; Hong, S.-J.; Choi, H.-G.; Byeon, S.-J.; Lim, H.-K.; Chung, S.-M.; Lee, J.-H.; Byun, S.-H. Bone Regeneration Capability of 3D Printed Ceramic Scaffolds. *Int. J. Mol. Sci.* **2020**, *21*, 4837. [CrossRef]
40. Jeong, J.; Kim, J.H.; Shim, J.H.; Hwang, N.S.; Heo, C.Y. Bioactive calcium phosphate materials and applications in bone regeneration. *Biomater. Res.* **2019**, *23*, 1–11. [CrossRef]
41. Eliaz, N.; Metoki, N. Calcium Phosphate Bioceramics: A Review of Their History, Structure, Properties, Coating Technologies and Biomedical Applications. *Materials* **2017**, *10*, 334. [CrossRef]
42. Dorozhkin, S.V. Calcium orthophosphate deposits: Preparation, properties and biomedical applications. *Mater. Sci. Eng. C* **2015**, *55*, 272–326. [CrossRef]
43. Barrère, F.; van Blitterswijk, C.A.; de Groot, K. Bone regeneration: Molecular and cellular interactions with calcium phosphate ceramics. *Int. J. Nanomed.* **2006**, *1*, 317–332.
44. Fernandes, H.R.; Gaddam, A.; Rebelo, A.; Brazete, D.; Stan, G.E.; Ferreira, J.M.J.M. Bioactive glasses and glass-ceramics for healthcare applications in bone regeneration and tissue engineering. *Materials* **2018**, *11*, 2530. [CrossRef] [PubMed]
45. Seidenstuecker, M.; Kerr, L.; Bernstein, A.; Mayr, H.O.; Suedkamp, N.P.; Gadow, R.; Krieg, P.; Hernandez Latorre, S.; Thomann, R.; Syrowatka, F.J.M. 3D powder printed bioglass and β-tricalcium phosphate bone scaffolds. *Materials* **2017**, *11*, 13. [CrossRef]
46. Ma, Y.; Dai, H.; Huang, X.; Long, Y. 3D printing of bioglass-reinforced β-TCP porous bioceramic scaffolds. *J. Mater. Sci.* **2019**, *54*, 10437–10446. [CrossRef]
47. Mancuso, E.; Bretcanu, O.A.; Marshall, M.; Birch, M.A.; McCaskie, A.W.; Dalgarno, K.W. Novel bioglasses for bone tissue repair and regeneration: Effect of glass design on sintering ability, ion release and biocompatibility. *Mater. Des.* **2017**, *129*, 239–248. [CrossRef] [PubMed]
48. Negut, I.; Floroian, L.; Ristoscu, C.; Mihailescu, C.N.; Mirza Rosca, J.C.; Tozar, T.; Badea, M.; Grumezescu, V.; Hapenciuc, C.; Mihailescu, I.N. Functional bioglass—Biopolymer double nanostructure for natural antimicrobial drug extracts delivery. *Nanomaterials* **2020**, *10*, 385. [CrossRef] [PubMed]
49. Yang, J. Progress of bioceramic and bioglass bone scaffolds for load-bearing applications. In *Orthopedic Biomaterials*; Springer: Berlin/Heidelberg, Germany, 2018; pp. 453–486.
50. Rizwan, M.; Hamdi, M.; Basirun, W.J. Bioglass® 45S5-based composites for bone tissue engineering and functional applications. *J. Biomed. Mater. Res. Part A* **2017**, *105*, 3197–3223. [CrossRef]
51. Liu, X.; Chu, P.K.; Ding, C. Surface modification of titanium, titanium alloys, and related materials for biomedical applications. *Mater. Sci. Eng. R Rep.* **2004**, *47*, 49–121. [CrossRef]
52. Rack, H.; Qazi, J. Titanium alloys for biomedical applications. *Mater. Sci. Eng. C* **2006**, *26*, 1269–1277. [CrossRef]
53. Niinomi, M.; Nakai, M.; Hieda, J. Development of new metallic alloys for biomedical applications. *Acta Biomater.* **2012**, *8*, 3888–3903. [CrossRef]
54. Vandrovcová, M.; Bacakova, L. Adhesion, growth and differentiation of osteoblasts on surface-modified materials developed for bone implants. *Physiol. Res.* **2011**, *60*, 403–417. [CrossRef]
55. Huiskes, R.; Weinans, H.; Van Rietbergen, B. The Relationship Between Stress Shielding and Bone Resorption Around Total Hip Stems and the Effects of Flexible Materials. *Clin. Orthop. Relat. Res.* **1992**, *274*, 124–134. [CrossRef]
56. Ryan, G.; Pandit, A.; Apatsidis, D.P. Fabrication methods of porous metals for use in orthopaedic applications. *Biomaterials* **2006**, *27*, 2651–2670. [CrossRef]
57. Chen, Q.; Thouas, G.A. Metallic implant biomaterials. *Mater. Sci. Eng. R Rep.* **2015**, *87*, 1–57. [CrossRef]
58. Strnad, J.; Strnad, Z.; Šestak, J.; Urban, K.; Povýšil, C. Bio-activated titanium surface utilizable for mimetic bone implantation in dentistry—Part III: Surface characteristics and bone–implant contact formation. *J. Phys. Chem. Solids* **2007**, *68*, 841–845. [CrossRef]
59. Zhao, X.; Wang, T.; Qian, S.; Liu, X.; Sun, J.; Li, B. Silicon-doped titanium dioxide nanotubes promoted bone formation on titanium implants. *Int. J. Mol. Sci.* **2016**, *17*, 292. [CrossRef] [PubMed]
60. Wiria, F.E.; Maleksaeedi, S.; He, Z. Manufacturing and characterization of porous titanium components. *Prog. Cryst. Growth Charact. Mater.* **2014**, *60*, 94–98. [CrossRef]
61. Zhang, L.C.; Klemm, D.; Eckert, J.; Hao, Y.L.; Sercombe, T.B. Manufacture by selective laser melting and mechanical behavior of a biomedical Ti–24Nb–4Zr–8Sn alloy. *Scr. Mater.* **2011**, *65*, 21–24. [CrossRef]

62. Zaharin, H.A.; Rani, A.; Majdi, A.; Azam, F.I.; Ginta, T.L.; Sallih, N.; Ahmad, A.; Yunus, N.A.; Zulkifli, T.Z.A. Effect of unit cell type and pore size on porosity and mechanical behavior of additively manufactured Ti6Al4V scaffolds. *Materials* **2018**, *11*, 2402. [CrossRef]
63. Bobyn, J.; Pilliar, R.; Cameron, H.; Weatherly, G.J.C.O.; Research®, R. The optimum pore size for the fixation of porous-surfaced metal implants by the ingrowth of bone. *Clin. Orthop. Relat. Res.* **1980**, *150*, 263–270. [CrossRef]
64. John, A.A.; Jaganathan, S.K.; Supriyanto, E.; Manikandan, A. Surface modification of titanium and its alloys for the enhancement of osseointegration in orthopaedics. *Curr. Sci.* **2016**, *111*, 1003–1015. [CrossRef]
65. Zhang, L.-C.; Chen, L.-Y.; Wang, L. Surface Modification of Titanium and Titanium Alloys: Technologies, Developments, and Future Interests. *Adv. Eng. Mater.* **2020**, *22*, 1901258. [CrossRef]
66. Sasikumar, Y.; Indira, K.; Rajendran, N. Surface Modification Methods for Titanium and its alloys and their corrosion behavior in biological environment: A review. *J. Bio. Tribo-Corros.* **2019**, *5*, 36. [CrossRef]
67. Civantos, A.; Martinez-Campos, E.; Ramos, V.; Elvira, C.; Gallardo, A.; Abarrategi, A. Titanium coatings and surface modifications: Toward clinically useful bioactive implants. *ACS Biomater. Sci. Eng.* **2017**, *3*, 1245–1261. [CrossRef]
68. Zhang, K.; Fan, Y.; Dunne, N.; Li, X. Effect of microporosity on scaffolds for bone tissue engineering. *Regen. Biomater.* **2018**, *5*, 115–124. [CrossRef] [PubMed]
69. Staruch, R.; Griffin, M.F.; Butler, P. Nanoscale Surface Modifications of Orthopaedic Implants: State of the Art and Perspectives. *Open Orthop. J.* **2016**, *10*, 920–938. [CrossRef] [PubMed]
70. Zemtsova, E.; Arbenin, A.; Valiev, R.; Smirnov, V. Modern techniques of surface geometry modification for the implants based on titanium and its alloys used for improvement of the biomedical characteristics. In *Titanium in Medical and Dental Applications*; Elsevier: Amsterdam, The Netherlands, 2018; pp. 115–145.
71. Gulati, K.; Prideaux, M.; Kogawa, M.; Lima-Marques, L.; Atkins, G.J.; Findlay, D.M.; Losic, D. Anodized 3D–printed titanium implants with dual micro- and nano-scale topography promote interaction with human osteoblasts and osteocyte-like cells. *J. Tissue Eng. Regen. Med.* **2016**, *11*, 3313–3325. [CrossRef] [PubMed]
72. Zafar, M.S.; Fareed, M.A.; Riaz, S.; Latif, M.; Habib, S.R.; Khurshid, Z. Customized Therapeutic Surface Coatings for Dental Implants. *Coatings* **2020**, *10*, 568. [CrossRef]
73. Kunrath, M.F.; Leal, B.F.; Hubler, R.; de Oliveira, S.D.; Teixeira, E.R. Antibacterial potential associated with drug-delivery built TiO2 nanotubes in biomedical implants. *AMB Express* **2019**, *9*, 1–13. [CrossRef] [PubMed]
74. Martinez-Marquez, D.; Mirnajafizadeh, A.; Carty, C.P.; Stewart, R.A. Application of quality by design for 3D printed bone prostheses and scaffolds. *PLoS ONE* **2018**, *13*, e0195291. [CrossRef]
75. Kasanen, E.; Lukka, K.; Siitonen, A. The constructive approach in management accounting research. *J. Manag. Account. Res.* **2003**, *5*, 243–264.
76. Oyegoke, A. The constructive research approach in project management research. *Int. J. Manag. Proj. Bus.* **2011**, *4*, 573–595. [CrossRef]
77. Wen, C.E.; Yamada, Y.; Shimojima, K.; Chino, Y.; Hosokawa, H.; Mabuchi, M. Novel titanium foam for bone tissue engineering. *J. Mater. Res.* **2002**, *17*, 2633–2639. [CrossRef]
78. Liberati, A.; Altman, D.G.; Tetzlaff, J.; Mulrow, C.; Gøtzsche, P.C.; Ioannidis, J.P.; Clarke, M.; Devereaux, P.J.; Kleijnen, J.; Moher, D. The PRISMA statement for reporting systematic reviews and meta-analyses of studies that evaluate health care interventions: Explanation and elaboration. *J. Clin. Epidemiol.* **2009**, *62*, e1–e34. [CrossRef] [PubMed]
79. Bobbert, F.S.L.; Lietaert, K.; Eftekhari, A.A.; Pouran, B.; Ahmadi, S.M.; Weinans, H.; Zadpoor, A.A. Additively manufactured metallic porous biomaterials based on minimal surfaces: A unique combination of topological, mechanical, and mass transport properties. *Acta Biomater.* **2017**, *53*, 572–584. [CrossRef] [PubMed]
80. Wu, S.; Liu, X.; Yeung, K.W.K.; Liu, C.; Yang, X. Biomimetic porous scaffolds for bone tissue engineering. *Mater. Sci. Eng. R Rep.* **2014**, *80*, 1–36. [CrossRef]
81. Wang, X.; Xu, S.; Zhou, S.; Xu, W.; Leary, M.; Choong, P.; Qian, M.; Brandt, M.; Xie, Y.M. Topological design and additive manufacturing of porous metals for bone scaffolds and orthopaedic implants: A review. *Biomaterials* **2016**, *83*, 127–141. [CrossRef] [PubMed]
82. Heinl, P.; Muller, L.; Korner, C.; Singer, R.F.; Muller, F.A. Cellular Ti-6Al-4V structures with interconnected macro porosity for bone implants fabricated by selective electron beam melting. *Acta Biomater.* **2008**, *4*, 1536–1544. [CrossRef] [PubMed]

83. Brett, E.; Flacco, J.; Blackshear, C.; Longaker, M.T.; Wan, D.C. Biomimetics of Bone Implants: The Regenerative Road. *BioRes. Open Access* **2017**, *6*, 1–6. [CrossRef]
84. Wang, X.; Li, Y.; Xiong, J.; Hodgson, P.D.; Wen, C. Porous TiNbZr alloy scaffolds for biomedical applications. *Acta Biomater.* **2009**, *5*, 3616–3624. [CrossRef]
85. Bose, S.; Darsell, J.; Hosick, H.L.; Yang, L.; Sarkar, D.; Bandyopadhyay, A. Processing and Characterization of Porous Alumina Scaffolds. *J. Mater. Sci. Mater. Med.* **2002**, *13*, 23–28. [CrossRef]
86. Karageorgiou, V.; Kaplan, D. Porosity of 3D biomaterial scaffolds and osteogenesis. *Biomaterials* **2005**, *26*, 5474–5491. [CrossRef] [PubMed]
87. Burg, K.J.L.; Porter, S.; Kellam, J.F. Biomaterial developments for bone tissue engineering. *Biomaterials* **2000**, *21*, 2347–2359. [CrossRef]
88. Wang, X.; Ni, Q. Determination of cortical bone porosity and pore size distribution using a low field pulsed NMR approach. *J. Orthop. Res.* **2003**, *21*, 312–319. [CrossRef]
89. Lee, H.; Jang, T.-S.; Song, J.; Kim, H.-E.; Jung, H.-D. Multi-scale porous Ti6Al4V scaffolds with enhanced strength and biocompatibility formed via dynamic freeze-casting coupled with micro-arc oxidation. *Mater. Lett.* **2016**, *185*, 21–24. [CrossRef]
90. Parthasarathy, J.; Starly, B.; Raman, S.; Christensen, A. Mechanical evaluation of porous titanium (Ti6Al4V) structures with electron beam melting (EBM). *J. Mech. Behav. Biomed. Mater.* **2010**, *3*, 249–259. [CrossRef]
91. Suska, F.; Kjeller, G.; Tarnow, P.; Hryha, E.; Nyborg, L.; Snis, A.; Palmquist, A. Electron Beam Melting Manufacturing Technology for Individually Manufactured Jaw Prosthesis: A Case Report. *J. Oral Maxillofac. Surg.* **2016**, *74*, 1706.e1–1706.e15. [CrossRef]
92. Andreasen, C.M.; Delaisse, J.-M.; van der Eerden, B.C.; van Leeuwen, J.P.; Ding, M.; Andersen, T.L. Understanding age-induced cortical porosity in women: Is a negative BMU balance in quiescent osteons a major contributor? *Bone* **2018**, *117*, 70–82. [CrossRef]
93. Morgan, E.F.; Unnikrisnan, G.U.; Hussein, A.I. Bone Mechanical Properties in Healthy and Diseased States. *Annu. Rev. Biomed. Eng.* **2018**, *20*, 119–143. [CrossRef]
94. Lopez-Heredia, M.A.; Sohier, J.; Gaillard, C.; Quillard, S.; Dorget, M.; Layrolle, P. Rapid prototyped porous titanium coated with calcium phosphate as a scaffold for bone tissue engineering. *Biomaterials* **2008**, *29*, 2608–2615. [CrossRef]
95. Hollander, D.A.; von Walter, M.; Wirtz, T.; Sellei, R.; Schmidt-Rohlfing, B.; Paar, O.; Erli, H.J. Structural, mechanical and in vitro characterization of individually structured Ti-6Al-4V produced by direct laser forming. *Biomaterials* **2006**, *27*, 955–963. [CrossRef]
96. Hara, D.; Nakashima, Y.; Sato, T.; Hirata, M.; Kanazawa, M.; Kohno, Y.; Yoshimoto, K.; Yoshihara, Y.; Nakamura, A.; Nakao, Y.; et al. Bone bonding strength of diamond-structured porous titanium-alloy implants manufactured using the electron beam-melting technique. *Mater. Sci. Eng. C Mater. Biol. Appl.* **2016**, *59*, 1047–1052. [CrossRef] [PubMed]
97. Wieding, J.; Jonitz, A.; Bader, R. The Effect of Structural Design on Mechanical Properties and Cellular Response of Additive Manufactured Titanium Scaffolds. *Materials* **2012**, *5*, 1336–1347. [CrossRef]
98. Xue, W.; Krishna, B.V.; Bandyopadhyay, A.; Bose, S. Processing and biocompatibility evaluation of laser processed porous titanium. *Acta Biomater.* **2007**, *3*, 1007–1018. [CrossRef]
99. De Peppo, G.M.; Palmquist, A.; Borchardt, P.; Lennerås, M.; Hyllner, J.; Snis, A.; Lausmaa, J.; Thomsen, P.; Karlsson, C. Free-form-fabricated commercially pure Ti and Ti6Al4V porous scaffolds support the growth of human embryonic stem cell-derived mesodermal progenitors. *Sci. World J.* **2012**, *2012*, 1–14. [CrossRef] [PubMed]
100. Zhang, X.-Y.; Fang, G.; Leeflang, S.; Zadpoor, A.A.; Zhou, J. Topological design, permeability and mechanical behavior of additively manufactured functionally graded porous metallic biomaterials. *Acta Biomater.* **2019**, *84*, 437–452. [CrossRef] [PubMed]
101. Taniguchi, N.; Fujibayashi, S.; Takemoto, M.; Sasaki, K.; Otsuki, B.; Nakamura, T.; Matsushita, T.; Kokubo, T.; Matsuda, S. Effect of pore size on bone ingrowth into porous titanium implants fabricated by additive manufacturing: An in vivo experiment. *Mater. Sci. Eng. C Mater. Biol. Appl.* **2016**, *59*, 690–701. [CrossRef]
102. Zou, C.; Zhang, E.; Li, M.; Zeng, S. Preparation, microstructure and mechanical properties of porous titanium sintered by Ti fibres. *J. Mater. Sci. Mater. Med.* **2007**, *19*, 401–405. [CrossRef]
103. Takemoto, M.; Fujibayashi, S.; Neo, M.; Suzuki, J.; Kokubo, T.; Nakamura, T. Mechanical properties and osteoconductivity of porous bioactive titanium. *Biomaterials* **2005**, *26*, 6014–6023. [CrossRef]

104. Van Hooreweder, B.; Apers, Y.; Lietaert, K.; Kruth, J.-P. Improving the fatigue performance of porous metallic biomaterials produced by Selective Laser Melting. *Acta Biomater.* **2017**, *47*, 193–202. [CrossRef]
105. Ahmadi, S.M.; Yavari, S.A.; Wauthle, R.; Pouran, B.; Schrooten, J.; Weinans, H.; Zadpoor, A.A. Additively Manufactured Open-Cell Porous Biomaterials Made from Six Different Space-Filling Unit Cells: The Mechanical and Morphological Properties. *Materials* **2015**, *8*, 1871–1896. [CrossRef]
106. Amin Yavari, S.; Ahmadi, S.M.; Wauthle, R.; Pouran, B.; Schrooten, J.; Weinans, H.; Zadpoor, A.A. Relationship between unit cell type and porosity and the fatigue behavior of selective laser melted meta-biomaterials. *J. Mech. Behav. Biomed. Mater.* **2015**, *43*, 91–100. [CrossRef] [PubMed]
107. Hrabe, N.W.; Heinl, P.; Flinn, B.; Körner, C.; Bordia, R.K. Compression-compression fatigue of selective electron beam melted cellular titanium (Ti-6Al-4V). *J. Biomed. Mater.* **2011**, *99B*, 313–320. [CrossRef]
108. Ponader, S.; von Wilmowsky, C.; Widenmayer, M.; Lutz, R.; Heinl, P.; Korner, C.; Singer, R.F.; Nkenke, E.; Neukam, F.W.; Schlegel, K.A. In vivo performance of selective electron beam-melted Ti-6Al-4V structures. *J. Biomed. Mater. Res. Part A* **2010**, *92*, 56–62. [CrossRef] [PubMed]
109. Yan, C.; Hao, L.; Hussein, A.; Young, P. Ti-6Al-4V triply periodic minimal surface structures for bone implants fabricated via selective laser melting. *J. Mech. Behav. Biomed. Mater.* **2015**, *51*, 61–73. [CrossRef] [PubMed]
110. Wysocki, B.; Idaszek, J.; Szlazak, K.; Strzelczyk, K.; Brynk, T.; Kurzydlowski, K.J.; Swieszkowski, W. Post Processing and Biological Evaluation of the Titanium Scaffolds for Bone Tissue Engineering. *Materials* **2016**, *9*, 197. [CrossRef]
111. Attar, H.; Löber, L.; Funk, A.; Calin, M.; Zhang, L.; Prashanth, K.G.; Scudino, S.; Zhang, Y.S.; Eckert, J. Mechanical behavior of porous commercially pure Ti and Ti-TiB composite materials Manufactured by selective laser melting. *Mater. Sci. Eng. A* **2015**, *625*, 350–356. [CrossRef]
112. Wauthle, R.; van der Stok, J.; Amin Yavari, S.; Van Humbeeck, J.; Kruth, J.P.; Zadpoor, A.A.; Weinans, H.; Mulier, M.; Schrooten, J. Additively manufactured porous tantalum implants. *Acta Biomater.* **2015**, *14*, 217–225. [CrossRef]
113. Zhang, B.; Pei, X.; Zhou, C.; Fan, Y.; Jiang, Q.; Ronca, A.; D'Amora, U.; Chen, Y.; Li, H.; Sun, Y. The biomimetic design and 3D printing of customized mechanical properties porous Ti6Al4V scaffold for load-bearing bone reconstruction. *Mater. Des.* **2018**, *152*, 30–39. [CrossRef]
114. Kim, T.B.; Yue, S.; Zhang, Z.; Jones, E.; Jones, J.R.; Lee, P.D. Additive manufactured porous titanium structures: Through-process quantification of pore and strut networks. *J. Mater. Process. Technol.* **2014**, *214*, 2706–2715. [CrossRef]
115. Ataee, A.; Li, Y.C.; Brandt, M.; Wen, C. Ultrahigh-strength titanium gyroid scaffolds manufactured by selective laser melting (SLM) for bone implant applications. *Acta Mater.* **2018**, *158*, 354–368. [CrossRef]
116. Zhao, D.; Huang, Y.; Ao, Y.; Han, C.; Wang, Q.; Li, Y.; Liu, J.; Wei, Q.; Zhang, Z. Effect of pore geometry on the fatigue properties and cell affinity of porous titanium scaffolds fabricated by selective laser melting. *J. Mech. Behav. Biomed. Mater.* **2018**, *88*, 478–487. [CrossRef] [PubMed]
117. Ahmadi, S.; Kumar, R.; Borisov, E.; Petrov, R.; Leeflang, S.; Li, Y.; Tümer, N.; Huizenga, R.; Ayas, C.; Zadpoor, A. From microstructural design to surface engineering: A tailored approach for improving fatigue life of additively manufactured meta-biomaterials. *Acta Biomater.* **2019**, *83*, 153–166. [CrossRef] [PubMed]
118. Xia, Y.; Feng, C.; Xiong, Y.; Luo, Y.; Li, X. Mechanical properties of porous titanium alloy scaffold fabricated using additive manufacturing technology. *Int. J. Appl. Electromagn. Mech.* **2019**, *59*, 1087–1095. [CrossRef]
119. Li, G.; Wang, L.; Pan, W.; Yang, F.; Jiang, W.; Wu, X.; Kong, X.; Dai, K.; Hao, Y. In vitro and in vivo study of additive manufactured porous Ti6Al4V scaffolds for repairing bone defects. *Sci. Rep.* **2016**, *6*, 34072. [CrossRef]
120. Yánez, A.; Cuadrado, A.; Martel, O.; Afonso, H.; Monopoli, D. Gyroid porous titanium structures: A versatile solution to be used as scaffolds in bone defect reconstruction. *Mater. Des.* **2018**, *140*, 21–29. [CrossRef]
121. Surmeneva, M.; Surmenev, R.; Chudinova, E.; Koptyug, A.; Tkachev, M.S.; Shkarina, S.; Rännar, L.-E. Fabrication of multiple-layered gradient cellular metal scaffold via electron beam melting for segmental bone reconstruction. *Mater. Des.* **2017**, *133*. [CrossRef]
122. Zhang, S.; Wei, Q.; Cheng, L.; Li, S.; Shi, Y. Effects of scan line spacing on pore characteristics and mechanical properties of porous Ti6Al4V implants fabricated by selective laser melting. *Mater. Des.* **2014**, *63*, 185–193. [CrossRef]

123. Van der Stok, J.; Van der Jagt, O.P.; Amin Yavari, S.; De Haas, M.F.; Waarsing, J.H.; Jahr, H.; Van Lieshout, E.M.; Patka, P.; Verhaar, J.A.; Zadpoor, A.A.; et al. Selective laser melting-produced porous titanium scaffolds regenerate bone in critical size cortical bone defects. *J. Orthop. Res.* **2013**, *31*, 792–799. [CrossRef]
124. Lin, C.Y.; Wirtz, T.; LaMarca, F.; Hollister, S.J. Structural and mechanical evaluations of a topology optimized titanium interbody fusion cage fabricated by selective laser melting process. *J. Biomed. Mater. Res. Part A* **2007**, *83*, 272–279. [CrossRef]
125. Yavari, S.A.; Wauthle, R.; van der Stok, J.; Riemslag, A.C.; Janssen, M.; Mulier, M.; Kruth, J.P.; Schrooten, J.; Weinans, H.; Zadpoor, A.A. Fatigue behavior of porous biomaterials manufactured using selective laser melting. *Mater. Sci. Eng. C Mater. Biol. Appl.* **2013**, *33*, 4849–4858. [CrossRef]
126. Liu, Y.; Li, S.; Hou, W.; Wang, S.; Hao, Y.L.; Yang, R.; Sercombe, T.; Zhang, L. Electron Beam Melted Beta-Type Ti-24Nb-4Zr-8Sn Porous Structures with High Strength-to-Modulus Ratio. *J. Mater. Sci. Technol.* **2016**, *32*, 505–508. [CrossRef]
127. Li, S.J.; Xu, Q.S.; Wang, Z.; Hou, W.T.; Hao, Y.L.; Yang, R.; Murr, L.E. Influence of cell shape on mechanical properties of Ti-6Al-4V meshes fabricated by electron beam melting method. *Acta Biomater.* **2014**, *10*, 4537–4547. [CrossRef] [PubMed]
128. Sallica-Leva, E.; Jardini, A.L.; Fogagnolo, J.B. Microstructure and mechanical behavior of porous Ti-6Al-4V parts obtained by selective laser melting. *J. Mech. Behav. Biomed. Mater.* **2013**, *26*, 98–108. [CrossRef]
129. Van Bael, S.; Kerckhofs, G.; Moesen, M.; Pyka, G.; Schrooten, J.; Kruth, J.P. Micro-CT-based improvement of geometrical and mechanical controllability of selective laser melted Ti6Al4V porous structures. *Mater. Sci. Eng. A* **2011**, *528*, 7423–7431. [CrossRef]
130. Cheng, A.; Humayun, A.; Cohen, D.J.; Boyan, B.D.; Schwartz, Z. Additively manufactured 3D porous Ti-6Al-4V constructs mimic trabecular bone structure and regulate osteoblast proliferation, differentiation and local factor production in a porosity and surface roughness dependent manner. *Biofabrication* **2014**, *6*, 045007. [CrossRef] [PubMed]
131. Roy, S.; Khutia, N.; Das, D.; Das, M.; Balla, V.K.; Bandyopadhyay, A.; Chowdhury, A.R. Understanding compressive deformation behavior of porous Ti using finite element analysis. *Mater. Sci. Eng. C Mater. Biol. Appl.* **2016**, *64*, 436–443. [CrossRef]
132. Mullen, L.; Stamp, R.C.; Brooks, W.K.; Jones, E.; Sutcliffe, C.J. Selective Laser Melting: A regular unit cell approach for the manufacture of porous, titanium, bone in-growth constructs, suitable for orthopedic applications. *J. Biomed. Mater. Res. Part B Appl. Biomater.* **2009**, *89*, 325–334. [CrossRef]
133. Van Grunsven, W.; Hernandez-Nava, E.; Reilly, G.C.; Goodall, R. Fabrication and Mechanical Characterisation of Titanium Lattices with Graded Porosity. *Metals* **2014**, *4*, 401–409. [CrossRef]
134. Andani, M.T.; Saedi, S.; Turabi, A.S.; Karamooz-Ravari, M.R.; Haberland, C.; Karaca, H.E.; Elahinia, M. Mechanical and shape memory properties of porous Ni 50.1 Ti 49.9 alloys manufactured by selective laser melting. *J. Mech. Behav. Biomed. Mater.* **2017**, *68*, 224–231. [CrossRef]
135. Lietaert, K.; Cutolo, A.; Boey, D.; Van Hooreweder, B. Fatigue life of additively manufactured Ti6Al4V scaffolds under tension-tension, tension-compression and compression-compression fatigue load. *Sci. Rep.* **2018**, *8*, 1–9. [CrossRef]
136. Barui, S.; Chatterjee, S.; Mandal, S.; Kumar, A.; Basu, B. Microstructure and compression properties of 3D powder printed Ti-6Al-4V scaffolds with designed porosity: Experimental and computational analysis. *Mater. Sci. Eng. C Mater. Biol. Appl.* **2017**, *70*, 812–823. [CrossRef]
137. Yang, L.; Yan, C.; Cao, W.; Liu, Z.; Song, B.; Wen, S.; Zhang, C.; Shi, Y.; Yang, S. Compression-compression fatigue behaviour of triply periodic minimal surface porous structures fabricated by selective laser melting. *Acta Mater.* **2019**, *181*, 49–66. [CrossRef]
138. Melancon, D.; Bagheri, Z.S.; Johnston, R.B.; Liu, L.; Tanzer, M.; Pasini, D. Mechanical characterization of structurally porous biomaterials built via additive manufacturing: Experiments, predictive models, and design maps for load-bearing bone replacement implants. *Acta Biomater.* **2017**, *63*, 350–368. [CrossRef]
139. Eldesouky, I.; Harrysson, O.; West, H.; Elhofy, H. Electron beam melted scaffolds for orthopedic applications. *Addit. Manuf.* **2017**, *17*. [CrossRef]
140. Vasireddi, R.; Basu, B. Conceptual design of three-dimensional scaffolds of powder-based materials for Bone Tissue Engineering Applications. *Rapid Prototyp. J.* **2015**, *21*. [CrossRef]

141. Rahyussalim, A.J.; Marsetio, A.F.; Saleh, I.; Kurniawati, T.; Whulanza, Y. The Needs of Current Implant Technology in Orthopaedic Prosthesis Biomaterials Application to Reduce Prosthesis Failure Rate. *J. Nanomater.* **2016**, *2016*, 1–9. [CrossRef]
142. Matassi, F.; Botti, A.; Sirleo, L.; Carulli, C.; Innocenti, M. Porous metal for orthopedics implants. *Clin. Cases Miner. Bone Metab.* **2013**, *10*, 111–115.
143. Frohlich, M.; Grayson, W.L.; Wan, L.Q.; Marolt, D.; Drobnic, M.; Vunjak-Novakovic, G. Tissue engineered bone grafts: Biological requirements, tissue culture and clinical relevance. *Curr. Stem Cell Res. Ther.* **2008**, *3*, 254–264. [CrossRef] [PubMed]
144. Nouri, A.; Hodgson, P.D.; Wen, C.E. *Biomimetic Porous Titanium Scaffolds for Orthopaedic and Dental Applications*; InTech: London, UK, 2010.
145. Rho, J.-Y.; Kuhn-Spearing, L.; Zioupos, P. Mechanical properties and the hierarchical structure of bone. *Med. Eng. Phys.* **1998**, *20*, 92–102. [CrossRef]
146. Currey, J.D. *Bones: Structure and Mechanics*; Princeton University Press: Princeton, NJ, USA, 2013.
147. Lee, S.; Porter, M.; Wasko, S.; Lau, G.; Chen, P.-Y.; Novitskaya, E.E.; Tomsia, A.P.; Almutairi, A.; Meyers, M.A.; McKittrick, J. Potential Bone Replacement Materials Prepared by Two Methods. *MRS Proc.* **2012**, *1418*. [CrossRef]
148. Rodan, G.A.; Raisz, L.G.; Bilezikian, J.P. *Principles of Bone Biology*; Academic Press: San Diego, CA, USA, 2002; Volume 2.
149. Buenzli, P.; Pivonka, P.; Gardiner, B.; Smith, D.; Dunstan, C.; Mundy, G. Theoretical analysis of the spatio-temporal structure of bone multicellular units. In *Proceedings of IOP Conference Series: Materials Science and Engineering*; IOP Publishing: Bristol, UK, 2010; Volume 10, p. 012132.
150. Wagermaier, W.; Gupta, H.S.; Gourrier, A.; Burghammer, M.; Roschger, P.; Fratzl, P. Spiral twisting of fiber orientation inside bone lamellae. *Biointerphases* **2006**, *1*, 1–5. [CrossRef]
151. Ekwaro-Osire, S.; Wanki, G.; Dias, J.P.; Science, P. Healthcare—Probabilistic Techniques for Bone as a Natural Composite. *J. Integr. Des. Process. Sci.* **2017**, *21*, 7–22. [CrossRef]
152. Fratzl, P. Collagen: Structure and mechanics, an introduction. In *Collagen*; Springer: Boston, MA, USA, 2008; pp. 1–13.
153. Arnett, T. Osteoclast Resorption #3. Available online: https://boneresearchsociety.org/resources/gallery/39/#top (accessed on 29 April 2020).
154. Wang, G.; Lu, Z.; Zhao, X.; Kondyurin, A.; Zreiqat, H. Ordered HAp nanoarchitecture formed on HAp–TCP bioceramics by "nanocarving" and mineralization deposition and its potential use for guiding cell behaviors. *J. Mater. Chem. B* **2013**, *1*, 2455–2462. [CrossRef]
155. Seeman, E.; Delmas, P.D. Bone quality–the material and structural basis of bone strength and fragility. *N. Engl. J. Med.* **2006**, *354*, 2250–2261. [CrossRef]
156. Boyde, A.; Haroon, Y.; Jones, S.; Riggs, C. Three dimensional structure of the distal condyles of the third metacarpal bone of the horse. *Equine Vet. J.* **1999**, *31*, 122–129. [CrossRef] [PubMed]
157. Ascenzi, M.-G.; Hetzer, N.; Lomovtsev, A.; Rude, R.; Nattiv, A.; Favia, A. Variation of trabecular architecture in proximal femur of postmenopausal women. *J. Biomech.* **2011**, *44*, 248–256. [CrossRef] [PubMed]
158. Boyde, A. Normal Bone. Available online: https://boneresearchsociety.org/resources/gallery/22/#top (accessed on 29 April 2020).
159. Schwarcz, H.P.; Abueidda, D.; Jasiuk, I. The Ultrastructure of Bone and Its Relevance to Mechanical Properties. *Front. Phys.* **2017**, *5*. [CrossRef]
160. Wolfram, U.; Schwiedrzik, J. Post-yield and failure properties of cortical bone. *Bonekey Rep.* **2016**, *5*, 829. [CrossRef] [PubMed]
161. Morgan, E.F.; Keaveny, T.M. Dependence of yield strain of human trabecular bone on anatomic site. *J. Biomech.* **2001**, *34*, 569–577. [CrossRef]
162. Bayraktar, H.H.; Morgan, E.F.; Niebur, G.L.; Morris, G.E.; Wong, E.K.; Keaveny, T.M. Comparison of the elastic and yield properties of human femoral trabecular and cortical bone tissue. *J. Biomech.* **2004**, *37*, 27–35. [CrossRef]
163. Yeni, Y.N.; Fyhrie, D.P. A rate-dependent microcrack-bridging model that can explain the strain rate dependency of cortical bone apparent yield strength. *J. Biomech.* **2003**, *36*, 1343–1353. [CrossRef]
164. Ebacher, V.; Wang, R. A Unique Microcracking Process Associated with the Inelastic Deformation of Haversian Bone. *Adv. Funct. Mater.* **2009**, *19*, 57–66. [CrossRef]

165. Dong, X.N.; Acuna, R.L.; Luo, Q.; Wang, X. Orientation dependence of progressive post-yield behavior of human cortical bone in compression. *J. Biomech.* **2012**, *45*, 2829–2834. [CrossRef] [PubMed]
166. Leng, H.; Dong, X.N.; Wang, X. Progressive post-yield behavior of human cortical bone in compression for middle-aged and elderly groups. *J. Biomech.* **2009**, *42*, 491–497. [CrossRef] [PubMed]
167. Mirzaalimazandarani, M.; Schwiedrzik, J.J.; Thaiwichai, S.; Best, J.P.; Michler, J.; Zysset, P.K.; Wolfram, U. Mechanical properties of cortical bone and their relationships with age, gender, composition and microindentation properties in the elderly. *Bone* **2016**, *93*, 196–211. [CrossRef]
168. Henkel, J.; Woodruff, M.A.; Epari, D.R.; Steck, R.; Glatt, V.; Dickinson, I.C.; Choong, P.F.; Schuetz, M.A.; Hutmacher, D.W. Bone Regeneration Based on Tissue Engineering Conceptions - A 21st Century Perspective. *Bone Res.* **2013**, *1*, 216–248. [CrossRef] [PubMed]
169. Caparros, C.; Guillem-Marti, J.; Molmeneu, M.; Punset, M.; Calero, J.A.; Gil, F.J. Mechanical properties and in vitro biological response to porous titanium alloys prepared for use in intervertebral implants. *J. Mech. Behav. Biomed. Mater.* **2014**, *39*, 79–86. [CrossRef]
170. Dendorfer, S.; Maier, H.J.; Taylor, D.; Hammer, J. Anisotropy of the fatigue behaviour of cancellous bone. *J. Biomech.* **2008**, *41*, 636–641. [CrossRef]
171. Zioupos, P.; Gresle, M.; Winwood, K. Fatigue strength of human cortical bone: Age, physical, and material heterogeneity effects. *J. Biomed. Mater. Res. Part A* **2008**, *86A*, 627–636. [CrossRef]
172. Zadpoor, A.A. Meta-biomaterials. *Biomater. Sci.* **2020**, *8*, 18–38. [CrossRef]
173. Cheah, C.; Chua, C.; Leong, K.; Chua, S. Development of a tissue engineering scaffold structure library for rapid prototyping. Part 1: Investigation and classification. *Int. J. Adv. Manuf. Technol.* **2003**, *21*, 291–301. [CrossRef]
174. Sun, W.; Starly, B.; Nam, J.; Darling, A. Bio-CAD modeling and its applications in computer-aided tissue engineering. *Comput. Des.* **2005**, *37*, 1097–1114. [CrossRef]
175. An, J.; Teoh, J.E.M.; Suntornnond, R.; Chua, C.K. Design and 3D printing of scaffolds and tissues. *Engineering* **2015**, *1*, 261–268. [CrossRef]
176. Chantarapanich, N.; Puttawibul, P.; Sucharitpwatskul, S.; Jeamwatthanachai, P.; Inglam, S.; Sitthiseripratip, K. Scaffold library for tissue engineering: A geometric evaluation. *Comput. Math. Methods Med.* **2012**, *2012*, 1–14. [CrossRef] [PubMed]
177. Wettergreen, M.; Bucklen, B.; Starly, B.; Yuksel, E.; Sun, W.; Liebschner, M. Creation of a unit block library of architectures for use in assembled scaffold engineering. *Comput. Des.* **2005**, *37*, 1141–1149. [CrossRef]
178. Yoo, D. New paradigms in hierarchical porous scaffold design for tissue engineering. *Mater. Sci. Eng. C* **2013**, *33*, 1759–1772. [CrossRef] [PubMed]
179. Feng, J.; Fu, J.; Lin, Z.; Shang, C.; Li, B. A review of the design methods of complex topology structures for 3D printing. *Vis. Comput. Ind. Biomed. Art* **2018**, *1*, 1–16. [CrossRef]
180. Rodríguez-Montaño, Ó.L.; Cortés-Rodríguez, C.J.; Naddeo, F.; Uva, A.E.; Fiorentino, M.; Naddeo, A.; Cappetti, N.; Gattullo, M.; Monno, G.; Boccaccio, A. Irregular Load Adapted Scaffold Optimization: A Computational Framework Based on Mechanobiological Criteria. *ACS Biomater. Sci. Eng.* **2019**, *5*, 5392–5411. [CrossRef]
181. Wally, Z.J.; Van Grunsven, W.; Claeyssens, F.; Goodall, R.; Reilly, G.C. Porous titanium for dental implant applications. *Metals* **2015**, *5*, 1902–1920. [CrossRef]
182. Liu, F.; Zhang, D.Z.; Zhang, P.; Zhao, M.; Jafar, S. Mechanical properties of optimized diamond lattice structure for bone scaffolds fabricated via selective laser melting. *Materials* **2018**, *11*, 374. [CrossRef]
183. Ungersböck, S.-E. *Advanced Modeling of Strained CMOS Technology*; Technical University of Vienna: Vienna, Austria, 2007.
184. Yang, E.; Leary, M.; Lozanovski, B.; Downing, D.; Mazur, M.; Sarker, A.; Khorasani, A.; Jones, A.; Maconachie, T.; Bateman, S. Effect of geometry on the mechanical properties of Ti-6Al-4V Gyroid structures fabricated via SLM: A numerical study. *Mater. Des.* **2019**, *184*, 108705. [CrossRef]
185. Kou, X.; Tan, S.T. A simple and effective geometric representation for irregular porous structure modeling. *Comput. Des.* **2010**, *42*, 930–941. [CrossRef]
186. Stamp, R.; Fox, P.; O'Neill, W.; Jones, E.; Sutcliffe, C. The development of a scanning strategy for the manufacture of porous biomaterials by selective laser melting. *J. Mater. Sci. Mater. Med.* **2009**, *20*, 1839–1848. [CrossRef]

187. Dorozhkin, S.V. Calcium Orthophosphate-Based Bioceramics. *Materials* **2013**, *6*, 3840–3942. [CrossRef] [PubMed]
188. Tang, D.; Tare, R.S.; Yang, L.Y.; Williams, D.F.; Ou, K.L.; Oreffo, R.O. Biofabrication of bone tissue: Approaches, challenges and translation for bone regeneration. *Biomaterials* **2016**, *83*, 363–382. [CrossRef] [PubMed]
189. Ghanaati, S.; Barbeck, M.; Orth, C.; Willershausen, I.; Thimm, B.W.; Hoffmann, C.; Rasic, A.; Sader, R.A.; Unger, R.E.; Peters, F.; et al. Influence of beta-tricalcium phosphate granule size and morphology on tissue reaction in vivo. *Acta Biomater.* **2010**, *6*, 4476–4487. [CrossRef]
190. Sarhadi, F.; Afarani, M.S.; Mohebbi-Kalhori, D.; Shayesteh, M. Fabrication of alumina porous scaffolds with aligned oriented pores for bone tissue engineering applications. *Appl. Phys. A* **2016**, *122*, 1–8. [CrossRef]
191. Schiefer, H.; Bram, M.; Buchkremer, H.P.; Stover, D. Mechanical examinations on dental implants with porous titanium coating. *J. Mater. Sci. Mater. Med.* **2009**, *20*, 1763–1770. [CrossRef]
192. Will, J.; Melcher, R.; Treul, C.; Travitzky, N.; Kneser, U.; Polykandriotis, E.; Horch, R.; Greil, P. Porous ceramic bone scaffolds for vascularized bone tissue regeneration. *J. Mater. Sci. Mater. Med.* **2008**, *19*, 2781–2790. [CrossRef]
193. Pattanayak, D.K.; Fukuda, A.; Matsushita, T.; Takemoto, M.; Fujibayashi, S.; Sasaki, K.; Nishida, N.; Nakamura, T.; Kokubo, T. Bioactive Ti metal analogous to human cancellous bone: Fabrication by selective laser melting and chemical treatments. *Acta Biomater.* **2011**, *7*, 1398–1406. [CrossRef]
194. U.S Food and Drug Administration Department of Health and Human Services FDA, Hip Joint Metal/Polymer/Metal Semi-Constrained Porous-Coated Uncemented Prosthesis. Available online: https://www.accessdata.fda.gov/scripts/cdrh/cfdocs/cfcfr/CFRSearch.cfm?fr=888.3358 (accessed on 29 April 2020).
195. Roosa, S.M.; Kemppainen, J.M.; Moffitt, E.N.; Krebsbach, P.H.; Hollister, S.J. The pore size of polycaprolactone scaffolds has limited influence on bone regeneration in an in vivo model. *J. Biomed. Mater. Res. Part A* **2010**, *92*, 359–368. [CrossRef]
196. Gray, C.; Boyde, A.; Jones, S.J.B. Topographically induced bone formation in vitro: Implications for bone implants and bone grafts. *Bone* **1996**, *18*, 115–123. [CrossRef]
197. Oh, I.-H.; Nomura, N.; Masahashi, N.; Hanada, S. Mechanical properties of porous titanium compacts prepared by powder sintering. *Scr. Mater.* **2003**, *49*, 1197–1202. [CrossRef]
198. Otsuki, B.; Takemoto, M.; Fujibayashi, S.; Neo, M.; Kokubo, T.; Nakamura, T. Pore throat size and connectivity determine bone and tissue ingrowth into porous implants: Three-dimensional micro-CT based structural analyses of porous bioactive titanium implants. *Biomaterials* **2006**, *27*, 5892–5900. [CrossRef]
199. Fukuda, A.; Takemoto, M.; Saito, T.; Fujibayashi, S.; Neo, M.; Pattanayak, D.K.; Matsushita, T.; Sasaki, K.; Nishida, N.; Kokubo, T.; et al. Osteoinduction of porous Ti implants with a channel structure fabricated by selective laser melting. *Acta Biomater.* **2011**, *7*, 2327–2336. [CrossRef]
200. Lemos, A.F.; Ferreira, J. Combining foaming and starch consolidation methods to develop macroporous hydroxyapatite implants. In *Key Engineering Materials*; Trans Tech Publications Ltd.: Freienbach, Switzerland, 2003; pp. 1041–1044.
201. Nyberg, E.L.; Farris, A.L.; Hung, B.P.; Dias, M.; Garcia, J.R.; Dorafshar, A.H.; Grayson, W.L. 3D-Printing Technologies for Craniofacial Rehabilitation, Reconstruction, and Regeneration. *Ann. Biomed. Eng.* **2017**, *45*, 45–57. [CrossRef]
202. Dhandayuthapani, B.; Yoshida, Y.; Maekawa, T.; Kumar, S. Polymeric Scaffolds in Tissue Engineering Application: A Review. *Int. J. Polym. Sci.* **2011**, *2011*. [CrossRef]
203. Li, J.P.; Li, S.H.; Van Blitterswijk, C.A.; de Groot, K. A novel porous Ti6Al4V: Characterization and cell attachment. *J. Biomed. Mater. Res. Part A* **2005**, *73*, 223–233. [CrossRef]
204. Chen, Y.; Frith, J.E.; Dehghan-Manshadi, A.; Attar, H.; Kent, D.; Soro, N.D.M.; Bermingham, M.J.; Dargusch, M.S. Mechanical properties and biocompatibility of porous titanium scaffolds for bone tissue engineering. *J. Mech. Behav. Biomed. Mater.* **2017**, *75*, 169–174. [CrossRef]
205. Hart, N.H.; Nimphius, S.; Rantalainen, T.; Ireland, A.; Siafarikas, A.; Newton, R.U. Mechanical basis of bone strength: Influence of bone material, bone structure and muscle action. *J. Musculoskelet. Neuronal Interact.* **2017**, *17*, 114–139.
206. Zadpoor, A.A. Mechanics of additively manufactured biomaterials. *J. Mech. Behav. Biomed. Mater.* **2017**, *70*, 1–6. [CrossRef] [PubMed]
207. Zadpoor, A.A. Mechanical performance of additively manufactured meta-biomaterials. *Acta Biomater.* **2019**, *85*, 41–59. [CrossRef]

208. Misch, C.E. *Stress Treatment Theorem for Implant Dentistry*, 3rd ed.; Elsevier Mosby: Chennai, India, 2015; pp. 159–192.
209. Saini, M.; Singh, Y.; Arora, P.; Arora, V.; Jain, K. Implant biomaterials: A comprehensive review. *World J. Clin. Cases* **2015**, *3*, 52–57. [CrossRef]
210. Alvarez, K.; Nakajima, H. Metallic scaffolds for bone regeneration. *Materials* **2009**, *2*, 790–832. [CrossRef]
211. Brammer, K.S.; Oh, S.; Cobb, C.J.; Bjursten, L.M.; van der Heyde, H.; Jin, S. Improved bone-forming functionality on diameter-controlled TiO 2 nanotube surface. *Acta Biomater.* **2009**, *5*, 3215–3223. [CrossRef]
212. Smith, T. The effect of plasma-sprayed coatings on the fatigue of titanium alloy implants. *JOM* **1994**, *46*, 54–56. [CrossRef]
213. Zanetti, E.M.; Aldieri, A.; Terzini, M.; Calì, M.; Franceschini, G.; Bignardi, C. Additively manufactured custom load-bearing implantable devices: Grounds for caution. *Australas Med. J.* **2017**, *10*, 694. [CrossRef]
214. Silva, M.; Shepherd, E.F.; Jackson, W.O.; Dorey, F.J.; Schmalzried, T.P. Average patient walking activity approaches 2 million cycles per year: Pedometers under-record walking activity. *J. Arthroplast.* **2002**, *17*, 693–697. [CrossRef]
215. Ahmadi, S.; Hedayati, R.; Li, Y.; Lietaert, K.; Tümer, N.; Fatemi, A.; Rans, C.; Pouran, B.; Weinans, H.; Zadpoor, A. Fatigue performance of additively manufactured meta-biomaterials: The effects of topology and material type. *Acta Biomater.* **2018**, *65*, 292–304. [CrossRef]
216. FDA. *Guidance Document for Testing Orthopedic Implants with Modified Metallic Surfaces Apposing Bone or Bone Cement*; Branch, O.D., Ed.; U.S. Food and Drug Administration: Rockville, MD, USA, 1994.
217. Aitchison, G.; Hukins, D.; Parry, J.; Shepherd, D.; Trotman, S. A review of the design process for implantable orthopedic medical devices. *Open Biomed. Eng. J.* **2009**, *3*, 21–27. [CrossRef]
218. Fratila-Apachitei, L.E.; Leoni, A.; Riemslag, A.; Fratila-Apachitei, L.; Duszczyk, J. Enhanced fatigue performance of porous coated Ti6Al4V biomedical alloy. *Appl. Surf. Sci.* **2011**, *257*, 6941–6944. [CrossRef]
219. Gong, H.; Rafi, K.; Gu, H.; Ram, G.J.; Starr, T.; Stucker, B. Influence of defects on mechanical properties of Ti–6Al–4 V components produced by selective laser melting and electron beam melting. *Mater. Des.* **2015**, *86*, 545–554. [CrossRef]
220. Liu, H. *Numerical Analysis of Thermal Stress and Deformation in Multi-Layer Laser Metal Deposition Process*; Missouri University of Science and Technology: Rolla, MO, USA, 2014.
221. Li, S.J.; Murr, L.E.; Cheng, X.Y.; Zhang, Z.B.; Hao, Y.L.; Yang, R.; Medina, F.; Wicker, R.B. Compression fatigue behavior of Ti–6Al–4V mesh arrays fabricated by electron beam melting. *Acta Mater.* **2012**, *60*, 793–802. [CrossRef]
222. Martinez-Marquez, D.; Gulati, K.; Carty, C.P.; Stewart, R.A.; Ivanovski, S. Determining the relative importance of titania nanotubes characteristics on bone implant surface performance: A quality by design study with a fuzzy approach. *Mater. Sci. Eng. C* **2020**, *114*, 110995. [CrossRef] [PubMed]
223. Yu, L.X. Pharmaceutical quality by design: Product and process development, understanding, and control. *Pharm. Res.* **2008**, *25*, 781–791. [CrossRef]

Publisher's Note: MDPI stays neutral with regard to jurisdictional claims in published maps and institutional affiliations.

© 2020 by the authors. Licensee MDPI, Basel, Switzerland. This article is an open access article distributed under the terms and conditions of the Creative Commons Attribution (CC BY) license (http://creativecommons.org/licenses/by/4.0/).

Review

Bone Morphogenetic Proteins, Carriers, and Animal Models in the Development of Novel Bone Regenerative Therapies

Nikola Stokovic [1], Natalia Ivanjko [1], Drazen Maticic [2], Frank P. Luyten [3] and Slobodan Vukicevic [1,*]

1. Laboratory for Mineralized Tissues, School of Medicine, University of Zagreb, 10000 Zagreb, Croatia; nikola.stokovic@mef.hr (N.S.); natalia.ivanjko@mef.hr (N.I.)
2. Clinics for Surgery, Orthopedics and Ophthalmology, Faculty of Veterinary Medicine, University of Zagreb, 10000 Zagreb, Croatia; drazen.maticic@vef.hr
3. Skeletal Biology & Engineering Research Center, KU Leuven, Herestraat 49, 3000 Leuven, Belgium; frank.luyten@kuleuven.be
* Correspondence: slobodan.vukicevic@mef.hr

Abstract: Bone morphogenetic proteins (BMPs) possess a unique ability to induce new bone formation. Numerous preclinical studies have been conducted to develop novel, BMP-based osteoinductive devices for the management of segmental bone defects and posterolateral spinal fusion (PLF). In these studies, BMPs were combined with a broad range of carriers (natural and synthetic polymers, inorganic materials, and their combinations) and tested in various models in mice, rats, rabbits, dogs, sheep, and non-human primates. In this review, we summarized bone regeneration strategies and animal models used for the initial, intermediate, and advanced evaluation of promising therapeutical solutions for new bone formation and repair. Moreover, in this review, we discuss basic aspects to be considered when planning animal experiments, including anatomical characteristics of the species used, appropriate BMP dosing, duration of the observation period, and sample size.

Keywords: animal model; bone fracture; bone healing; posterolateral spinal fusion; regenerative medicine; bone morphogenetic proteins

1. Introduction

Bone tissue possesses unique regenerative properties, and bone fractures regularly heal under physiological conditions. However, large segmental bone defects resulting from severe trauma or extensive tumor resection cannot be restored by endogenous self-repair mechanisms, decrease quality of life, and may sometimes lead to limb amputation. Indeed, the management of large segmental defects is one of the most challenging issues in orthopedic medicine, typically due to the biologically hampered microenvironment [1,2]. The standard of care for the healing of large bone defects requires the use of an autologous bone graft (ABG), which is usually harvested from the iliac crest. ABG is also used as a gold standard to achieve spinal fusions, including posterolateral spinal fusion (PLF). PLF is a commonly performed surgical procedure used for the treatment of degenerative diseases of the spine, including degenerative disc disease, spondylolisthesis, spinal instability, and symptomatic scoliosis [3–6].

However, ABG possesses several disadvantages, including a limited amount of bone that might be harvested, the potential transfer of contaminating agents, acute and chronic pain, skin scarring, and deformity at the donor site [4,7]. In addition, the use of ABG increases the blood loss, duration, and cost of surgical procedures. Therefore, there remains an imminent need for the development of novel bone regeneration strategies to enrich or replace ABG. Among these, osteoinductive devices are under investigation for clinical use in PLF and healing of large segmental long bone defects.

In the last few decades, numerous preclinical studies using animal models have been conducted to test novel bone bridging or fusion strategies [8,9]. The principles of

the rational use of animal models in the evaluation of novel bone regenerative therapies have been previously described [8]. Hence, we further investigated the use of animal models in the development of osteoinductive therapies of large segmental bone defects and PLF procedures, in particular the selection of a proper anatomical model, treatment dose, observation period, and sample size. We specifically analyzed published in vivo studies looking into the development of bone morphogenetic protein (BMP)-based bone inducing implants.

2. Bone Regeneration by Bone Morphogenetic Protein Devices

BMPs are well-known osteoinductive molecules, required and sufficient for ectopic bone induction, and powerful agents for the restoration of large orthotopic bone defects [10]. BMP2 is the most widely used osteoinductive BMP in preclinical testing, and it is a part of an osteoinductive device (Infuse™, Medtronic, Dublin, Ireland), currently approved for anterior lumbar interbody fusion (ALIF), acute tibial fractures, and maxillofacial reconstructions [11–15]. However, BMPs have been used off- label in various spinal indications, including cervical spine fusion, posterior lumbar interbody fusion (PLIF), transforaminal lumbar interbody fusion (TLIF), posterolateral spinal fusion (PLF), and thoracolumbar fusions [16,17]. Reported side effects in patients included implant displacement, infection, swelling of the adjacent tissue and dysphagia, formations of seroma, radiculitis and nerve root compression, ectopic bone formation, osteolysis, and retrograde ejaculation [11,12,16–21]. These side effects eventually resulted from the use of suprphysiological doses as registered BMP2-based devices contain 4–12 mg recombinant protein, while the human body contains only a total of 2 mg of BMPs [22].

Other commonly used osteoinductive BMPs are BMP7, which is no longer in clinical use, and more recently, BMP6 [23]. We demonstrated that BMP6 appears to be superior to BMP2 and BMP7 in promoting osteoblast differentiation in vitro and inducing bone formation in vivo [23,24]. The superiority of BMP6 may arise from its resistance to noggin inhibition and affinity across the BMP type I receptors. Therefore, BMP6-based devices are expected to be more efficacious at lower doses compared to BMP2 and BMP7.

BMPs require a carrier/delivery system that will sustain the BMP concentration and allow prolonged BMP release [25–28]. Moreover, the ideal BMP carrier should be biocompatible, enable vascular and cellular infiltration, resist compression, and define the contours of the resulting bone [25,26,29]. BMP carriers can be divided into four major groups: natural polymers, synthetic polymers, inorganic materials, and combinations between these groups [25,26,30].

Natural polymers include collagen, hyaluronic acid, gelatin, fibrin, chitosan, alginate, and silk and have been extensively evaluated in preclinical studies [6,31–38]. The advantages of natural polymers are biocompatibility, biodegradability, and resorbability in the physiological environment [25–27]. The most commonly used is bovine tendon collagen which delivers BMP2 in the clinically approved Infuse™ device. However, collagen has significant disadvantages, including a low affinity for BMPs, immunogenicity due to its animal origin, and weak biomechanical properties resulting in compression by surrounding tissues [25–27].

Biocompatible and biodegradable synthetic polymers, such as polylactic acid (PLA), polyglycolic acid (PGA), poly(D, L-lactide-co-glycolide) (PLGA), polyethylene glycol (PEG), poly-E-caprolactone (PCL), and polypropylene fumarate (PPF), as well as their block polymers have been evaluated as potential BMP carriers to overcome the disadvantages of natural polymers, including immunogenicity and disease transmission risk [39–42]. They are also moldable into highly porous three-dimensional scaffolds, linearly oriented scaffolds, fibers, sheets, blocks, or microspheres [26]. Apart from these advantages, synthetic polymers decrease local pH as a result of acidic breakdown byproducts, have poor clearance, cause bulk degradation, and cause chronic inflammation associated with high-molecular weight polymers, resulting in substantial disadvantages [26]. They have also been tested with other potentially osteogenic molecules, such as PGE2 and PGE4 prostaglandin receptor analogs [43], and materials such as calcium silicate (CaSi) and dicalcium phosphate dihydrate (DCPD) [44–47].

Inorganic materials as potential BMP carriers include calcium phosphate (CaP) ceramics, calcium phosphate and calcium sulfate cement, and bioglass [2,5,29,32,38,42,48–72]. The most commonly used inorganic preclinical materials are CaP ceramics, further subdivided into hydroxyapatite (HA), tricalcium phosphate (TCP), and biphasic calcium phosphate (BCP) containing both HA and TCP at various ratios. We have recently shown that the chemical composition of ceramics does not affect the amount of newly formed bone induced by the osteoinductive device [73,74]. However, HA and TCP significantly differ in resorbability (HA is very stable, while TCP is more resorbable), which would eventually result in different residual ceramic volumes. The resorbability might be adjusted by varying HA/TCP ratios in BCP ceramics [75]. Moreover, CaP ceramics might be formulated into particles or blocks in a broad range of sizes and geometrical shapes while porosity, pore size, and interconnectivity are adjusted during the sintering process [73,75,76]. We demonstrated that particle size affects the volume of newly formed bone; smaller particles (74–420 µm) combined with rhBMP6 resulted in higher bone volume than larger particles (1000–4000 µm) [73]. Another important determinant of ceramics is the pore size since pores from 300 to 400 µm promoted the formation of the largest bone volume [51].

The fourth group of BMP carriers are composites of the aforementioned materials which have been introduced to overcome the encountered limitations of a single component. The most typical combinations are composites containing either natural or synthetic polymers with CaP ceramics [39,77–85]. In these combinations, ceramics increase the biomechanical properties of the implants and are used to address compressibility issues. Less frequent, natural, and synthetic polymers might be combined.

We have recently developed an autologous bone graft substitute (ABGS) comprised of BMP6 delivered within an autologous blood coagulum to which a compression-resistant matrix, such as allograft or synthetic ceramics, can be added [22,73,74,76,86–92]. Moreover, the volume of newly induced bone increased with the elevation of the CRM amount, which might be attributed to the enlargement in an overall surface area [73].

3. Animal Models

Animal models are routinely used in the development of novel bone regenerative therapies [8]. Models might be categorized according to the species (mouse, rat, rabbit, sheep, non-human primate) and tested indication (ectopic model, critical-size defect, PLF). In this review, we suggested classification based on the stage of preclinical development, namely as initial, intermediate, and advanced testing of osteoinductive devices (Figure 1). Initial testing includes rodent ectopic and rodent critical-size defect models for rapid comparison of different osteoinductive responses. Intermediate evaluation includes adequate rabbit models (segmental defect and PLF model), while advanced testing uses canine, sheep, and non-human primates as a final step before clinical trials.

3.1. Initial Evaluation in Rodents

3.1.1. Ectopic Models

Rodent ectopic models have been extensively used for the initial evaluation of novel osteoinductive therapies. They might be also used for investigating the biology of ectopic bone induction and the formation of a bone organ or ossicle, including bone and bone marrow [31,32,39,48–57,71,73,76,86,87,93–105]. Rodent ectopic models (Tables 1 and 2) are further subdivided according to the species (mouse, rat) and the implantation site (subcutaneous or intramuscular). Implantation under the skin (Figure 2A–D) or into the muscle does not affect the bone formation outcome, and the bone formation occurs in the first two weeks following implantation of an osteoinductive device [76,86,87]. The later time points are needed for the evaluation of the bone longevity and maintenance of the ectopic bone structure. Molecular and cellular events during the cascade of bone formation can be evaluated using microCT/nanoCT and histological analyses. Immunohistochemistry, flow cytometry, gene profile microarrays, and single-cell RNA sequencing are among other analytical techniques used for unraveling the mechanism of ectopic bone formation.

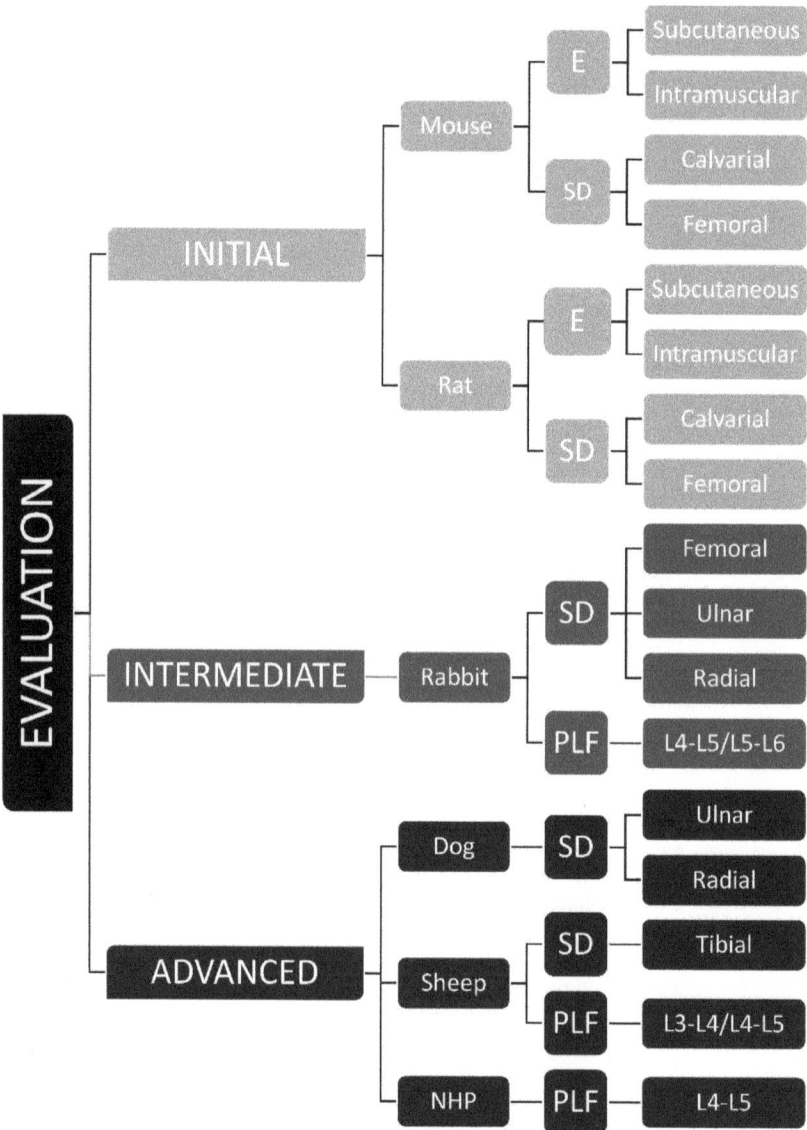

Figure 1. Classification of animal models from mouse to NHP based on the stage of preclinical development, namely as initial, intermediate, and advanced evaluation of osteoinductive devices. E—ectopic, SD—segmental defect, PLF—posterolateral spinal fusion models, and L—level of lumbar transverse processes.

Table 1. Mouse ectopic and bone defect models.

Mouse Ectopic Model				
Author (Year)	Carrier	BMP Dose (µg)	Time (Weeks or Days)	Sample Size (n)
Kato et al. (2006)	PLA-DX-PEG, PLA-DX-PEG/TCP, TCP	2 and 5	3 and 6 weeks	6
Roldan et al. (2010)	BCP		12 weeks	8
Liang et al. (2014)	TCP	50	3, 7, 14, 21, and 28 days	5
Bolander et al. (2016)	CaP granules/Collagen	1.06 and 1.77	5 weeks	4
Ji et al. (2018)	CaP-based materials	0.81, 3.24, and 5.67	2 and 5 weeks	3
Hashimoto et al. (2020)	Collagen	2.5	7, 10, 14, and 21 days	

Mouse Calvarial Defect Model					
Author (Year)	Calvarial Defect Size (mm)	Carrier	BMP Dose (µg)	Time (Weeks)	Sample Size (n)
La et al. (2012)	4	TCP, Heparin—conjugated fibrin	0.3	8	10
Yang et al. (2012)	4	Collagen, Apatite—coated collagen	0.5, 0.75, 1, 2, and 3	8	6
Fan et al. (2015)	3	PLGA/Apatite layer	0.3, 0.6, and 1	6	8, 12
Gronowitz et al. (2017)	3.5	Collagen/HA	2	3	4
Herberg et al. (2017)	5	Acellular dermis	0.542	4	10
Huang et al. (2017)	3.5	PLA	50	2, 4, 6, and 8	16
Seo et al. (2017)	5	Poly(phosphazene) hydrogels Poly(phosphazene) hydrogels/BCP	5 and 10	8	3
Terauchi et al. (2017)	3.5	Sulphopropyl ether—modified polyrotaxanes/Collagen	0.1	4	5, 6
Maisani et al. (2018)	3.5	Hydrogel	1	8	6
Reyes et al. (2018)	4	PLGA	0.1, 0.3, and 0.6	4 and 8	4

Mouse Femoral Defect Model					
Author (Year)	Femoral Defect Size (mm)	Carrier	BMP Dose (µg)	Time (Weeks)	Sample Size (n)
Alaee et al. (2014)	2	Collagen	5	4 days, 1, 2, 3, 4, and 8	6
Bougrouli et al. (2016)	2	Collagen	5	1, 2, 4, and 8	6
Zwingenbergen et al. (2016)	3	Heparin/functionalized mineralized collagen matrix	2.5 and 15	6	11

Table 2. Rat ectopic and bone defect models.

Rat Ectopic Model				
Author (Year)	Carrier	BMP Dose (µg)	Time (Weeks)	Sample Size (n)
Kuboki et al. (1998)	HA		1, 2, 3, and 4	
Tsuruga et al. (1998)	HA	4	1, 2, 3, and 4	3
Alam et al. (2000)	TCP, HA, BCP	1.5 and 10	2 and 4	3
Vehof et al. (2002)	HA	8	3, 5, 7, and 9	3
Kim Chang-Sung et al. (2004)	TCP, Collagen	5	2 and 8	10
Tazaki et al. (2006)	HA	0.5, 1, and 5	3	
Tazaki et al. (2008)	HA, TCP	0.5, 1, and 5	3	3
Luca et al. (2010)	Chitosan/Hyaluronan hydrogel	150	3	3, 6
Reves et al. (2011)	Chitosan-nano-HA	36	4	6
Park et al. (2011)	BCP	2.5	2 and 8	5–8
Bhakta et al. (2012)	Hyaluronan-based hydrogel	5	8	6
Strobel et al. (2012)	BCP	1.6	2, 4, and 6	6
Kisiel et al. (2013)	Hyaluronan hydrogel/Fibronectin fragments	4	7	6
Ma et al. (2014)	BCP	20	8	6
Mumcuoglu et al. (2018)	Collagen-based microspheres/Alginate	0.3, 1, and 10	10	8
Lin et al. (2019)	Coralline HA	20	5	6
Rat Calvarial Defect Model				
Author (Year)	Carrier	BMP Dose (µg)	Time (Weeks)	Sample Size (n)
Jung et al. (2006)	TCP	2.5	2 and 8	20
Kim et al. (2011)	BCP	50 and 250	2 and 8	20
Park et al. (2011)	BCP	2.5	2 and 8	5–8
Notodihardjo et al. (2011)	HA	10	4	5
Jang et al. (2012)	BCP	2.5, 5, 10, and 20	2 and 8	8
Lee JH et al. (2013)	TCP, HA, BCP	5	4 and 8	13
Bae et al. (2017)	PCL/TCP	5	4	7
Rat Femoral Defect Model				
Author (Year)	Carrier	BMP Dose (µg)	Time (Weeks)	Sample Size (n)
Chu et al. (2006)	Poly(propylene fumarate)/TCP/DCP	10	6 and 15	4, 7
Johnson et al. (2011)	Collagen, Collagen/Heparin, Heparin	3	12	7, 9
Diab et al. (2011)	PCL/Silk fibroin hydrogel	5	12	10
Lee et al. (2012)	BCP	1000	4 and 8	6
Rodriguez-Evora et al. (2013)	Segmented polyurethane/PLGA/ TCP ceramics	1.6 and 6.5	12	9
Wai-Ching et al. (2014)	Bioactive glass/DCP	10	15	8, 9
Williams et al. (2015)	Collagen	25, 50, 75 and 100	8	8, 11
Krishnan et al. (2015)	Nanofiber mesh alginate	5 µg	12	14

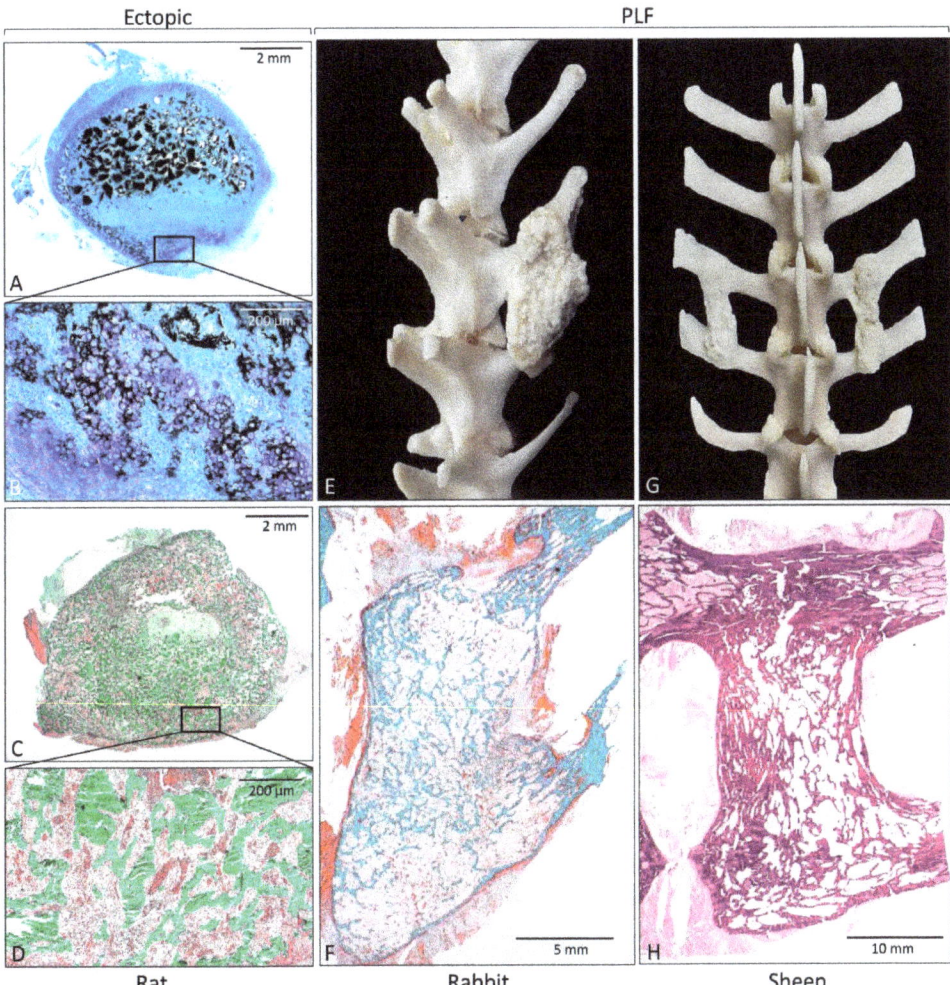

Figure 2. Histological sections and gross anatomy of newly formed bone induced by rhBMP6 in a rat subcutaneous assay (**A–D**); rabbit (**E,F**) and sheep (**G,H**) posterolateral spinal fusion model. (**A,B**) On day 7 following implantation, endochondral bone formation occurs at the peripheral site of ABGS, while 14 days (**C,D**) after implantation, newly formed bone is present throughout the implant containing rhBMP6 lyophilized on allograft mixed with ABC. Gross anatomy of newly formed bone between transverse processes in rabbit (**E**) and sheep (**G**) PLF model. Histological sections through fusion mass in rabbit (**F**) and sheep (**H**) PLF model. Histological sections were processed undecalcified and stained by Von Kossa (**A,B**), Goldner (**C,D,F**), and hematoxylin and eosin (**H**). (Modified from [74,88], respectively.)

Murine models are initially used to unravel the potential mechanism of action of various signaling pathways and genes or proteins in bone induction or enhancement. However, due to limited translation of mouse to human bone regeneration and disease outcome, the rat is therefore a more suitable model when testing functional outcomes [106–108].

3.1.2. Bone Defect Models

Mouse or rat bone defects are the initial orthotopic models to evaluate the osteoinductive properties of novel therapies and the osseointegration of newly formed bone with adjacent native bone. There are two main bone defect models in rodents: a calvarial critical-size defect and segmental femoral defect. In the calvarial critical-size defect, circular bone defects are created in the mouse (3–5 mm) [109–118] and rat (4–8 mm) [29,55,58–60,66,77] calvaria (Figure 3A; Tables 1 and 2), while segmental defects of the long bone are typically created in the femur, both in mice (2–3 mm) [119–121] and rats (6–10 mm) [5,33–35,67,78,79] (Figure 3B; Tables 1 and 2) and filled with tested osteoinductive material. The development of a reproducible non-union model in the mouse is demanding, and, in contrast to rat non-union models, mouse non-union models are sparse [122]. The main shortcoming of this model is a relatively small defect size compared to clinically relevant proportions. Moreover, it is difficult to obtain a full stabilization of the fracture, therefore resulting in increased callus formation. Methods of evaluation include analyses of radiological images (CT/microCT), histological and histomorphometric analyses, and biomechanical testing, which might be conducted at the end of the observation period (Figure 4) [5,29,33–35,55,58–60,66,67,77–79,109–121,123].

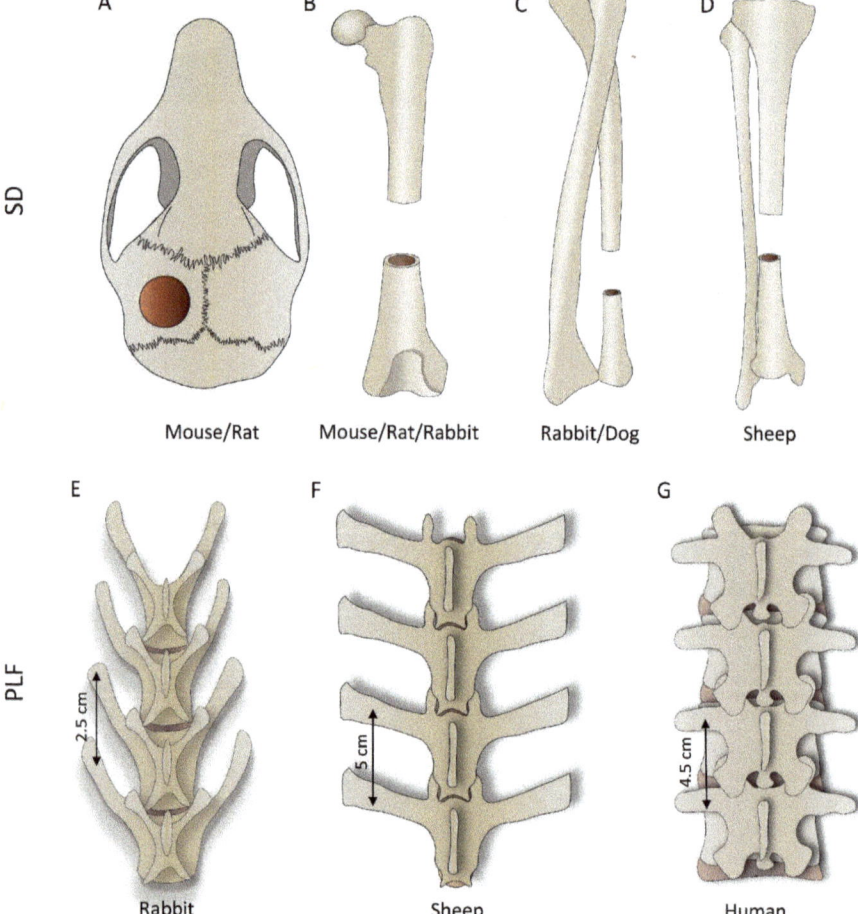

Figure 3. First row: segmental defect (SD) models performed on various bones depending on the species; (**A**) calvarial, (**B**) femoral, (**C**) ulnar (or radial), and (**D**) tibial defect. Second row: posterolateral lumbar fusion (PLF) is conducted between adjacent transverse processes; the figure shows differences in the anatomy of rabbit (**E**), sheep (**F**), and human (**G**) lumbar spine.

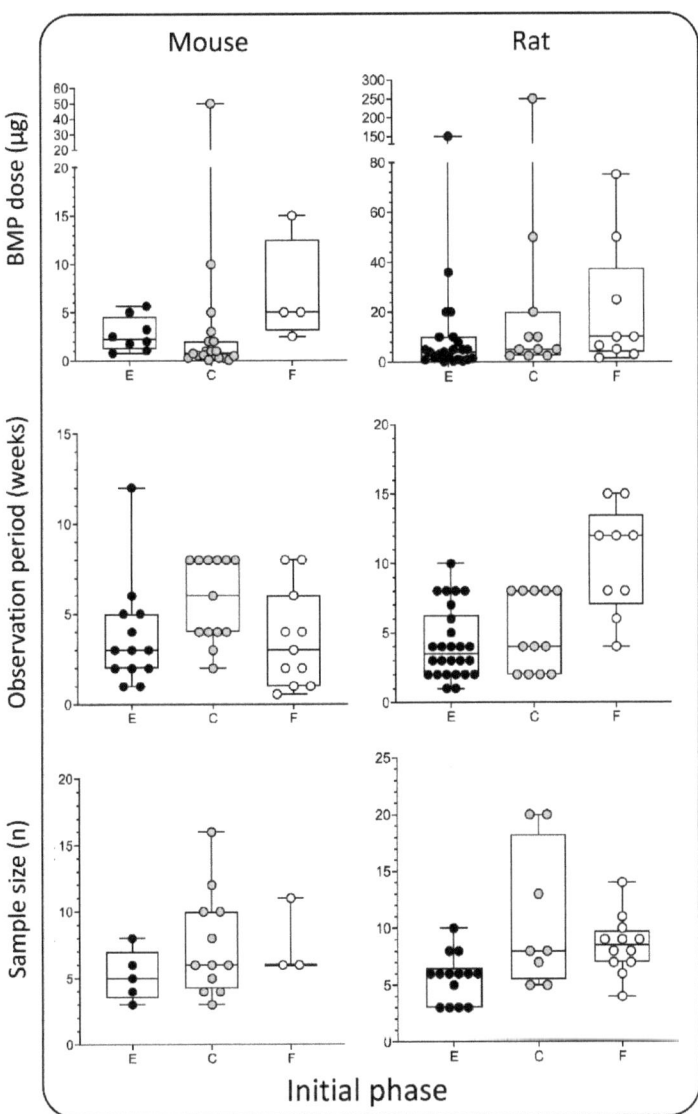

Figure 4. BMP doses (μg), observation period (weeks), and sample size (*n*) used in the initial phase —rodent models. Data is presented through box and whisker plots in which dots represent all individual values from studies listed in Tables 1 and 2. E—ectopic model, C—calvarial, and F—femoral defect.

3.2. Intermediate Evaluation in Rabbits

3.2.1. Segmental Defect Model

Potential therapeutical solutions for segmental bone defect restoration have been extensively evaluated in rabbits [36,41,42,61,68–70,80–84,87,124–126] (Table 3), and a defect can be created in the femur, radius, or ulna (Figure 3B,C). Regardless of the chosen anatomical site, the typical defect size is 15–20 mm. In published work, the defect was bridged with a broad range of delivery systems containing up to 150 μg of BMPs. The observation period was typically 6–12 weeks. Few studies evaluated bone formation at earlier time points (2 and 4 weeks) or for a prolonged period (24 weeks) (Figure 5, 1st column).

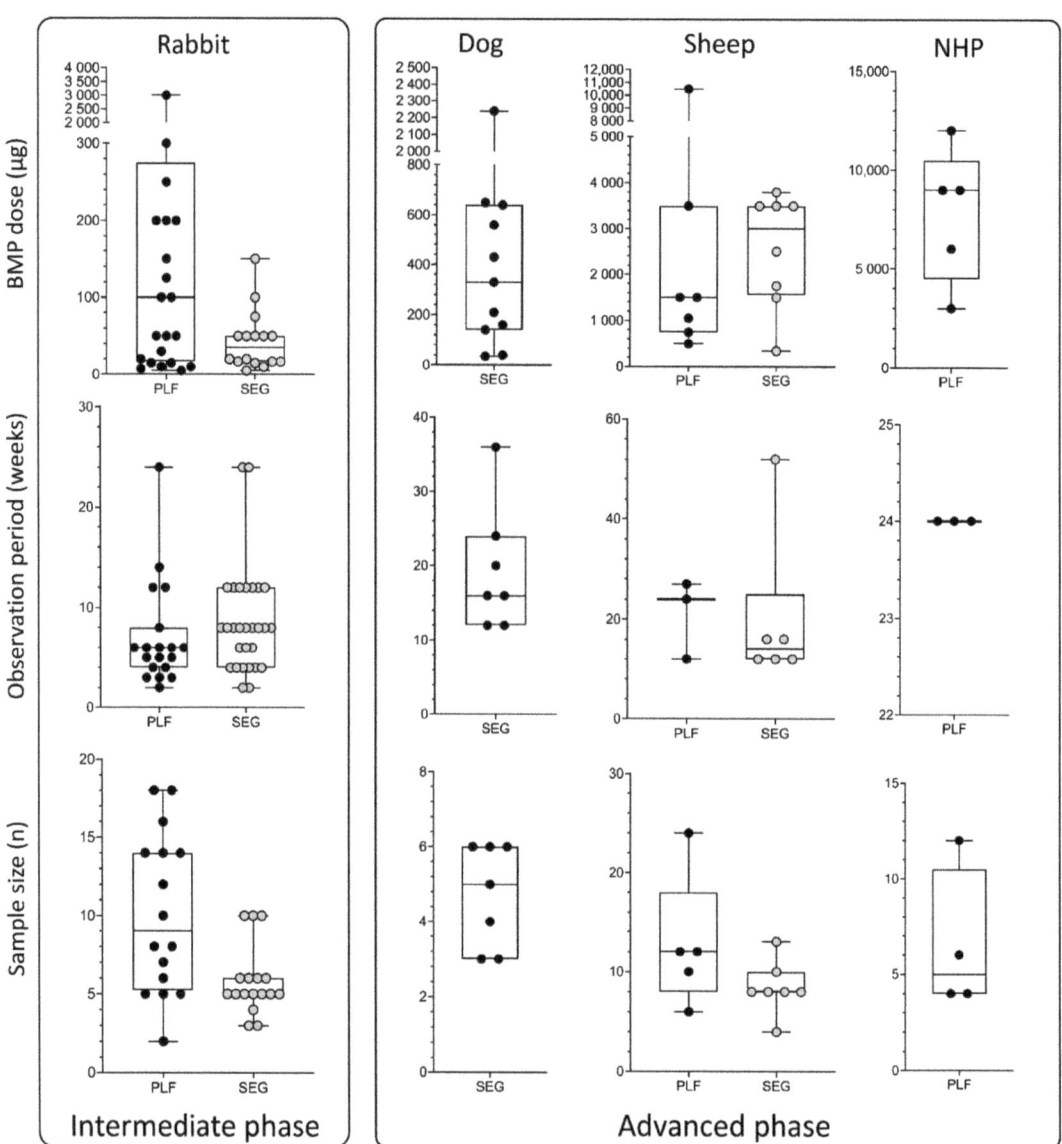

Figure 5. BMP doses (µg), observation period (weeks), and sample size (*n*) used in the intermediate and advanced phases, which include rabbit, dog, sheep, and non-human primate models. Data are presented through box and whisker plots in which dots represent all individual values from studies listed in Tables 3–6. SEG—segmental defect models; PLF—posterolateral spinal fusion model.

Table 3. Rabbit segmental defect model.

Author (Year)	Model	Carrier	BMP Dose (μg)	Time (Weeks)	Sample Size (n)
Yoneda et al. (2004)	Femur (1.5 cm)	PLA-DX-PEG/TCP	50	24	5
Yamamoto et al. (2006)	Ulna (2 cm)	Gelatin hydrogel	17	6	3
Liu et al. (2009)	Radius (1.5 cm)	Gelatin/nanoHA/Fibrin	100	4, 8, and 12	5
Luca et al. (2010)	Radius (1.5 cm)	Chitosan hydrogel/TCP	150	8	1 (pilot)
Zhu et al. (2010)	Radius	nanoHA		4, 8, and 12	10
Bae et al. (2011)	Ulna (1.5 cm)	PCL/fibrin	75	8	5
Fujita et al. (2011)	Ulna (2 cm)	Gelatin/TCP	17	4 and 8	6, 10
Sun-Woong et al. (2012)	Ulna (2 cm)	PCL	15	12	6
Hou et al. (2012)	Radius (1.5 cm)	Collagen, Collagen/Chitosan	50	2, 4, 8, and 12	3, 5
Choi et al. (2014)	Radius (2 cm)	Collagen, Fibrin glue	50	6 and 12	4
Wu et al. (2014)	Radius (1.5 cm)	CaP cement, Hydroxypropylmethyl cellulose/CaP cement	50	2, 4, 8, and 12	5
Yamamoto et al. (2015)	Ulna (2 cm)	Gelatin/TCP	17	6	6
Peng et al. (2016)	Femur (1 cm)	PEG-PLGA hydrogel	5, 10, and 20	12	6
Pan et al. (2017)	Femur (2 cm)	Bioglass/TCP	20	4 and 8	5
Kuroiwa et al. (2018)	Femur (2 cm)	TCP	50	12 and 24	10
Grgurevic et al. (2019)	Ulna (1.5 cm)	Autologous blood coagulum	25, 50, and 100	23	5
Huang et al. (2021)	Ulna (2 cm)	TCP	20	8	5

Table 4. Rabbit PLF model.

Author (Year)	Carrier	BMP Dose (μg)	Time (Weeks)	Sample Size (n)
Boden et al. (1995)	DBM, Biocoral/Collagen	100 and 300	5	14–16
Itoh et al. (1999)	Collagen	10, 50, and 200	24	6
Louis-Ugbo et al. (2001)	BCP, Collagen/BCP	3000/mL	5	18
Jenis et al. (2002)	Collagen	-	3 and 12	8
Konishi et al. (2002)	Autograft/HA	200	2, 4, and 6	2–7
Suh et al. (2002)	Collagen/BCP, BCP	860	5	14
Minamide et al. (2003)	TCP cement, True bone ceramics, Collagen	100	3 and 6	5–10
Namikawa et al. (2005)	TCP/PLA-DX-PEG	7.5, 15, and 30	6	5

Table 4. Cont.

Rabbit PLF Model				
Author (Year)	Carrier	BMP Dose (μg)	Time (Weeks)	Sample Size (n)
Valdes et al. (2007)	-		6	18
Dohzono et al. (2009)	TCP	5, 15, 50, and 150	4 and 8	5–8
Lee JW et al. (2011)	Heparin—conjugated PLGA nanospheres, PLGA nanospheres	20	12	12
Lee JH et al. (2012)	HA	10, 50, 200, and 500	3 and 6	14
Vukicevic et al. (2019)	Autologous blood coagulum, Autologous blood coagulum/Allograft	125, 250, 500, and 1000	14	4

Table 5. Dog and sheep segmental defect model.

Dog Segmental Defect Model					
Author (Year)	Model	Carrier	BMP Dose (μg)	Time (Weeks)	Sample Size (n)
Itoh (1998)	Ulna (2 cm)	PLGA/Gelatin	40, 160, and 640	16	4
Tuominen (2000)	Ulna (2 cm)	Coral	-	16 and 36	3, 6
Hu (2003)	Radius (2 cm)	HA/Collagen/PLA	-	24	6
Jones (2008)	Ulna (2.5 cm)	Collagen/Allograft, Collagen/BCP ceramics	210, 430, and 650	12	6
Harada (2012)	Ulna (2.5 cm)	TCP	35, 140, 560, and 2240	12	3
Minier (2014)	Ulna (2 cm)	CaP/Hydrogel	330	20	5
Sheep Segmental Defect Model					
Author (Year)	Model	Carrier	BMP Dose (μg)	Time (Weeks)	Sample Size (n)
Den Boer et al. (2003)	Tibia (3 cm)	HA	2500	12	8
Pluhar et al. (2006)	Tibia (5 cm)	Carboxymethylcellulose/Bovine collagen, Collagen	3500	16	10
Reichert et al. (2012)	Tibia (3 cm)	mPCL-TCP	3500	12 and 52	8
Cipitria et al. (2013)	Tibia (3 cm)	mPCL-TCP	1750 and 3500	12	8
Lammens et al. (2020)	Tibia (3 and 4.5 cm)	CaP ceramics	344, 1500, and 3800	16	4, 8, 13

Table 6. Sheep and non-humane primate PLF models.

Sheep PLF Model				
Author (Year)	Carrier	BMP Dose (μg)	Time (Weeks)	Sample Size (n)
Pelletier et al. (2014)	TCP	1050, 3500, and 10,500	12	12
Toth et al. (2016)	Collagen/BCP, Collagen-ceramic sponge	750 and 1500/cm^3	24	12–24
Grgurevic et al. (2020)	Autologous blood coagulum, Autologous blood coagulum/Allograft	500 and 1500	27	6–10

Table 6. *Cont.*

NHP PLF Model				
Author (Year)	Carrier	BMP Dose (µg)	Time (Weeks)	Sample Size (*n*)
Boden et al. (1999)	BCP	6000, 9000, and 12,000	24	4–12
Suh et al. (2002)	Ceramic/Collagen	9000	24	4
Akamaru et al. (2003)	Collagen/BCP Collagen/Allograft	3000	24	6

3.2.2. Posterolateral Spinal Fusion (PLF) Model

Rabbit is the most commonly used species for the evaluation of the efficacy and safety of promising therapeutical solutions for achieving PLF [5,6,37,38,40,62–64,74,85,86,127–129]. The transverse processes of lumbar vertebrae are exposed, and an osteoinductive device is implanted bilaterally between adjacent transverse processes (L4-L5 or L5-L6) [127]. Transverse processes should be decorticated before the implantation to promote osseointegration of newly formed bone with native bone [86]. In the majority of previous studies, the BMP dose was up to 1000 µg and was delivered on various carriers (Table 4). The spinal fusion outcome was evaluated 6 weeks following surgery, and the majority of rabbit PLF studies had an observation period of fewer than 10 weeks. Few studies had a prolonged observation period (>10 weeks), but later time points might be important to determine the survival and long-term maintenance of newly induced bone [6,86], which is clinically relevant in patients undergoing PLF surgery (Figure 5, 1st column).

Methods of evaluation in rabbit segmental defect and PLF models are similar and include segmental mobility testing, radiological methods (x-ray and CT/microCT), histological (Figure 2F), histomorphometric analyses, and biomechanical testing [5,6,37,38,40,62–64,74,85,86,127–129].

3.3. Advanced Evaluation of Bone Regeneration Therapies

3.3.1. Dog and Sheep Segmental Defect Model

Dog and sheep segmental defect models are used for advanced evaluation of bone regeneration therapies. In dogs, the defect (20–25 mm) is created in the radius or ulna (Figure 3C) [130–135]. Applied doses of BMPs were in the range between 100 and 650 µg, which is higher compared to the rabbit model. Moreover, the typical observation period (12–24 weeks) was also prolonged compared to the rabbit model (Figure 5, 2nd and 3rd columns).

Tibial segmental bone defects in sheep (Figure 3D) were recently developed to evaluate novel bone regeneration therapies in conditions mimicking the size and biology of segmental bone defects in the clinics [2,136–140]. Moreover, there are two subtypes of this model: a fresh defect (FD) and biologically exhausted defect (BED), the latter mimicking a patient with a non-union [2]. Following the creation of a large defect (30 or 45 mm) in the sheep tibia in the FD model, a polymethyl-methacrylate spacer is inserted to induce the formation of the Masquelet membrane. Six weeks following the creation of the defect, an osteoinductive device was inserted after the removal of the spacer (FD model). In the BED model, the defect is in the first instance left untreated leading to a non-union. Subsequently, debridement of the non-union or fibrotic tissue ingrowth (BED model) is performed, followed by implantation of a spacer for 6 weeks, and then, finally, after removal of the spacer, an implant is inserted. BMP doses applied in this model ranged from 344 to 3800 µg, while the typical observation period was up to 16 weeks [2]. Although the osteoinductive device containing BMP6 on a carrier achieved bridging in FD (30 mm), it was found that larger, biologically exhausted defects appear to require a cell-based implant together with BMP to achieve proper clinically relevant bridging (Table 5) [2]. Importantly, defects were mechanically well stabilized with a circular external fixator according to the Ilizarov technique.

3.3.2. Sheep PLF Model

The sheep PLF model is highly translatable to clinics because the size of the lumbar vertebrae of the sheep is comparable to humans. However, only a few preclinical studies have been conducted on this model [3,65,88] (Table 6). Sheep PLF may be conducted at a single level or as a multisegmental procedure. Moreover, it may be performed with or without instrumentation [88]. The observation period and applied BMP doses in this model were typically significantly longer/higher than in studies on small animals: the follow-up period was up to 6 months with a BMP amount up to 10 mg (Figure 5, 3rd column). Methods of evaluation included X-ray monitoring, microCT evaluation, histological analyses (Figure 2H), and biomechanical testing [3,65,88].

3.3.3. Non-human Primate (NHP) PLF Model

Non-human primates are the most similar animal species to humans, both anatomically and genetically. However, only a few studies were conducted using NHP PLF [14,63,141] (Table 6), primarily due to ethical and economic reasons. In these studies, the goal was to achieve a single-level fusion between adjacent lumbar transverse processes, which are anatomically similar to humans. The applied BMP2 doses (3–12 mg), as well as observation period (24 weeks), were comparable to the sheep PLF model (Figure 5, 4th column).

3.4. Anatomical Characteristics of the Species

3.4.1. Segmental Bone Defect

The general anatomy of long bones (Figure 3B–D) of species discussed in this review is similar, and obviously, the greatest difference among them is size. Differences in the bone size reflect the segmental defect created in each species. The length of the long bone segmental defect in mice (3 mm) or rats (5–8 mm) is small and created in the femur, the largest bone in rodents [5,33–35,67,78,79,119–121,123,142]. Long bones are significantly larger in rabbits/dogs, and segmental defects (15–20 mm in rabbits and 20–25 mm in dogs) are created in the femur, radius, or ulna [36,41,42,61,68–70,80–84,87,124–126,130–135]. Sheep segmental bone defects are usually created in the tibia and, due to their size (30–45 mm), are considered clinically relevant because the defect size compares to those in patients [2].

3.4.2. PLF

Posterolateral spinal fusion (PLF) in preclinical studies is conducted in the lumbar portion of the spine. Although the basic anatomical features of lumbar vertebrae are similar among species discussed in this review, they differ in size and proportions of the different parts of the vertebrae. Rabbits (Figure 3E) and sheep (Figure 3F) have long transverse processes compared to the size of the vertebral body, while humans (Figure 3G), as an adaptation to erect posture and bipedal locomotion, have large bodies and short transverse processes. Importantly, transverse processes in rabbits are slanted and oriented anteriorly (Figure 2E or Figure 3E). On the other hand, the transverse processes in sheep (Figure 2G or Figure 3F) and humans (Figure 3G) are horizontal. The distances between the transverse processes are relatively short in rabbits (20–30 mm), while they are comparable in sheep and humans (40–50 mm).

4. Appropriate Bone Morphogenetic Proteins Dosing

The selection of the appropriate dose for each indication is one of the most challenging steps in the design of experiments. In this review, we compared doses used in various models in mice, rats, rabbits, sheep, and NHP. In the mouse and rat ectopic models, BMP doses were typically up to 25 µg [31,32,39,48–57,71,73,76,86,87] (Figure 4, 1st column, 1st row) per implant, while they were increased in the rat bone defects (up to 50 µg) [5,29,33–35,55,58–60,66,67,77–79,123] (Figure 4, 2nd column, 1st row). BMP doses used in rabbits were up to 150 µg in segmental bone defects [36,41,42,61,68–70,80–84,87,124–126,143] (Figure 5, 1st column, 1st row) and up to 300 µg in rabbit PLF procedures [5,6,37,38,40,62–64,74,85,86,127–129] (Figure 5, 1st column, 1st row). Tested BMP doses in dogs were higher and were typically in the range between 100 and 650 µg [130–135] (Figure 5, 2nd column, 1st row). Moreover, in the sheep and NHP models, BMP doses were significantly higher: in the sheep, the doses were between 500 µg and up to 4 mg [3,65,88] (Figure 5, 3rd column, 1st row), while in the NHP, the doses were up to 12 mg [14,63,141] (Figure 5, 4th column, 1st row).

5. Duration of the Observation Period

The bone induction process is significantly faster in small animals compared to higher species animals and humans. Therefore, observation periods were significantly shorter in rodent ectopic and bone defect models than in studies on sheep and non-human primates (Figures 4 and 5, 2nd row). The average observation period in mouse and rat ectopic models was 3–4 weeks [31,32,39,48–57,71,73,76,86,87] (Figure 4, 1st and 2nd columns, 2nd row). However, depending on the purpose of the study, observation periods in these studies might vary from a few days (studies on the mechanism of bone induction) to several months (bone longevity). Typical observation periods in the rat calvarial defect and femoral segmental defect models are slightly prolonged and last 5 and 10 weeks, respectively [5,29,33–35,55,58–60,66,77–79,123] (Figure 4, 2nd column, 2nd row). The observation period in segmental defect studies was up to 12 weeks in rabbits. [36,41,42,61,68–70,80–84,87,124–126,130–135,143] (Figure 5, 1st column. 2nd row). On the other hand, spinal fusion outcome in the rabbit PLF model was typically evaluated after 6 weeks [5,6,37,38,40,62–64,74,85,86,127–129] (Figure 5, 1st column, 2nd row). However, to study longevity or resorbability of compression-resistant matrices, the follow-up period might be prolonged. As expected, a longer observation period in dogs and sheep was up to 12 months [3,65,88,130–135] (Figure 5, 2nd and 3rd columns, 2nd row), while in NHP studies, it was 6 months [14,63,141] (Figure 5, 4th column, 2nd row).

6. Sample Size

Defining an appropriate sample size is a prerequisite for obtaining valid conclusions from each study. Moreover, the appropriate size of the sample is affected by several parameters, including experimental design and purpose of the study as well as expected differences among experimental groups. The sample size in the majority of reviewed studies here was 5–10 per group regardless of the animal species or model (Figures 4 and 5, 3rd row). Moreover, there is a consensus in published work that the minimal number of animals per experimental group is four. However, a few animals might die during surgery or follow-up periods due to reasons non-related to the tested osteoinductive therapy; therefore, at least five animals per experimental group should be included.

7. Study Outcomes

In Tables 1–6, it was not possible to describe the study outcomes due to non-comparable scoring grades for healing or spinal fusion experiments. The prerequisite in reporting the outcome of bone defect and spinal fusion studies is a clearly described success rate as the percentage of successfully rebridged defects or fused spine segments, respectively. Moreover, the method (radiological images, mobility testing) used to determine rebridgment/fusion should be clearly described. Surprisingly, in a large number of published studies, the success rate was not explicitly described. Several authors used their own scoring grades instead of standardized binary outcomes (successful or unsuccessful rebridgment/fusion). However, even when the binary outcome was used, the determination of successful rebridgment/fusion differed among authors. For example, a few authors determined success rate only on X-ray images without microCT, histology, and biomechanical testing. We suggest that successful rebridgment/fusion should also be determined with microCT, histological sectioning, and biomechanical testing.

The experimental outcome of osteoinductive therapies using rodent ectopic models should be determined by microCT and histology. MicroCT analyses provide information on newly formed bone volume expressed as bone volume (BV) or bone volume/tissue volume ratio (BV/TV). Additionally, if the tested osteoinductive device contains ceramics, microCT analyses might be used to determine the amount of residual ceramic matrix. Moreover, microCT analyses allow the determination of structural properties of newly formed bone by calculating trabecular parameters (trabecular number, trabecular thickness, trabecular separation). The structural properties of newly induced bone should also be analysed by histology and histomorphometry to determine the volume of the bone and remaining carrier/matrix.

8. Conclusions

Due to the large socioeconomic burden of degenerative diseases of the spine and segmental defects of long bones, there is an imminent need for the development of novel osteoinductive therapeutic solutions [1,22]. However, until now, none of the osteoinductive devices have been approved for use in PLF and large segmental defects in patients. A broad range of bone regeneration strategies have been proposed and tested in different animal models. A vast majority of these studies have been conducted in rats and rabbits, leading only to the initial and intermediate steps of preclinical testing, and despite claiming positive results, only a few have been further tested in sheep and NHP models. Infuse™, a BMP2-containing osteoinductive device, has been approved for use in ALIF and acute tibial fractures but has also been used in various off-label indications. However, numerous side effects related to high BMP dose and a large release from the bovine collagen as a carrier have been reported. Therefore, there is a need for an osteoinductive device that would be efficacious at lower doses of BMP delivered on a carrier with a prolonged BMP release. There is some hope that novel engineered BMPs or innovative delivery systems for BMPs may reduce the required therapeutic doses. A novel ABGS containing rhBMP6 within autologous blood coagulum was evaluated in preclinical studies, and in exploratory clinical trials (high tibial osteotomy, distal radial fracture, and posterolateral interbody fusion), it was proven safe and efficacious at relatively low BMP6 doses [73,74,76,86–88,91,92].

Author Contributions: Conceptualization, N.S. and S.V.; Writing—Original Draft Preparation, N.S. and S.V.; Writing—Review and Editing, N.S., N.I., D.M., F.P.L., and S.V.; Visualization, N.I. and N.S.; Supervision, S.V., F.P.L., and D.M.; Funding Acquisition, S.V., F.P.L., and D.M. All authors have read and agreed to the published version of the manuscript.

Funding: This program was funded by the FP7 Health Program (FP7/2007-2013) under grant agreement HEALTH-F4-2011-279239 (Osteogrow), H2020 Health GA 779340 (OSTEOproSPINE), and European Regional Development Fund—Scientific Center of Excellence for Reproductive and Regenerative Medicine (project "Reproductive and regenerative medicine—exploration of new platforms and potentials," GA KK.01.1.1.01.0008 funded by the EU through the ERDF).

Institutional Review Board Statement: Not applicable.

Informed Consent Statement: Not applicable.

Acknowledgments: Special thanks to Jack Ratliff (Ratliff Histology Consultants LLC) for the excellent preparation of the undecalcified histology sections presented in this paper.

Conflicts of Interest: S.V. is a cofounder of perForm Biologics and a coordinator of the EU HORIZON 2020 grant OSTEOproSPINE funding clinical studies of the new drug for bone repair (patent WO2019076484A1).

References

1. Dumic-Cule, I.; Pecina, M.; Jelic, M.; Jankolija, M.; Popek, I.; Grgurevic, L.; Vukicevic, S. Biological aspects of segmental bone defects management. *Int. Orthop.* **2015**, *39*, 1005–1011. [CrossRef] [PubMed]
2. Lammens, J.; Marechal, M.; Delport, H.; Geris, L.; Oppermann, H.; Vukicevic, S.; Luyten, F.P. A cell-based combination product for the repair of large bone defects. *Bone* **2020**, *138*, 115511. [CrossRef] [PubMed]
3. Toth, J.M.; Wang, M.; Lawson, J.; Badura, J.M.; DuBose, K.B. Radiographic, biomechanical, and histological evaluation of rhBMP-2 in a 3-level intertransverse process spine fusion: An ovine study. *J. Neurosurg. Spine* **2016**, *25*, 733–739. [CrossRef] [PubMed]

4. Dimar, J.R.; Glassman, S.D.; Burkus, K.J.; Carreon, L.Y. Clinical outcomes and fusion success at 2 years of single-level instrumented posterolateral fusions with recombinant human bone morphogenetic protein-2/compression resistant matrix versus iliac crest bone graft. *Spine* **2006**, *31*, 2534–2539. [CrossRef] [PubMed]
5. Lee, J.H.; Yu, C.H.; Yang, J.J.; Baek, H.R.; Lee, K.M.; Koo, T.Y.; Chang, B.S.; Lee, C.K. Comparative study of fusion rate induced by different dosages of *Escherichia coli*-derived recombinant human bone morphogenetic protein-2 using hydroxyapatite carrier. *Spine J.* **2012**, *12*, 239–248. [CrossRef] [PubMed]
6. Itoh, H.; Ebara, S.; Kamimura, M.; Tateiwa, Y.; Kinoshita, T.; Yuzawa, Y.; Takaoka, K. Experimental spinal fusion with use of recombinant human bone morphogenetic protein 2. *Spine* **1999**, *24*, 1402–1405. [CrossRef]
7. Glassman, S.D.; Dimar, J.R.; Carreon, L.Y.; Campbell, M.J.; Puno, R.M.; Johnson, J.R. Initial fusion rates with recombinant human bone morphogenetic protein-2/compression resistant matrix and a hydroxyapatite and tricalcium phosphate/collagen carrier in posterolateral spinal fusion. *Spine* **2005**, *30*, 1694–1698. [CrossRef]
8. Peric, M.; Dumic-Cule, I.; Grcevic, D.; Matijasic, M.; Verbanac, D.; Paul, R.; Grgurevic, L.; Trkulja, V.; Bagi, C.M.; Vukicevic, S. The rational use of animal models in the evaluation of novel bone regenerative therapies. *Bone* **2015**, *70*, 73–86. [CrossRef]
9. McGovern, J.A.; Griffin, M.; Hutmacher, D.W. Animal models for bone tissue engineering and modelling disease. *Dis. Model. Mech.* **2018**, 11. [CrossRef]
10. Sampath, T.K.; Vukicevic, S. Biology of bone morphogenetic protein in bone repair and regeneration: A role for autologous blood coagulum as carrier. *Bone* **2020**, 115602. [CrossRef]
11. Carragee, E.J.; Hurwitz, E.L.; Weiner, B.K. A critical review of recombinant human bone morphogenetic protein-2 trials in spinal surgery: Emerging safety concerns and lessons learned. *Spine J.* **2011**, *11*, 471–491. [CrossRef]
12. James, A.W.; LaChaud, G.; Shen, J.; Asatrian, G.; Nguyen, V.; Zhang, X.; Ting, K.; Soo, C. A review of the clinical side effects of bone morphogenetic protein-2. *Tissue Eng. Part B Rev.* **2016**, *22*, 284–297. [CrossRef]
13. Govender, S.; Csimma, C.; Genant, H.K.; Valentin-Opran, A.; Amit, Y.; Arbel, R.; Aro, H.; Atar, D.; Bishay, M.; Borner, M.G.; et al. Recombinant human bone morphogenetic protein-2 for treatment of open tibial fractures: A prospective, controlled, randomized study of four hundred and fifty patients. *J. Bone Jt. Surg. Am.* **2002**, *84*, 2123–2134. [CrossRef] [PubMed]
14. Boden, S.D.; Zdeblick, T.A.; Sandhu, H.S.; Heim, S.E. The use of rhBMP-2 in interbody fusion cages. Definitive evidence of osteoinduction in humans: A preliminary report. *Spine* **2000**, *25*, 376–381. [CrossRef] [PubMed]
15. Wei, S.; Cai, X.; Huang, J.; Xu, F.; Liu, X.; Wang, Q. Recombinant human BMP-2 for the treatment of open tibial fractures. *Orthopedics* **2012**, *35*, e847–e854. [CrossRef]
16. Ong, K.L.; Villarraga, M.L.; Lau, E.; Carreon, L.Y.; Kurtz, S.M.; Glassman, S.D. Off-label use of bone morphogenetic proteins in the United States using administrative data. *Spine* **2010**, *35*, 1794–1800. [CrossRef]
17. Vincentelli, A.F.; Szadkowski, M.; Vardon, D.; Litrico, S.; Fuentes, S.; Steib, J.P.; le Huec, J.C.; Huppert, J.; Dubois, G.; Lenoir, T.; et al. rhBMP-2 (Recombinant Human Bone Morphogenetic Protein-2) in real world spine surgery. A phase IV, National, multicentre, retrospective study collecting data from patient medical files in French spinal centres. *Orthop. Traumatol. Surg. Res.* **2019**, *105*, 1157–1163. [CrossRef] [PubMed]
18. Tannoury, C.A.; An, H.S. Complications with the use of bone morphogenetic protein 2 (BMP-2) in spine surgery. *Spine J.* **2014**, *14*, 552–559. [CrossRef] [PubMed]
19. Fu, R.; Selph, S.; McDonagh, M.; Peterson, K.; Tiwari, A.; Chou, R.; Helfand, M. Effectiveness and harms of recombinant human bone morphogenetic protein-2 in spine fusion: A systematic review and meta-analysis. *Ann. Intern. Med.* **2013**, *158*, 890–902. [CrossRef]
20. Simmonds, M.C.; Brown, J.V.; Heirs, M.K.; Higgins, J.P.; Mannion, R.J.; Rodgers, M.A.; Stewart, L.A. Safety and effectiveness of recombinant human bone morphogenetic protein-2 for spinal fusion: A meta-analysis of individual-participant data. *Ann. Intern. Med.* **2013**, *158*, 877–889. [CrossRef] [PubMed]
21. Hiremath, G.K.; Steinmetz, M.P.; Krishnaney, A.A. Is it safe to use recombinant human bone morphogenetic protein in posterior cervical fusion? *Spine* **2009**, *34*, 885–889. [CrossRef]
22. Vukicevic, S.; Oppermann, H.; Verbanac, D.; Jankolija, M.; Popek, I.; Curak, J.; Brkljacic, J.; Pauk, M.; Erjavec, I.; Francetic, I.; et al. The clinical use of bone morphogenetic proteins revisited: A novel biocompatible carrier device OSTEOGROW for bone healing. *Int. Orthop.* **2014**, *38*, 635–647. [CrossRef] [PubMed]
23. Vukicevic, S.; Grgurevic, L. BMP-6 and mesenchymal stem cell differentiation. *Cytokine Growth Factor Rev.* **2009**, *20*, 441–448. [CrossRef]
24. Song, K.; Krause, C.; Shi, S.; Patterson, M.; Suto, R.; Grgurevic, L.; Vukicevic, S.; van Dinther, M.; Falb, D.; Ten Dijke, P.; et al. Identification of a key residue mediating bone morphogenetic protein (BMP)-6 resistance to noggin inhibition allows for engineered BMPs with superior agonist activity. *J. Biol. Chem.* **2010**, *285*, 12169–12180. [CrossRef]
25. El Bialy, I.; Jiskoot, W.; Nejadnik, M.R. Formulation, delivery and stability of bone morphogenetic proteins for effective bone regeneration. *Pharm. Res.* **2017**, *34*, 1152–1170. [CrossRef]
26. Seeherman, H.; Wozney, J.M. Delivery of bone morphogenetic proteins for orthopedic tissue regeneration. *Cytokine Growth Factor Rev.* **2005**, *16*, 329–345. [CrossRef]
27. Haidar, Z.S.; Hamdy, R.C.; Tabrizian, M. Delivery of recombinant bone morphogenetic proteins for bone regeneration and repair. Part B: Delivery systems for BMPs in orthopaedic and craniofacial tissue engineering. *Biotechnol. Lett.* **2009**, *31*, 1825–1835. [CrossRef] [PubMed]

28. Agrawal, V.; Sinha, M. A review on carrier systems for bone morphogenetic protein-2. *J. Biomed. Mater. Res. B Appl. Biomater.* **2017**, *105*, 904–925. [CrossRef]
29. Jung, U.W.; Choi, S.Y.; Pang, E.K.; Kim, C.S.; Choi, S.H.; Cho, K.S. The effect of varying the particle size of beta tricalcium phosphate carrier of recombinant human bone morphogenetic protein-4 on bone formation in rat calvarial defects. *J. Periodontol.* **2006**, *77*, 765–772. [CrossRef]
30. Lee, S.H.; Shin, H. Matrices and scaffolds for delivery of bioactive molecules in bone and cartilage tissue engineering. *Adv. Drug Deliv. Rev.* **2007**, *59*, 339–359. [CrossRef] [PubMed]
31. Hashimoto, K.; Kaito, T.; Furuya, M.; Seno, S.; Okuzaki, D.; Kikuta, J.; Tsukazaki, H.; Matsuda, H.; Yoshikawa, H.; Ishii, M. In vivo dynamic analysis of BMP-2-induced ectopic bone formation. *Sci. Rep.* **2020**, *10*, 4751. [CrossRef]
32. Kim, C.S.; Kim, J.I.; Kim, J.; Choi, S.H.; Chai, J.K.; Kim, C.K.; Cho, K.S. Ectopic bone formation associated with recombinant human bone morphogenetic proteins-2 using absorbable collagen sponge and beta tricalcium phosphate as carriers. *Biomaterials* **2005**, *26*, 2501–2507. [CrossRef]
33. Johnson, M.R.; Boerckel, J.D.; Dupont, K.M.; Guldberg, R.E. Functional restoration of critically sized segmental defects with bone morphogenetic protein-2 and heparin treatment. *Clin. Orthop. Relat. Res.* **2011**, *469*, 3111–3117. [CrossRef]
34. Williams, J.C.; Maitra, S.; Anderson, M.J.; Christiansen, B.A.; Reddi, A.H.; Lee, M.A. BMP-7 and bone regeneration: Evaluation of dose-response in a rodent segmental defect model. *J. Orthop. Trauma* **2015**, *29*, e336–e341. [CrossRef]
35. Krishnan, L.; Priddy, L.B.; Esancy, C.; Li, M.T.; Stevens, H.Y.; Jiang, X.; Tran, L.; Rowe, D.W.; Guldberg, R.E. Hydrogel-based Delivery of rhBMP-2 Improves Healing of Large Bone Defects Compared with Autograft. *Clin. Orthop. Relat. Res.* **2015**, *473*, 2885–2897. [CrossRef] [PubMed]
36. Hou, J.; Wang, J.; Cao, L.; Qian, X.; Xing, W.; Lu, J.; Liu, C. Segmental bone regeneration using rhBMP-2-loaded collagen/chitosan microspheres composite scaffold in a rabbit model. *Biomed. Mater.* **2012**, *7*, 035002. [CrossRef] [PubMed]
37. Jenis, L.G.; Wheeler, D.; Parazin, S.J.; Connolly, R.J. The effect of osteogenic protein-1 in instrumented and noninstrumented posterolateral fusion in rabbits. *Spine J.* **2002**, *2*, 173–178. [CrossRef]
38. Minamide, A.; Kawakami, M.; Hashizume, H.; Sakata, R.; Yoshida, M.; Tamaki, T. Experimental study of carriers of bone morphogenetic protein used for spinal fusion. *J. Orthop. Sci.* **2004**, *9*, 142–151. [CrossRef] [PubMed]
39. Kato, M.; Namikawa, T.; Terai, H.; Hoshino, M.; Miyamoto, S.; Takaoka, K. Ectopic bone formation in mice associated with a lactic acid/dioxanone/ethylene glycol copolymer-tricalcium phosphate composite with added recombinant human bone morphogenetic protein-2. *Biomaterials* **2006**, *27*, 3927–3933. [CrossRef]
40. Lee, J.W.; Lee, S.; Lee, S.H.; Yang, H.S.; Im, G.I.; Kim, C.S.; Park, J.H.; Kim, B.S. Improved spinal fusion efficacy by long-term delivery of bone morphogenetic protein-2 in a rabbit model. *Acta Orthop.* **2011**, *82*, 756–760. [CrossRef]
41. Kang, S.W.; Bae, J.H.; Park, S.A.; Kim, W.D.; Park, M.S.; Ko, Y.J.; Jang, H.S.; Park, J.H. Combination therapy with BMP-2 and BMSCs enhances bone healing efficacy of PCL scaffold fabricated using the 3D plotting system in a large segmental defect model. *Biotechnol. Lett.* **2012**, *34*, 1375–1384. [CrossRef] [PubMed]
42. Pan, Z.; Jiang, P.; Xue, S.; Wang, T.; Li, H.; Wang, J. Repair of a critical-size segmental rabbit femur defect using bioglass-beta-TCP monoblock, a vascularized periosteal flap and BMP-2. *J. Biomed. Mater. Res. B Appl. Biomater.* **2018**, *106*, 2148–2156. [CrossRef] [PubMed]
43. Paralkar, V.M.; Borovecki, F.; Ke, H.Z.; Cameron, K.O.; Lefker, B.; Grasser, W.A.; Owen, T.A.; Li, M.; DaSilva-Jardine, P.; Zhou, M.; et al. An EP2 receptor-selective prostaglandin E2 agonist induces bone healing. *Proc. Natl. Acad. Sci. USA* **2003**, *100*, 6736–6740. [CrossRef]
44. Gandolfi, M.G.; Zamparini, F.; Esposti, M.D.; Chiellini, F.; Fava, F.; Fabbri, P.; Taddei, P.; Prati, C. Highly porous polycaprolactone scaffolds doped with calcium silicate and dicalcium phosphate dihydrate designed for bone regeneration. *Mater. Sci. Eng. C Mater. Biol. Appl.* **2019**, *102*, 341–361. [CrossRef] [PubMed]
45. Gandolfi, M.G.; Zamparini, F.; Esposti, M.D.; Chiellini, F.; Aparicio, C.; Fava, F.; Fabbri, P.; Taddei, P.; Prati, C. Polylactic acid-based porous scaffolds doped with calcium silicate and dicalcium phosphate dihydrate designed for biomedical application. *Mater. Sci. Eng. C Mater. Biol. Appl.* **2018**, *82*, 163–181. [CrossRef]
46. Qin, Y.; Sun, R.; Wu, C.; Wang, L.; Zhang, C. Exosome: A novel approach to stimulate bone regeneration through regulation of osteogenesis and angiogenesis. *Int. J. Mol. Sci.* **2016**, *17*, 712. [CrossRef]
47. Gandolfi, M.G.; Gardin, C.; Zamparini, F.; Ferroni, L.; Esposti, M.D.; Parchi, G.; Ercan, B.; Manzoli, L.; Fava, F.; Fabbri, P.; et al. Mineral-doped poly(L-lactide) acid scaffolds enriched with exosomes improve osteogenic commitment of human adipose-derived mesenchymal stem cells. *Nanomaterials* **2020**, *10*, 432. [CrossRef]
48. Roldan, J.C.; Detsch, R.; Schaefer, S.; Chang, E.; Kelantan, M.; Waiss, W.; Reichert, T.E.; Gurtner, G.C.; Deisinger, U. Bone formation and degradation of a highly porous biphasic calcium phosphate ceramic in presence of BMP-7, VEGF and mesenchymal stem cells in an ectopic mouse model. *J. Craniomaxillofac. Surg.* **2010**, *38*, 423–430. [CrossRef]
49. Liang, G.; Yang, Y.; Oh, S.; Ong, J.L.; Zheng, C.; Ran, J.; Yin, G.; Zhou, D. Ectopic osteoinduction and early degradation of recombinant human bone morphogenetic protein-2-loaded porous beta-tricalcium phosphate in mice. *Biomaterials* **2005**, *26*, 4265–4271. [CrossRef]
50. Kuboki, Y.; Takita, H.; Kobayashi, D.; Tsuruga, E.; Inoue, M.; Murata, M.; Nagai, N.; Dohi, Y.; Ohgushi, H. BMP-induced osteogenesis on the surface of hydroxyapatite with geometrically feasible and nonfeasible structures: Topology of osteogenesis. *J. Biomed. Mater. Res.* **1998**, *39*, 190–199. [CrossRef]

51. Tsuruga, E.; Takita, H.; Itoh, H.; Wakisaka, Y.; Kuboki, Y. Pore size of porous hydroxyapatite as the cell-substratum controls BMP-induced osteogenesis. *J. Biochem.* **1997**, *121*, 317–324. [CrossRef] [PubMed]
52. Alam, M.I.; Asahina, I.; Ohmamiuda, K.; Takahashi, K.; Yokota, S.; Enomoto, S. Evaluation of ceramics composed of different hydroxyapatite to tricalcium phosphate ratios as carriers for rhBMP-2. *Biomaterials* **2001**, *22*, 1643–1651. [CrossRef]
53. Vehof, J.W.; Takita, H.; Kuboki, Y.; Spauwen, P.H.; Jansen, J.A. Histological characterization of the early stages of bone morphogenetic protein-induced osteogenesis. *J. Biomed. Mater. Res.* **2002**, *61*, 440–449. [CrossRef]
54. Tazaki, J.; Murata, M.; Akazawa, T.; Yamamoto, M.; Ito, K.; Arisue, M.; Shibata, T.; Tabata, Y. BMP-2 release and dose-response studies in hydroxyapatite and beta-tricalcium phosphate. *Biomed. Mater. Eng.* **2009**, *19*, 141–146. [PubMed]
55. Park, J.C.; So, S.S.; Jung, I.H.; Yun, J.H.; Choi, S.H.; Cho, K.S.; Kim, C.S. Induction of bone formation by Escherichia coli-expressed recombinant human bone morphogenetic protein-2 using block-type macroporous biphasic calcium phosphate in orthotopic and ectopic rat models. *J. Periodontal Res.* **2011**, *46*, 682–690. [CrossRef] [PubMed]
56. Strobel, L.A.; Rath, S.N.; Maier, A.K.; Beier, J.P.; Arkudas, A.; Greil, P.; Horch, R.E.; Kneser, U. Induction of bone formation in biphasic calcium phosphate scaffolds by bone morphogenetic protein-2 and primary osteoblasts. *J. Tissue Eng. Regen. Med.* **2014**, *8*, 176–185. [CrossRef]
57. Ma, J.; Yang, F.; Both, S.K.; Prins, H.J.; Helder, M.N.; Pan, J.; Cui, F.Z.; Jansen, J.A.; van den Beucken, J.J. Bone forming capacity of cell- and growth factor-based constructs at different ectopic implantation sites. *J. Biomed. Mater. Res. A* **2015**, *103*, 439–450. [CrossRef]
58. Kim, J.W.; Choi, K.H.; Yun, J.H.; Jung, U.W.; Kim, C.S.; Choi, S.H.; Cho, K.S. Bone formation of block and particulated biphasic calcium phosphate lyophilized with Escherichia coli-derived recombinant human bone morphogenetic protein 2 in rat calvarial defects. *Oral Surg. Oral Med. Oral Pathol. Oral Radiol. Endodontol.* **2011**, *112*, 298–306. [CrossRef]
59. Jang, J.W.; Yun, J.H.; Lee, K.I.; Jang, J.W.; Jung, U.W.; Kim, C.S.; Choi, S.H.; Cho, K.S. Osteoinductive activity of biphasic calcium phosphate with different rhBMP-2 doses in rats. *Oral Surg. Oral Med. Oral Pathol. Oral Radiol.* **2012**, *113*, 480–487. [CrossRef]
60. Lee, J.H.; Ryu, M.Y.; Baek, H.R.; Lee, K.M.; Seo, J.H.; Lee, H.K.; Ryu, H.S. Effects of porous beta-tricalcium phosphate-based ceramics used as an E. coli-derived rhBMP-2 carrier for bone regeneration. *J. Mater. Sci. Mater. Med.* **2013**, *24*, 2117–2127. [CrossRef]
61. Kuroiwa, Y.; Niikura, T.; Lee, S.Y.; Oe, K.; Iwakura, T.; Fukui, T.; Matsumoto, T.; Matsushita, T.; Nishida, K.; Kuroda, R. Escherichia coli-derived BMP-2-absorbed beta-TCP granules induce bone regeneration in rabbit critical-sized femoral segmental defects. *Int. Orthop.* **2019**, *43*, 1247–1253. [CrossRef]
62. Louis-Ugbo, J.; Kim, H.S.; Boden, S.D.; Mayr, M.T.; Li, R.C.; Seeherman, H.; D'Augusta, D.; Blake, C.; Jiao, A.; Peckham, S. Retention of 125I-labeled recombinant human bone morphogenetic protein-2 by biphasic calcium phosphate or a composite sponge in a rabbit posterolateral spine arthrodesis model. *J. Orthop. Res.* **2002**, *20*, 1050–1059. [CrossRef]
63. Suh, D.Y.; Boden, S.D.; Louis-Ugbo, J.; Mayr, M.; Murakami, H.; Kim, H.S.; Minamide, A.; Hutton, W.C. Delivery of recombinant human bone morphogenetic protein-2 using a compression-resistant matrix in posterolateral spine fusion in the rabbit and in the non-human primate. *Spine* **2002**, *27*, 353–360. [CrossRef] [PubMed]
64. Dohzono, S.; Imai, Y.; Nakamura, H.; Wakitani, S.; Takaoka, K. Successful spinal fusion by E. coli-derived BMP-2-adsorbed porous beta-TCP granules: A pilot study. *Clin. Orthop. Relat. Res.* **2009**, *467*, 3206–3212. [CrossRef] [PubMed]
65. Pelletier, M.H.; Oliver, R.A.; Christou, C.; Yu, Y.; Bertollo, N.; Irie, H.; Walsh, W.R. Lumbar spinal fusion with beta-TCP granules and variable Escherichia coli-derived rhBMP-2 dose. *Spine J.* **2014**, *14*, 1758–1768. [CrossRef] [PubMed]
66. Notodihardjo, F.Z.; Kakudo, N.; Kushida, S.; Suzuki, K.; Kusumoto, K. Bone regeneration with BMP-2 and hydroxyapatite in critical-size calvarial defects in rats. *J. Craniomaxillofac. Surg.* **2012**, *40*, 287–291. [CrossRef] [PubMed]
67. Liu, W.C.; Robu, I.S.; Patel, R.; Leu, M.C.; Velez, M.; Chu, T.M. The effects of 3D bioactive glass scaffolds and BMP-2 on bone formation in rat femoral critical size defects and adjacent bones. *Biomed. Mater.* **2014**, *9*, 045013. [CrossRef] [PubMed]
68. Zhu, W.; Wang, D.; Zhang, X.; Lu, W.; Han, Y.; Ou, Y.; Zhou, K.; Fen, W.; Liu, J.; Peng, L.; et al. Experimental study of nano-hydroxyapatite/recombinant human bone morphogenetic protein-2 composite artificial bone. *Artif. Cells Blood Substit. Immobil. Biotechnol.* **2010**, *38*, 150–156. [CrossRef]
69. Huang, T.Y.; Wu, C.C.; Weng, P.W.; Chen, J.M.; Yeh, W.L. Effect of ErhBMP-2-loaded beta-tricalcium phosphate on ulna defects in the osteoporosis rabbit model. *Bone Rep.* **2021**, *14*, 100739. [CrossRef]
70. Wu, Y.; Hou, J.; Yin, M.; Wang, J.; Liu, C. Enhanced healing of rabbit segmental radius defects with surface-coated calcium phosphate cement/bone morphogenetic protein 2 scaffolds. *Mater. Sci. Eng. C Mater. Biol. Appl.* **2014**, *44*, 326–335. [CrossRef]
71. Tazaki, J.; Akazawa, T.; Murata, M.; Yamamoto, M.; Tabata, Y.; Yoshimoto, R.; Arisue, M. BMP-2 Dose-response and release studies in functionally graded HAp. *Key Eng. Mater.* **2006**, *309–311*, 965–968. [CrossRef]
72. Vukicevic, S.; Stokovic, N.; Pecina, M. Is ceramics an appropriate bone morphogenetic protein delivery system for clinical use? *Int. Orthop.* **2019**, *43*, 1275–1276. [CrossRef]
73. Stokovic, N.; Ivanjko, N.; Erjavec, I.; Milosevic, M.; Oppermann, H.; Shimp, L.; Sampath, K.T.; Vukicevic, S. Autologous bone graft substitute containing rhBMP6 within autologous blood coagulum and synthetic ceramics of different particle size determines the quantity and structural pattern of bone formed in a rat sucutaneous assay. *Bone* **2020**, 115654. [CrossRef]

74. Stokovic, N.; Ivanjko, N.; Pecin, M.; Erjavec, I.; Karlovic, S.; Smajlovic, A.; Capak, H.; Milosevic, M.; Bubic Spoljar, J.; Vnuk, D.; et al. Evaluation of synthetic ceramics as compression resistant matrix to promote osteogenesis of autologous blood coagulum containing recombinant human bone morphogenetic protein 6 in rabbit posterolateral lumbar fusion model. *Bone* **2020**, *140*, 115544. [CrossRef]
75. Dorozhkin, S.V. Bioceramics of calcium orthophosphates. *Biomaterials* **2010**, *31*, 1465–1485. [CrossRef]
76. Stokovic, N.; Ivanjko, N.; Milesevic, M.; Matic Jelic, I.; Bakic, K.; Rumenovic, V.; Oppermann, H.; Shimp, L.; Sampath, T.K.; Pecina, M.; et al. Synthetic ceramic macroporous blocks as a scaffold in ectopic bone formation induced by recombinant human bone morphogenetic protein 6 within autologous blood coagulum in rats. *Int. Orthop.* **2020**, *45*, 1097–1107. [CrossRef] [PubMed]
77. Bae, E.B.; Park, K.H.; Shim, J.H.; Chung, H.Y.; Choi, J.W.; Lee, J.J.; Kim, C.H.; Jeon, H.J.; Kang, S.S.; Huh, J.B. Efficacy of rhBMP-2 Loaded PCL/beta-TCP/bdECM Scaffold Fabricated by 3D Printing Technology on Bone Regeneration. *Biomed. Res. Int.* **2018**, *2018*, 2876135. [CrossRef]
78. Chu, T.M.; Warden, S.J.; Turner, C.H.; Stewart, R.L. Segmental bone regeneration using a load-bearing biodegradable carrier of bone morphogenetic protein-2. *Biomaterials* **2007**, *28*, 459–467. [CrossRef]
79. Rodriguez-Evora, M.; Delgado, A.; Reyes, R.; Hernandez-Daranas, A.; Soriano, I.; San Roman, J.; Evora, C. Osteogenic effect of local, long versus short term BMP-2 delivery from a novel SPU-PLGA-betaTCP concentric system in a critical size defect in rats. *Eur. J. Pharm. Sci.* **2013**, *49*, 873–884. [CrossRef]
80. Yoneda, M.; Terai, H.; Imai, Y.; Okada, T.; Nozaki, K.; Inoue, H.; Miyamoto, S.; Takaoka, K. Repair of an intercalated long bone defect with a synthetic biodegradable bone-inducing implant. *Biomaterials* **2005**, *26*, 5145–5152. [CrossRef] [PubMed]
81. Liu, Y.; Lu, Y.; Tian, X.; Cui, G.; Zhao, Y.; Yang, Q.; Yu, S.; Xing, G.; Zhang, B. Segmental bone regeneration using an rhBMP-2-loaded gelatin/nanohydroxyapatite/fibrin scaffold in a rabbit model. *Biomaterials* **2009**, *30*, 6276–6285. [CrossRef] [PubMed]
82. Luca, L.; Rougemont, A.L.; Walpoth, B.H.; Boure, L.; Tami, A.; Anderson, J.M.; Jordan, O.; Gurny, R. Injectable rhBMP-2-loaded chitosan hydrogel composite: Osteoinduction at ectopic site and in segmental long bone defect. *J. Biomed. Mater. Res. A* **2011**, *96*, 66–74. [CrossRef] [PubMed]
83. Fujita, N.; Matsushita, T.; Ishida, K.; Sasaki, K.; Kubo, S.; Matsumoto, T.; Kurosaka, M.; Tabata, Y.; Kuroda, R. An analysis of bone regeneration at a segmental bone defect by controlled release of bone morphogenetic protein 2 from a biodegradable sponge composed of gelatin and beta-tricalcium phosphate. *J. Tissue Eng. Regen. Med.* **2012**, *6*, 291–298. [CrossRef] [PubMed]
84. Yamamoto, M.; Hokugo, A.; Takahashi, Y.; Nakano, T.; Hiraoka, M.; Tabata, Y. Combination of BMP-2-releasing gelatin/beta-TCP sponges with autologous bone marrow for bone regeneration of X-ray-irradiated rabbit ulnar defects. *Biomaterials* **2015**, *56*, 18–25. [CrossRef] [PubMed]
85. Namikawa, T.; Terai, H.; Suzuki, E.; Hoshino, M.; Toyoda, H.; Nakamura, H.; Miyamoto, S.; Takahashi, N.; Ninomiya, T.; Takaoka, K. Experimental spinal fusion with recombinant human bone morphogenetic protein-2 delivered by a synthetic polymer and beta-tricalcium phosphate in a rabbit model. *Spine* **2005**, *30*, 1717–1722. [CrossRef]
86. Vukicevic, S.; Grgurevic, L.; Erjavec, I.; Pecin, M.; Bordukalo-Niksic, T.; Stokovic, N.; Lipar, M.; Capak, H.; Maticic, D.; Windhager, R.; et al. Autologous blood coagulum is a physiological carrier for BMP6 to induce new bone formation and promote posterolateral lumbar spine fusion in rabbits. *J. Tissue Eng. Regen. Med.* **2020**, *14*, 147–159. [CrossRef]
87. Grgurevic, L.; Oppermann, H.; Pecin, M.; Erjavec, I.; Capak, H.; Pauk, M.; Karlovic, S.; Kufner, V.; Lipar, M.; Bubic Spoljar, J.; et al. Recombinant human bone morphogenetic protein 6 delivered within autologous blood coagulum restores critical size segmental defects of ulna in rabbits. *JBMR Plus* **2019**, *3*, e10085. [CrossRef]
88. Grgurevic, L.; Erjavec, I.; Gupta, M.; Pecin, M.; Bordukalo-Niksic, T.; Stokovic, N.; Vnuk, D.; Farkas, V.; Capak, H.; Milosevic, M.; et al. Autologous blood coagulum containing rhBMP6 induces new bone formation to promote anterior lumbar interbody fusion (ALIF) and posterolateral lumbar fusion (PLF) of spine in sheep. *Bone* **2020**. [CrossRef]
89. Vukicevic, S.; Peric, M.; Oppermann, H.; Stokovic, N.; Ivanjko, N.; Erjavec, I.; Kufner, V.; Vnuk, D.; Bubic Spoljar, J.; Pecin, M.; et al. Bone morphogenetic proteins: From discovery to development of a novel autologous bone graft substitute consisting of recombinant human BMP6 delivered in autologous blood coagulum carrier. *Rad CASA Med. Sci.* **2020**, *544*, 26–41.
90. Pecin, M.; Stokovic, N.; Ivanjko, N.; Smajlovic, A.; Kreszinger, M.; Capak, H.; Vrbanac, Z.; Oppermann, H.; Maticic, D.; Vukicevic, S. A novel autologous bone graft substitute containing rhBMP6 in autologous blood coagulum with synthetic ceramics for reconstruction of a large humerus segmental gunshot defect in a dog: The first veterinary patient to receive a novel osteoinductive therapy. *Bone Rep.* **2021**, *14*, 100759. [CrossRef]
91. Chiari, C.; Grgurevic, L.; Bordukalo-Niksic, T.; Oppermann, H.; Valentinitsch, A.; Nemecek, E.; Staats, K.; Schreiner, M.; Trost, C.; Kolb, A.; et al. Recombinant human BMP6 applied within autologous blood coagulum accelerates bone healing: Randomized controlled trial in high tibial osteotomy patients. *J. Bone Miner. Res.* **2020**, *35*, 1893–1903. [CrossRef]
92. Durdevic, D.; Vlahovic, T.; Pehar, S.; Miklic, D.; Oppermann, H.; Bordukalo-Niksic, T.; Gavrankapetanovic, I.; Jamakosmanovic, M.; Milosevic, M.; Martinovic, S.; et al. A novel autologous bone graft substitute comprised of rhBMP6 blood coagulum as carrier tested in a randomized and controlled Phase I trial in patients with distal radial fractures. *Bone* **2020**, *140*, 115551. [CrossRef]
93. Ji, W.; Kerckhofs, G.; Geeroms, C.; Marechal, M.; Geris, L.; Luyten, F.P. Deciphering the combined effect of bone morphogenetic protein 6 and calcium phosphate on bone formation capacity of periosteum derived cells-based tissue engineering constructs. *Acta Biomater.* **2018**, *80*, 97–107. [CrossRef]
94. Eyckmans, J.; Roberts, S.J.; Schrooten, J.; Luyten, F.P. A clinically relevant model of osteoinduction: A process requiring calcium phosphate and BMP/Wnt signalling. *J. Cell. Mol. Med.* **2010**, *14*, 1845–1856. [CrossRef] [PubMed]

95. Eyckmans, J.; Roberts, S.J.; Bolander, J.; Schrooten, J.; Chen, C.S.; Luyten, F.P. Mapping calcium phosphate activated gene networks as a strategy for targeted osteoinduction of human progenitors. *Biomaterials* **2013**, *34*, 4612–4621. [CrossRef] [PubMed]
96. Bolander, J.; Chai, Y.C.; Geris, L.; Schrooten, J.; Lambrechts, D.; Roberts, S.J.; Luyten, F.P. Early BMP, Wnt and Ca(2+)/PKC pathway activation predicts the bone forming capacity of periosteal cells in combination with calcium phosphates. *Biomaterials* **2016**, *86*, 106–118. [CrossRef]
97. Bolander, J.; Ji, W.; Geris, L.; Bloemen, V.; Chai, Y.C.; Schrooten, J.; Luyten, F.P. The combined mechanism of bone morphogenetic protein- and calcium phosphate-induced skeletal tissue formation by human periosteum derived cells. *Eur. Cell Mater.* **2016**, *31*, 11–25. [CrossRef] [PubMed]
98. Katagiri, H.; Mendes, L.F.; Luyten, F.P. Reduction of BMP6-induced bone formation by calcium phosphate in wild-type compared with nude mice. *J. Tissue Eng. Regen. Med.* **2019**, *13*, 846–856. [CrossRef]
99. Chai, Y.C.; Roberts, S.J.; Desmet, E.; Kerckhofs, G.; van Gastel, N.; Geris, L.; Carmeliet, G.; Schrooten, J.; Luyten, F.P. Mechanisms of ectopic bone formation by human osteoprogenitor cells on CaP biomaterial carriers. *Biomaterials* **2012**, *33*, 3127–3142. [CrossRef]
100. Mumcuoglu, D.; Fahmy-Garcia, S.; Ridwan, Y.; Nicke, J.; Farrell, E.; Kluijtmans, S.G.; van Osch, G.J. Injectable BMP-2 delivery system based on collagen-derived microspheres and alginate induced bone formation in a time- and dose-dependent manner. *Eur. Cell Mater.* **2018**, *35*, 242–254. [CrossRef]
101. Lin, X.; Hunziker, E.B.; Liu, T.; Hu, Q.; Liu, Y. Enhanced biocompatibility and improved osteogenesis of coralline hydroxyapatite modified by bone morphogenetic protein 2 incorporated into a biomimetic coating. *Mater. Sci. Eng. C Mater. Biol. Appl.* **2019**, *96*, 329–336. [CrossRef] [PubMed]
102. Reves, B.T.; Jennings, J.A.; Bumgardner, J.D.; Haggard, W.O. Osteoinductivity Assessment of BMP-2 Loaded Composite Chitosan-Nano-Hydroxyapatite Scaffolds in a Rat Muscle Pouch. *Materials* **2011**, *4*, 1360–1374. [CrossRef]
103. Bhakta, G.; Rai, B.; Lim, Z.X.; Hui, J.H.; Stein, G.S.; van Wijnen, A.J.; Nurcombe, V.; Prestwich, G.D.; Cool, S.M. Hyaluronic acid-based hydrogels functionalized with heparin that support controlled release of bioactive BMP-2. *Biomaterials* **2012**, *33*, 6113–6122. [CrossRef]
104. Kisiel, M.; Martino, M.M.; Ventura, M.; Hubbell, J.A.; Hilborn, J.; Ossipov, D.A. Improving the osteogenic potential of BMP-2 with hyaluronic acid hydrogel modified with integrin-specific fibronectin fragment. *Biomaterials* **2013**, *34*, 704–712. [CrossRef]
105. Luca, L.; Rougemont, A.L.; Walpoth, B.H.; Gurny, R.; Jordan, O. The effects of carrier nature and pH on rhBMP-2-induced ectopic bone formation. *J. Control. Release* **2010**, *147*, 38–44. [CrossRef] [PubMed]
106. Sommer, N.G.; Hahn, D.; Okutan, B.; Marek, R.; Weinberg, A.M. Animal models in orthopedic research: The proper animal model to answer fundamental questions on bone healing depending on pathology and implant material. In *Animal Models in Medicine and Biology*; Tvrdá, E., Yenisetti, C.S., Eds.; IntechOpen: London, UK, 2020.
107. Simpson, A.H.; Murray, I.R. Osteoporotic fracture models. *Curr. Osteoporos Rep.* **2015**, *13*, 9–15. [CrossRef] [PubMed]
108. Jacenko, O.; Olsen, B.R. Transgenic mouse models in studies of skeletal disorders. *J. Rheumatol. Suppl.* **1995**, *43*, 39–41.
109. Fan, J.; Im, C.S.; Cui, Z.K.; Guo, M.; Bezouglaia, O.; Fartash, A.; Lee, J.Y.; Nguyen, J.; Wu, B.M.; Aghaloo, T.; et al. Delivery of phenamil enhances BMP-2-induced osteogenic differentiation of adipose-derived stem cells and bone formation in calvarial defects. *Tissue Eng. Part A* **2015**, *21*, 2053–2065. [CrossRef]
110. Gronowicz, G.; Jacobs, E.; Peng, T.; Zhu, L.; Hurley, M.; Kuhn, L.T. Calvarial bone regeneration is enhanced by sequential delivery of FGF-2 and BMP-2 from layer-by-layer coatings with a biomimetic calcium phosphate barrier layer. *Tissue Eng. Part A* **2017**, *23*, 1490–1501. [CrossRef] [PubMed]
111. Herberg, S.; Aguilar-Perez, A.; Howie, R.N.; Kondrikova, G.; Periyasamy-Thandavan, S.; Elsalanty, M.E.; Shi, X.; Hill, W.D.; Cray, J.J. Mesenchymal stem cell expression of SDF-1beta synergizes with BMP-2 to augment cell-mediated healing of critical-sized mouse calvarial defects. *J. Tissue Eng. Regen. Med.* **2017**, *11*, 1806–1819. [CrossRef]
112. Huang, K.C.; Yano, F.; Murahashi, Y.; Takano, S.; Kitaura, Y.; Chang, S.H.; Soma, K.; Ueng, S.W.N.; Tanaka, S.; Ishihara, K.; et al. Sandwich-type PLLA-nanosheets loaded with BMP-2 induce bone regeneration in critical-sized mouse calvarial defects. *Acta Biomater.* **2017**, *59*, 12–20. [CrossRef]
113. La, W.G.; Kwon, S.H.; Lee, T.J.; Yang, H.S.; Park, J.; Kim, B.S. The effect of the delivery carrier on the quality of bone formed via bone morphogenetic protein-2. *Artif. Organs* **2012**, *36*, 642–647. [CrossRef]
114. Maisani, M.; Sindhu, K.R.; Fenelon, M.; Siadous, R.; Rey, S.; Mantovani, D.; Chassande, O. Prolonged delivery of BMP-2 by a non-polymer hydrogel for bone defect regeneration. *Drug Deliv. Transl. Res.* **2018**, *8*, 178–190. [CrossRef]
115. Reyes, R.; Rodriguez, J.A.; Orbe, J.; Arnau, M.R.; Evora, C.; Delgado, A. Combined sustained release of BMP2 and MMP10 accelerates bone formation and mineralization of calvaria critical size defect in mice. *Drug Deliv.* **2018**, *25*, 750–756. [CrossRef]
116. Seo, B.B.; Koh, J.T.; Song, S.C. Tuning physical properties and BMP-2 release rates of injectable hydrogel systems for an optimal bone regeneration effect. *Biomaterials* **2017**, *122*, 91–104. [CrossRef]
117. Terauchi, M.; Inada, T.; Kanemaru, T.; Ikeda, G.; Tonegawa, A.; Nishida, K.; Arisaka, Y.; Tamura, A.; Yamaguchi, S.; Yui, N. Potentiating bioactivity of BMP-2 by polyelectrolyte complexation with sulfonated polyrotaxanes to induce rapid bone regeneration in a mouse calvarial defect. *J. Biomed. Mater. Res. A* **2017**, *105*, 1355–1363. [CrossRef] [PubMed]
118. Yang, H.S.; La, W.G.; Park, J.; Kim, C.S.; Im, G.I.; Kim, B.S. Efficient bone regeneration induced by bone morphogenetic protein-2 released from apatite-coated collagen scaffolds. *J. Biomater. Sci. Polym. Ed.* **2012**, *23*, 1659–1671. [CrossRef] [PubMed]

119. Alaee, F.; Hong, S.H.; Dukas, A.G.; Pensak, M.J.; Rowe, D.W.; Lieberman, J.R. Evaluation of osteogenic cell differentiation in response to bone morphogenetic protein or demineralized bone matrix in a critical sized defect model using GFP reporter mice. *J. Orthop. Res.* **2014**, *32*, 1120–1128. [CrossRef]
120. Bougioukli, S.; Jain, A.; Sugiyama, O.; Tinsley, B.A.; Tang, A.H.; Tan, M.H.; Adams, D.J.; Kostenuik, P.J.; Lieberman, J.R. Combination therapy with BMP-2 and a systemic RANKL inhibitor enhances bone healing in a mouse critical-sized femoral defect. *Bone* **2016**, *84*, 93–103. [CrossRef]
121. Zwingenberger, S.; Langanke, R.; Vater, C.; Lee, G.; Niederlohmann, E.; Sensenschmidt, M.; Jacobi, A.; Bernhardt, R.; Muders, M.; Rammelt, S.; et al. The effect of SDF-1alpha on low dose BMP-2 mediated bone regeneration by release from heparinized mineralized collagen type I matrix scaffolds in a murine critical size bone defect model. *J. Biomed. Mater. Res. A* **2016**, *104*, 2126–2134. [CrossRef] [PubMed]
122. Garcia, P.; Histing, T.; Holstein, J.H.; Klein, M.; Laschke, M.W.; Matthys, R.; Ignatius, A.; Wildemann, B.; Lienau, J.; Peters, A.; et al. Rodent animal models of delayed bone healing and non-union formation: A comprehensive review. *Eur. Cell Mater.* **2013**, *26*, 1–12. [CrossRef] [PubMed]
123. Diab, T.; Pritchard, E.M.; Uhrig, B.A.; Boerckel, J.D.; Kaplan, D.L.; Guldberg, R.E. A silk hydrogel-based delivery system of bone morphogenetic protein for the treatment of large bone defects. *J. Mech. Behav. Biomed. Mater.* **2012**, *11*, 123–131. [CrossRef] [PubMed]
124. Peng, K.T.; Hsieh, M.Y.; Lin, C.T.; Chen, C.F.; Lee, M.S.; Huang, Y.Y.; Chang, P.J. Treatment of critically sized femoral defects with recombinant BMP-2 delivered by a modified mPEG-PLGA biodegradable thermosensitive hydrogel. *BMC Musculoskelet. Disord.* **2016**, *17*, 286. [CrossRef] [PubMed]
125. Bae, J.H.; Song, H.R.; Kim, H.J.; Lim, H.C.; Park, J.H.; Liu, Y.; Teoh, S.H. Discontinuous release of bone morphogenetic protein-2 loaded within interconnected pores of honeycomb-like polycaprolactone scaffold promotes bone healing in a large bone defect of rabbit ulna. *Tissue Eng. Part A* **2011**, *17*, 2389–2397. [CrossRef] [PubMed]
126. Choi, E.J.; Kang, S.H.; Kwon, H.J.; Cho, S.W.; Kim, H.J. Bone healing properties of autoclaved autogenous bone grafts incorporating recombinant human bone morphogenetic protein-2 and comparison of two delivery systems in a segmental rabbit radius defect. *Maxillofac Plast. Reconstr. Surg.* **2014**, *36*, 94–102. [CrossRef]
127. Boden, S.D.; Schimandle, J.H.; Hutton, W.C. 1995 Volvo Award in basic sciences. The use of an osteoinductive growth factor for lumbar spinal fusion. Part II: Study of dose, carrier, and species. *Spine* **1995**, *20*, 2633–2644. [CrossRef]
128. Valdes, M.; Moore, D.C.; Palumbo, M.; Lucas, P.R.; Robertson, A.; Appel, J.; Ehrlich, M.G.; Keeping, H.S. rhBMP-6 stimulated osteoprogenitor cells enhance posterolateral spinal fusion in the New Zealand white rabbit. *Spine J.* **2007**, *7*, 318–325. [CrossRef] [PubMed]
129. Konishi, S.; Nakamura, H.; Seki, M.; Nagayama, R.; Yamano, Y. Hydroxyapatite granule graft combined with recombinant human bone morphogenic protein-2 for solid lumbar fusion. *J. Spinal Disord. Tech.* **2002**, *15*, 237–244. [CrossRef]
130. Itoh, T.; Mochizuki, M.; Nishimura, R.; Matsunaga, S.; Kadosawa, T.; Kokubo, S.; Yokota, S.; Sasaki, N. Repair of ulnar segmental defect by recombinant human bone morphogenetic protein-2 in dogs. *J. Vet. Med. Sci.* **1998**, *60*, 451–458. [CrossRef]
131. Tuominen, T.; Jamsa, T.; Tuukkanen, J.; Nieminen, P.; Lindholm, T.C.; Lindholm, T.S.; Jalovaara, P. Native bovine bone morphogenetic protein improves the potential of biocoral to heal segmental canine ulnar defects. *Int. Orthop.* **2000**, *24*, 289–294. [CrossRef]
132. Hu, Y.; Zhang, C.; Zhang, S.; Xiong, Z.; Xu, J. Development of a porous poly(L-lactic acid)/hydroxyapatite/collagen scaffold as a BMP delivery system and its use in healing canine segmental bone defect. *J. Biomed. Mater. Res. A* **2003**, *67*, 591–598. [CrossRef]
133. Jones, C.B.; Sabatino, C.T.; Badura, J.M.; Sietsema, D.L.; Marotta, J.S. Improved healing efficacy in canine ulnar segmental defects with increasing recombinant human bone morphogenetic protein-2/allograft ratios. *J. Orthop. Trauma* **2008**, *22*, 550–559. [CrossRef]
134. Harada, Y.; Itoi, T.; Wakitani, S.; Irie, H.; Sakamoto, M.; Zhao, D.; Nezu, Y.; Yogo, T.; Hara, Y.; Tagawa, M. Effect of Escherichia coli-produced recombinant human bone morphogenetic protein 2 on the regeneration of canine segmental ulnar defects. *J. Bone Miner. Metab.* **2012**, *30*, 388–399. [CrossRef]
135. Minier, K.; Toure, A.; Fusellier, M.; Fellah, B.; Bouvy, B.; Weiss, P.; Gauthier, O. BMP-2 delivered from a self-crosslinkable CaP/hydrogel construct promotes bone regeneration in a critical-size segmental defect model of non-union in dogs. *Vet. Comp. Orthop. Traumatol.* **2014**, *27*, 411–421. [PubMed]
136. Lammens, J.; Marechal, M.; Delport, H.; Geris, L.; Luyten, F.P. A flowchart for the translational research of cell-based therapy in the treatment of long bone defects. *J. Regen. Med.* **2021**, *10*, 1.
137. Cipitria, A.; Reichert, J.C.; Epari, D.R.; Saifzadeh, S.; Berner, A.; Schell, H.; Mehta, M.; Schuetz, M.A.; Duda, G.N.; Hutmacher, D.W. Polycaprolactone scaffold and reduced rhBMP-7 dose for the regeneration of critical-sized defects in sheep tibiae. *Biomaterials* **2013**, *34*, 9960–9968. [CrossRef] [PubMed]
138. den Boer, F.C.; Wippermann, B.W.; Blokhuis, T.J.; Patka, P.; Bakker, F.C.; Haarman, H.J. Healing of segmental bone defects with granular porous hydroxyapatite augmented with recombinant human osteogenic protein-1 or autologous bone marrow. *J. Orthop. Res.* **2003**, *21*, 521–528. [CrossRef]
139. Pluhar, G.E.; Turner, A.S.; Pierce, A.R.; Toth, C.A.; Wheeler, D.L. A comparison of two biomaterial carriers for osteogenic protein-1 (BMP-7) in an ovine critical defect model. *J. Bone Jt. Surg. Br.* **2006**, *88*, 960–966. [CrossRef]

140. Reichert, J.C.; Epari, D.R.; Wullschleger, M.E.; Berner, A.; Saifzadeh, S.; Noth, U.; Dickinson, I.C.; Schuetz, M.A.; Hutmacher, D.W. Bone tissue engineering. Reconstruction of critical sized segmental bone defects in the ovine tibia. *Orthopade* **2012**, *41*, 280–287. [CrossRef]
141. Akamaru, T.; Suh, D.; Boden, S.D.; Kim, H.S.; Minamide, A.; Louis-Ugbo, J. Simple carrier matrix modifications can enhance delivery of recombinant human bone morphogenetic protein-2 for posterolateral spine fusion. *Spine* **2003**, *28*, 429–434. [CrossRef]
142. Zwingenberger, S.; Yao, Z.; Jacobi, A.; Vater, C.; Valladares, R.D.; Li, C.; Nich, C.; Rao, A.J.; Christman, J.E.; Antonios, J.K.; et al. Enhancement of BMP-2 induced bone regeneration by SDF-1alpha mediated stem cell recruitment. *Tissue Eng. Part A* **2014**, *20*, 810–818. [PubMed]
143. Yamamoto, M.; Takahashi, Y.; Tabata, Y. Enhanced bone regeneration at a segmental bone defect by controlled release of bone morphogenetic protein-2 from a biodegradable hydrogel. *Tissue Eng.* **2006**, *12*, 1305–1311. [CrossRef] [PubMed]

Case Report

Design Techniques to Optimize the Scaffold Performance: Freeze-dried Bone Custom-Made Allografts for Maxillary Alveolar Horizontal Ridge Augmentation

Felice Roberto Grassi [1], Roberta Grassi [2], Leonardo Vivarelli [3], Dante Dallari [3], Marco Govoni [3], Gianna Maria Nardi [4], Zamira Kalemaj [5] and Andrea Ballini [1,6,7,*]

1. Department of Basic Medical Sciences, Neurosciences and Sense Organs, University of Bari "Aldo Moro", 70121 Bari, Italy; feliceroberto.grassi@uniba.it
2. Department of Surgical, Medical and Sperimental Sciences, University of Sassari, 07100 Sassari, Italy; grassi.roberta93@gmail.com
3. Reconstructive Orthopaedic Surgery and Innovative Techniques—Musculoskeletal Tissue Bank, IRCCS Istituto Ortopedico Rizzoli, 40124 Bologna, Italy; leonardo.vivarelli@ior.it (L.V.); dante.dallari@ior.it (D.D.); marco.govoni@ior.it (M.G.)
4. Department of Oral and Maxillo-Facial Sciences, University of Rome "La Sapienza", 00148 Rome, Italy; giannamaria.nardi@uniroma1.it
5. Private Practice, 20139 Milano, Italy; zamirakalemaj@hotmail.com
6. Department of Biosciences, Biotechnologies and Biopharmaceutics, Campus Universitario "Ernesto Quagliariello", University of Bari "Aldo Moro", 70121 Bari, Italy
7. Department of Precision Medicine, University of Campania "Luigi Vanvitelli", 80123 Naples, Italy
* Correspondence: andrea.ballini@uniba.it

Received: 29 January 2020; Accepted: 16 March 2020; Published: 19 March 2020

Abstract: The purpose of the current investigation was to evaluate the clinical success of horizontal ridge augmentation in severely atrophic maxilla (Cawood and Howell class IV) using freeze-dried custom made bone harvested from the tibial hemiplateau of cadaver donors, and to analyze the marginal bone level gain prior to dental implant placement at nine months subsequent to bone grafting and before prosthetic rehabilitation. A 52-year-old woman received custom made bone grafts. The patient underwent CT scans two weeks prior and nine months after surgery for graft volume and density analysis. The clinical and radiographic bone observations showed a very low rate of resorption after bone graft and implant placement. The custom-made allograft material was a highly effective modality for restoring the alveolar horizontal ridge, resulting in a reduction of the need to obtain autogenous bone from a secondary site with predictable procedure. Further studies are needed to investigate its behavior at longer time periods.

Keywords: geometry optimization of scaffolds; allograft; block bone grafts; custom made bone; design techniques for scaffold; precision and translational medicine

1. Introduction

Implant-supported rehabilitation of the edentulous ridge requires adequate volume and integrity of the alveolar bone [1].

Bone resorption in the maxillary ridge, due to trauma, pathology or tooth loss, frequently results in a knife-edged deformity, which complicates implant placement and stabilization, particularly in the posterior jaw [2–4]. Grafting with allograft bone has been documented to be a useful tool in reconstructing jaw anatomy, [5] restoring esthetics [6], and providing biomechanical support for the placement of dental implants [7].

Clinically, the most suitable banked bone allografts are fresh-frozen (FFBAs), freeze-dried bone allografts (FDBAs), and demineralized freeze-dried (DFDBAs) [8], although in oro-maxillofacial surgical interventions FDBAs and DFDBAs are the most used.

Frozen bone is accessible for human receivers after at least 6 months of quarantine at −80 °C [9] and no additional preparation is required. Moreover, the osteoinductive proteins are preserved [10].

Strict guidelines for tissue harvesting and storing at −80 °C make the risk of primary infections and antigenicity reasonably low [11]. In addition, the reduction of water content in frozen bone by the freeze-drying process further decreases potential microbial contaminations [12].

Frozen bone and relative freeze-dried/demineralized products are accessible as cancellous granules/blocks, corticocancellous granules/blocks, and cortical granules or chips. Once defrosted, frozen bone has handling qualities similar to fresh bone [8] while freeze-dried allografts have the additional advantage of storage at room temperature.

Bone density can be measured with high reproducibility by means of Cone-Beam-Computed-Tomography (CBCT) scans [13]. Further techniques, such as orthopantomography, do not assure fitting precision in density determination [13].

The aim of the present translational study was to evaluate the clinical success of horizontal ridge augmentation in severely atrophic maxilla (Cawood and Howell class IV) using custom made bone harvested from the tibial hemiplateau of cadaver donors.

2. Materials and Methods

A 52-year-old woman presented compromised anterior maxillary ridges, who presented for the placement of dental implants, was included in this pilot study. Written and verbal information was given to the patient before enrollment, and written informed consent was obtained. The study was conducted in full accordance with the World Medical Association Declaration of Helsinki on experimentation involving human subjects, as revised in 2008.

Inclusion criteria was a horizontal severely atrophic maxilla (Cawood and Howell class IV), needing a bone grafting procedure prior to implant placement. Exclusion criteria were established according to Venet et al. (2017) [12]. Plaque index score was maintained ≤25% throughout the study [14].

Graft Sample Blocks

The processing was performed on corticocancellous bone blocks obtained from a proximal tibial epiphysis, in the anatomical region between the articular surface of the tibial plateau and tuberosity.

Human allogeneic bone blocks were collected from cadaveric donors, stored at −80 °C and processed in the accredited public non-profit Musculoskeletal Tissue Bank of IRCCS Istituto Ortopedico Rizzoli (Bologna, Italy), authorized by the Italian National Transplant Center for the collection, processing, and distribution of human musculoskeletal tissue [15], and registered in the European Tissue Establishment list (code IT000096). The choice of the block to be machined involved considering dimensions slightly greater than the machining area defined with the graft design.

After thawing, each block was fixed with clamps on a special stainless-steel table in a GMP-Class A (Good Manufacturing Practice) Clean Room environment. Then, the table was fixed inside the CNC (Computer Numerical Control) milling machine (model Bright, Delta Macchine, Rieti, Italy). After tool fixing, a 3 mm diameter ball mill, the execution of the machining trajectory was started and accurately monitored by the operator.

At the end of the processing, the block was removed and finished, the supports were broken by hand and sharp edges, and rough corners were hand-refined using sterile rasps. Then, grafts were cleaned with organic solvents, washed with sterile water, and freeze-dried (VirTis Genesis 25, SP Scientific, Warminster, PA, USA). Finally, grafts were individually wrapped in triple pack. Microbiological sampling was performed during the processing of grafts and after the lyophilization protocol in order to exclude microbial contamination and declare tissues suitable for implantation.

According to the Guidelines of the Italian National Transplant Center, lyophilized bone grafts are preserved at room temperature for a maximum of 5 years.

Graft Design

Digital Imaging and Communications in Medicine (DICOM) data of the maxilla (Figure 1) were acquired by a Cone Beam Computerized Tomography (CBCT) scanner (Orthophos XG 3D Ceph, Dentsply Sirona Italia Srl, Roma, Italy) and imported into the 3DSlicer software (www.slicer.org) [16–18].

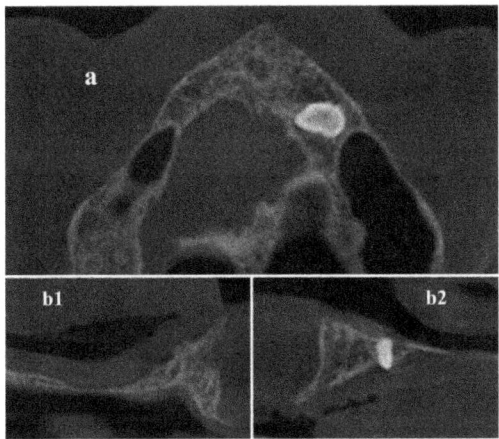

Figure 1. Cone-Beam-Computed-Tomography (CBCT) axial (**a**) and cross-sectional (**b1,b2**) images before surgical procedure.

After setting a threshold for the automatic selection of the areas delimiting the cortical bone, a manual analysis for each slice was performed to correct potential errors. Starting from the selected areas, an automatic procedure generated the 3D model as a stereolithography (STL) file.

The graft design was performed by Rhinoceros ver. 4 (www.rhino3d.com) following these steps: (a) placement of a parallelepiped in the position where the volume increase was required; (b) Boolean subtraction between the parallelepiped and the STL model of the patient's anatomy; (c) revision of the model for the manufacture by a 3-axis milling machine to obtain a L-shape section. This procedure was performed for each graft.

The design of the grafts has been repeatedly validated and subsequently corrected following the surgeon's indications, up to the final model (Figure 2).

Trajectory Planning and Graft Manufacturing

The planning of machining trajectories was performed using Rhinoceros by the Rhino-CAM 2 plug-in (MecSoft Corporation, Irvine, CA, USA). Each graft was positioned in the center of the machining CAM area with the larger flat area facing down to keep the cortical portion of the tissue intact, in order to enhance the graft resistance during processing and avoid breaking during implantation. Bone bridges have been added to the CAD design to maintain fixation during processing and removed by hand at the end of the graft manufacturing procedure. The working area was defined according to the design's dimensions and the diameter of the milling tool.

The tool-path trajectories have been programmed, subsequently simulated and modified to obtain the best result (Figure 3). Then, the definitive trajectories have been exported as G-Code coding specific for the milling machine used.

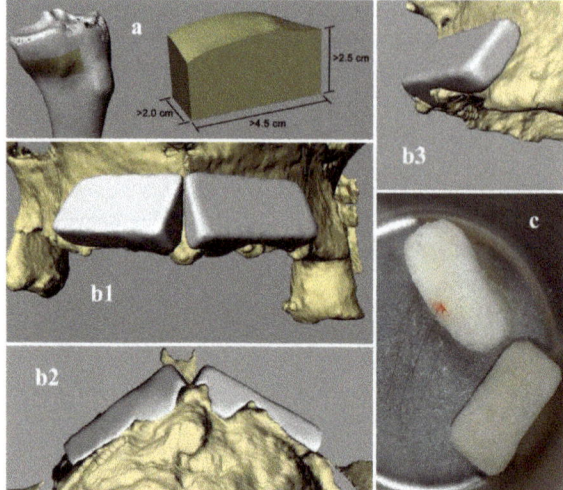

Figure 2. (a) Render of 3D bone donor site harvested from a cadaver tibial hemiplateau. (b1–b3) Render of 3D reconstruction of patient's anatomy (yellow) and designed bone grafts (white). (c) Final block for clinical use.

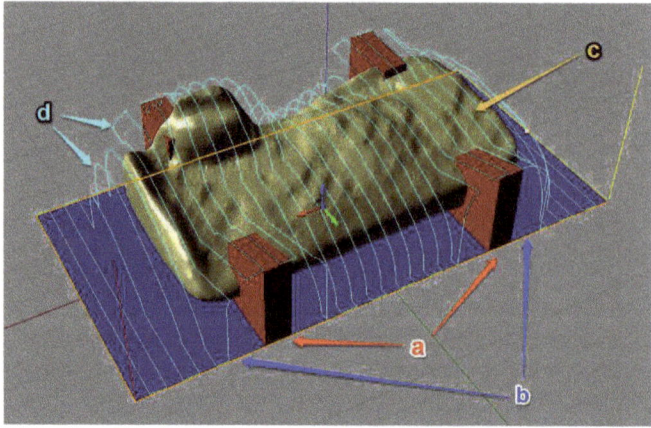

Figure 3. Tool-path trajectory planning: (a) added bone bridges; (b) work area; (c) graft design; (d) tool-path.

The machining operation was composed by two different phases: horizontal roughing and parallel finishing. The tool used for each phase was the same, a fluted 3 mm ball mill with two teeth, hard metal k10 material (tungsten carbide). In each phase rotational velocity of the tool was 5300 RPM and feed rate was 200 mm/min, that corresponds to a feed per tooth of 19 μm and a tangential velocity of 50 m/min (0.833 m/s). Horizontal roughing was performed as a full immersion vertical milling, with a machining depth of 3.5 mm, leaving a stock margin of 0.3 mm. Parallel finishing was a vertical milling, with a machining depth depending on stock margins and parallel steps of 0.7 mm (Figure 3d) leaving a variable stock margin between 0 mm and 0.15 mm (a sufficient dimensional accuracy for oral implant surgery). This machining strategy, performed by few passages, reduced the machining time and the bone heating during the process. Although temperature was not monitored during the fabrication of these two custom grafts, our machining parameters, especially the small feed per tooth, allow a local temperature between 40 °C and 60 °C for cortical bone and lower values for cancellous bone. According to Krause [19], these parameters do not cause the denaturation of proteins and enzymes that occurs for temperature above 70 °C.

As reported by Van Isacker et al. [20], frozen bone can be handled and reshaped like normal bone and is fully workable, but freeze-dried bone, unless rehydrated in saline, is brittle like ceramic, and is not fully workable.

Thus, to execute a better processing procedure and to obtain a further reduction in temperature, machining was performed on a frozen bone block. At the end of the machining operation, the custom allograft was freeze-dried.

Surgical Procedure

Local anesthesia was obtained by infiltrating articaine (4% containing 1:100,000 adrenaline). The exposure of the three-dimensional aspect of the bone defects was achieved through a full-thickness crestal incision with two vertical releasing incisions (Figure 4a).

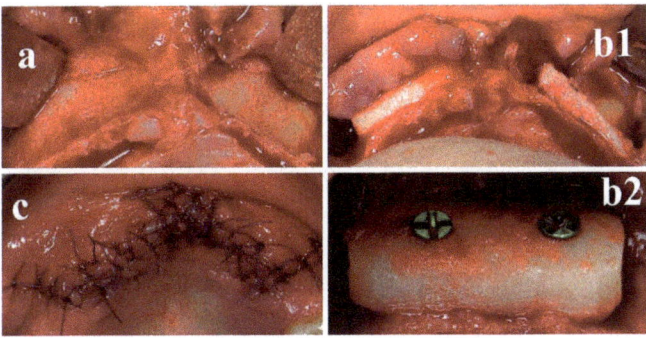

Figure 4. The custom-made freeze-dried bone allograft (FDBA) was positioned in the deficit area and stabilized with a screw. (**a**) Anatomical site prior surgical procedure and clinical view of the defect. (**b1,b2**) Screw positioning on a custom-made FDBA. (**c**) Sutures.

The clinically sized, anatomically shaped, custom-made bone block was placed in position strictly overlapping the underlying alveolar crest and fitted securely to the residual bone (Figure 4b1). The recipient site was weakened with multiple micro-holes to enhance bleeding from the trabecular bone [12,21,22]. Rigid fixation of the scaffold to the residual crest was obtained by means of a titanium mini-screw (1.5 mm width, 8 mm length) (Tekka by Global-D, Lyon, France) (Figure 4b2) [12].

The grafted area was closed with a pulley suture for proper flap adaptation and to avoid any tissue strangulation by an absorbable 4.0/5.0 suture material. Sutures were removed 14 days postoperatively (Figure 4c) and no prosthetic device was admitted in the following 90 days, with particular care in domiciliary oral hygiene procedures.

Preceding to the second phase, supplementary CBCT scans were performed to evaluate grafts gain (Figure 5).

After a 9 month healing period, micro-screws were removed (Figure 6a,b), and 3.7 mm in diameter for a 10 mm in length dental implants (iRES SAGL, Mendrisio, Switzerland) were placed, and sutures performed (Figure 6c). No supplementary grafting procedure was required in the current investigation [22].

Statistical Evaluation

The outcome values were analyzed using the *t-test* for paired samples for pre–post differences with time as the factor and IBM Statistical Package for the Social Sciences (SPSS Inc. Version 21.0,

Chicago, IL, USA) software to detect significant differences between pre-test and post-test scores obtained from bone volume analysis and measurements of variation in ridge thickness from CBCT images at 9 months after allograft insertion.

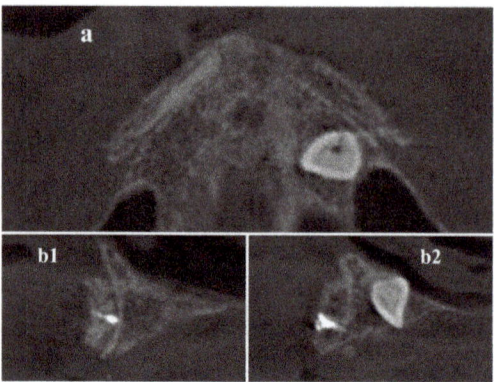

Figure 5. CBCT scans were done after 9 months to define implant placement width and length, on (**a**) axial and (**b1,b2**) cross-sectional images.

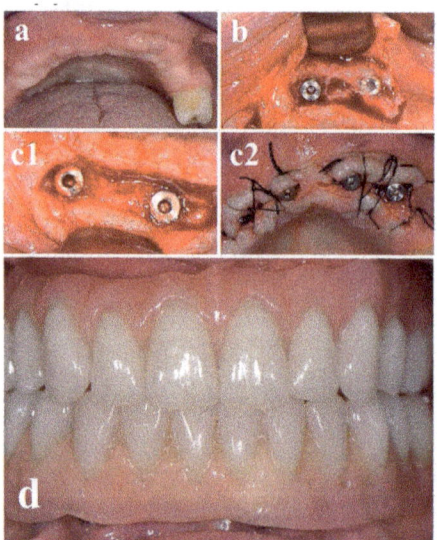

Figure 6. Nine months after the surgical procedure, the implant was positioned. (**a**) Soft tissues before surgical procedure. (**b**) Surgical flap is elevated. (**c1**) Implant was positioned and (**c2**) sutures performed. (**d**) Frontal view of the final rehabilitation.

3. Results

A patient presenting atrophic anterior maxillae was selected to participate in the study. The patient received 2 custom-made allografts. The healing period was uneventful.

At the crestal level, the patient met the inclusion criteria of ridge width around 2 mm [22]. The titanium fixation mini-screws detached and bleeding from the bone graft were detected, demonstrating revascularization of the site.

Moreover, the regenerative surgical procedure went well. In fact, the custom-made allograft scaffolds perfectly fitted in the bone anatomy and were therefore easily adapted to the bone defects during surgery, secured by titanium mini-screws (Figures 4–6). This excellent matching of the size/shape helped the surgeon to reduce the operation time [23]. Moreover, all implants were inserted with fitting primary stability (Figure 6d).

Paired comparisons showed a significant mean increase in ridge thickness of 5.0 ± 0.55 mm from the preoperative measurement to the postoperative measurement ($p < 0.01$).

The bone resorption that occurred during the incorporation period corresponds to 7.8% (95% confidence interval (CI): (6.4%, 9.2%)) of the measured postoperative ridge thickness.

4. Discussion

Maxillary ridge augmentation relates to procedures designed to correct a thin alveolar ridge. For dental implant placement, adequate bone volume is necessary, unfortunately, bone volume is not always adequate, particularly in elderly patients.

Through the current improvements in planning and manufacturing software and hardware, graft customization can be combined with the implant dentistry digital workflow with the potential patients to become part of daily clinical practice [24].

The reduction of risk factors such as infections caused by contamination of surgical site, or contamination of the graft during preoperative preparation, can be obtained with the manufacturing of a custom sterile graft [23]. The correct design of a custom graft, with also a reduction of surgical time, can be obtained with a proper planning of surgical procedure.

Nevertheless, the sterile manufacturing of bone in a GMP-Class A environment, without the need of terminal sterilization via gamma radiation, guarantees the safety of graft in terms of sterility and the quality of tissue in terms of biological properties [25,26].

Among bone replacement grafts, autograft is considered the gold standard [27]. However, owing to the low level of patient acceptance of autografts, allografts (grafts from the same species) have become the most popular bone grafts in oral surgery. Among allografts available for the periodontal regeneration, freeze-dried bone allograft (FDBA) and the demineralized freeze-dried allograft (DFDBA) are the allograft-forms most used [28].

Although FDBAs and DFDBAs were first used in periodontal therapy in early 1970s, to date they continue to be the most requested allografts in clinical practice because of their osteoconductive and osteoinductive properties, respectively [29]. Compared to autogenous bone, allografts feature several advantages such as adequate availability and the elimination of additional donor site surgery and morbidity [30]. Besides, the freeze-drying technique removes more than 95% of water content from bone, allowing further storage of tissues at room temperature and reducing antigenicity [31]. In addition, the application of stringent tissue banking guidelines together with bone processing and liophilization protocols performed in a GMP-Class A environment, minimize the risk of microbial contaminations [32], and reduce the need for terminal sterilization by gamma rays and consequent loss of mechanical strength of bone tissues [33]. As a matter of fact, although in freeze-dried bone tissues most of the water content has been removed, ionization has a direct effect on the collagen fiber bundles causing the breakage of protein chains and hence alters mechanical resistance [34].

Since only the demineralized bone matrix retains substantial amounts of endogenous osteoinductive proteins such as bone morphogenetic proteins (BMPs), known to promote bone growth and regeneration (i.e., BMP-2, BMP-4, BMP-7) [35,36], FDBAs are considered only as osteoconductive allografts. However, their structure can serve as a scaffold and support for the host cells that can migrate into the grafts and then differentiate into new osteogenic cells. In addition, in respect to DFDBAs, the superior mechanical properties of FDBAs make these allografts more suitable for restoring the alveolar horizontal ridge in athrophic maxillae.

Hence, the findings of our study show, with a high success rate, that FDBAs represent a reliable treatment option for extensive rehabilitation of atrophic maxillae, consistent with findings reported with the use of autologous bone.

The success rate of the block grafts was very effective and comparable with those reported by other authors [37–39]. Besides, the augmentation procedure permitted the insertion of implants in the grafted area 9 months after surgery. The clinical and radiographic observations showed a very low rate of bone resorption of the graft material, improving the ability to place endosseous implants.

A potential scaffold frontier for bone tissue engineering protocols seems to be the mesenchymal stem cells seeded in nanocomposite materials with antibacterial properties, but further studies in this direction are necessary prior to be consider it as a safe and useful tool for human implantation purposes [40].

Our data showed that custom-made FDBA can be used as successful graft material for the treatment of bone maxillary ridge defects. If adequate surgical techniques are adopted, this type of bone graft can be safely used in regions of implant placement as a suitable alternative to autogenous grafts. Furthermore, the personalized design of custom grafts and sterile machining in a GMP-Class A environment, as described above, can be applied to different surgical specialties, such as orthopedics, spinal surgery, and maxillofacial surgery.

5. Conclusions

This pilot study demonstrated the possibility of fabricating customized CAD/CAM grafts from the tibial hemiplateau of cadaver donors. The grafts were digitally designed based on CBCT scans of partially dentate patients using a set of 3D software, showing that grafts dimensions were correlated to the defect type. The sterile manufacturing of grafts resulted in the correct reproduction of the designed graft, leading to a correct matching with the bone surface of patient. Additional in vivo studies with custom-made bone allografts are required for the validation of this digital workflow for maxillary bone augmentation.

Author Contributions: Conception and design: F.R.G., R.G. and G.M.N.; data acquisition: Z.K.; writing and manuscript editing, analysis and interpretation of data: A.B., L.V., M.G.; 3D reconstructions and volume renderings: L.V., D.D., M.G.; supervised the manuscript final approval of the version to be published: F.R.G. and A.B. All authors read and approved the final manuscript.

Funding: This research received no external funding.

Conflicts of Interest: The authors declare no conflicts of interest.

References

1. Barone, A.; Covani, U. Maxillary alveolar ridge reconstruction with nonvascularized autogenous block bone: Clinical results. *J. Oral Maxillofac. Surg.* **2007**, *65*, 2039–2046. [CrossRef] [PubMed]
2. Cawood, J.I.; Howell, R.A. A classification of the edentulous jaws. *Int. J. Oral Maxillofac. Surg.* **1988**, *17*, 232–236. [CrossRef]
3. Cawood, J.I.; Howell, R.A. Reconstructive preprosthetic surgery. I. Anatomical considerations. *Int. J. Oral Maxillofac. Surg.* **1991**, *20*, 75–82. [CrossRef]
4. Leonetti, J.A.; Koup, R. Localized maxillary ridge augmentation with a block allograft for dental implant placement: Case reports. *Implant. Dent.* **2003**, *12*, 217–226. [CrossRef]
5. Burchardt, H. Biology of bone transplantation. *Orthop. Clin. N. Am.* **1987**, *18*, 187–196.
6. Schultze-Mosgau, S.; Schliephake, H.; Schultze-Mosgau, S.; Neukam, F.W. Soft tissue profile changes after autogenous iliac crest onlay grafting for the extremely atrophic maxilla. *J. Oral Maxillofac. Surg.* **2000**, *58*, 971–975. [CrossRef]
7. Misch, C.M.; Misch, C.E.; Resnik, R.R.; Ismail, Y.H. Reconstruction of maxillary alveolar defects with mandibular symphysis grafts for dental implants: A preliminary procedural report. *Int. J. Oral Maxillofac. Implant.* **1992**, *7*, 360–366.
8. Hardin, C.K. Banked bone. *Otolaryngol. Clin. N. Am.* **1994**, *27*, 911–925.

9. Simpson, D.; Kakarala, G.; Hampson, K.; Steele, N.; Ashton, B. Viable cells survive in fresh frozen human bone allografts. *Acta Orthop.* **2007**, *78*, 26–30. [CrossRef]
10. Perrott, D.H.; Smith, R.A.; Kaban, L.B. The use of fresh frozen allogeneic bone for maxillary and mandibular reconstruction. *Int. J. Oral Maxillofac. Surg.* **1992**, *21*, 260–265. [CrossRef]
11. Gocke, D.J. Tissue donor selection and safety. *Clin. Orthop. Relat. Res.* **2005**, 17–21. [CrossRef] [PubMed]
12. Venet, L.; Perriat, M.; Mangano, F.G.; Fortin, T. Horizontal ridge reconstruction of the anterior maxilla using customized allogeneic bone blocks with a minimally invasive technique—A case series. *BMC Oral Health* **2017**, *17*, 146. [CrossRef] [PubMed]
13. De Vos, W.; Casselman, J.; Swennen, G.R. Cone-beam computerized tomography (CBCT) imaging of the oral and maxillofacial region: A systematic review of the literature. *Int. J. Oral Maxillofac. Surg.* **2009**, *38*, 609–625. [CrossRef] [PubMed]
14. Lumetti, S.; Consolo, U.; Galli, C.; Multinu, A.; Piersanti, L.; Bellini, P.; Manfredi, E.; Corinaldesi, G.; Zaffe, D.; Macaluso, G.M.; et al. Fresh-frozen bone blocks for horizontal ridge augmentation in the upper maxilla: 6-month outcomes of a randomized controlled trial. *Clin. Implant. Dent. Relat. Res.* **2014**, *16*, 116–123. [CrossRef]
15. Tissues' Banks in Italy. Available online: http://www.trapianti.salute.gov.it/trapianti/dettaglioContenutiCnt.jsp?lingua=italiano&area=cnt&menu=chiSiamo&sottomenu=rete&id=237 (accessed on 27 January 2020).
16. Fedorov, A.; Beichel, R.; Kalpathy-Cramer, J.; Finet, J.; Fillion-Robin, J.C.; Pujol, S.; Bauer, C.; Jennings, D.; Fennessy, F.; Sonka, M.; et al. 3D Slicer as an image computing platform for the Quantitative Imaging Network. *Magn. Reson. Imaging* **2012**, *30*, 1323–1341. [CrossRef]
17. Kikinis, R.; Pieper, S.D.; Vosburgh, K.G. 3D Slicer: A Platform for Subject-Specific Image Analysis, Visualization, and Clinical Support. In *Intraoperative Imaging and Image-Guided Therapy*; Jolesz, F.A., Ed.; Springer: New York, NY, USA, 2014; pp. 277–289. [CrossRef]
18. Matsiushevich, K.; Belvedere, C.; Leardini, A.; Durante, S. Quantitative comparison of freeware software for bone mesh from DICOM files. *J. Biomech.* **2019**, *84*, 247–251. [CrossRef]
19. Krause, W.R. Orthogonal bone cutting: Saw design and operating characteristics. *J. Biomech. Eng.* **1987**, *109*, 263–271. [CrossRef]
20. Van Isacker, T.; Cornu, O.; Barbier, O.; Dufrane, D.; de Gheldere, A.; Delloye, C. Bone Autografting, Allografting and Banking. In *European Surgical Orthopaedics and Traumatology: The LEFORT Textbook*; Bentley, G., Ed.; Springer: Berlin/Heidelberg, Germany, 2014; pp. 77–90. [CrossRef]
21. Mangano, F.; Macchi, A.; Shibli, J.A.; Luongo, G.; Iezzi, G.; Piattelli, A.; Caprioglio, A.; Mangano, C. Maxillary ridge augmentation with custom-made CAD/CAM scaffolds. A 1-year prospective study on 10 patients. *J. Oral Implantol.* **2014**, *40*, 561–569. [CrossRef]
22. Pereira, E.; Messias, A.; Dias, R.; Judas, F.; Salvoni, A.; Guerra, F. Horizontal Resorption of Fresh-Frozen Corticocancellous Bone Blocks in the Reconstruction of the Atrophic Maxilla at 5 Months. *Clin. Implant. Dent. Relat. Res.* **2015**, *17* (Suppl. S2), e444–e458. [CrossRef]
23. Luongo, F.; Mangano, F.G.; Macchi, A.; Luongo, G.; Mangano, C. Custom-Made Synthetic Scaffolds for Bone Reconstruction: A Retrospective, Multicenter Clinical Study on 15 Patients. *Biomed. Res. Int.* **2016**, *2016*, 5862586. [CrossRef]
24. Yen, H.H.; Stathopoulou, P.G. CAD/CAM and 3D-Printing Applications for Alveolar Ridge Augmentation. *Curr. Oral Health Rep.* **2018**, *5*, 127–132. [CrossRef] [PubMed]
25. Nguyen, H.; Cassady, A.I.; Bennett, M.B.; Gineyts, E.; Wu, A.; Morgan, D.A.; Forwood, M.R. Reducing the radiation sterilization dose improves mechanical and biological quality while retaining sterility assurance levels of bone allografts. *Bone* **2013**, *57*, 194–200. [CrossRef] [PubMed]
26. Nguyen, H.; Morgan, D.A.; Forwood, M.R. Sterilization of allograft bone: Effects of gamma irradiation on allograft biology and biomechanics. *Cell Tissue Bank.* **2007**, *8*, 93–105. [CrossRef] [PubMed]
27. Kao, S.T.; Scott, D.D. A review of bone substitutes. *Oral Maxillofac. Surg. Clin. N. Am.* **2007**, *19*, 513–521. [CrossRef] [PubMed]
28. Naidu, P. Allografts in Periodontal Regeneration. *Madr. J. Case Rep. Stud.* **2019**, *3*, 121–125. [CrossRef]
29. Mellonig, J.T.; Bowers, G.M.; Bright, R.W.; Lawrence, J.J. Clinical evaluation of freeze-dried bone allografts in periodontal osseous defects. *J. Periodontol.* **1976**, *47*, 125–131. [CrossRef]
30. Grover, V.; Kapoor, A.; Malhotra, R.; Sachdeva, S. Bone allografts: A review of safety and efficacy. *Indian J. Dent. Res.* **2011**, *22*, 496. [CrossRef]

31. Delloye, C.; Cornu, O.; Druez, V.; Barbier, O. Bone allografts: What they can offer and what they cannot. *J. Bone Jt. Surg. Br.* **2007**, *89*, 574–579. [CrossRef]
32. Delloye, C. Tissue allografts and health risks. *Acta Orthop. Belg.* **1994**, *60* (Suppl. S1), 62–67.
33. Cornu, O.; Banse, X.; Docquier, P.L.; Luyckx, S.; Delloye, C. Effect of freeze-drying and gamma irradiation on the mechanical properties of human cancellous bone. *J. Orthop. Res.* **2000**, *18*, 426–431. [CrossRef]
34. Dziedzic-Goclawska, A.; Kaminski, A.; Uhrynowska-Tyszkiewicz, I.; Stachowicz, W. Irradiation as a safety procedure in tissue banking. *Cell Tissue Bank.* **2005**, *6*, 201–219. [CrossRef] [PubMed]
35. Holt, D.J.; Grainger, D.W. Demineralized bone matrix as a vehicle for delivering endogenous and exogenous therapeutics in bone repair. *Adv. Drug Deliv. Rev.* **2012**, *64*, 1123–1128. [CrossRef] [PubMed]
36. Urist, M.R.; Strates, B.S. Bone morphogenetic protein. *J. Dent. Res.* **1971**, *50*, 1392–1406. [CrossRef] [PubMed]
37. Barone, A.; Varanini, P.; Orlando, B.; Tonelli, P.; Covani, U. Deep-frozen allogeneic onlay bone grafts for reconstruction of atrophic maxillary alveolar ridges: A preliminary study. *J. Oral Maxillofac. Surg.* **2009**, *67*, 1300–1306. [CrossRef]
38. Blume, O.; Hoffmann, L.; Donkiewicz, P.; Wenisch, S.; Back, M.; Franke, J.; Schnettler, R.; Barbeck, M. Treatment of Severely Resorbed Maxilla Due to Peri-Implantitis by Guided Bone Regeneration Using a Customized Allogenic Bone Block: A Case Report. *Materials (Basel)* **2017**, *10*, 1213. [CrossRef]
39. Motamedian, S.R.; Khojaste, M.; Khojasteh, A. Success rate of implants placed in autogenous bone blocks versus allogenic bone blocks: A systematic literature review. *Ann. Maxillofac. Surg.* **2016**, *6*, 78–90. [CrossRef]
40. Makvandi, P.; Ali, G.W.; Della Sala, F.; Abdel-Fattah, W.I.; Borzacchiello, A. Hyaluronic acid/corn silk extract based injectable nanocomposite: A biomimetic antibacterial scaffold for bone tissue regeneration. *Mater. Sci. Eng. C Mater. Biol. Appl.* **2020**, *107*, 110195. [CrossRef]

© 2020 by the authors. Licensee MDPI, Basel, Switzerland. This article is an open access article distributed under the terms and conditions of the Creative Commons Attribution (CC BY) license (http://creativecommons.org/licenses/by/4.0/).

Article

Evaluation of the Use of an Inorganic Bone Matrix in the Repair of Bone Defects in Rats Submitted to Experimental Alcoholism

Iris Jasmin Santos German [1,2,3], Karina Torres Pomini [1], Ana Carolina Cestari Bighetti [1], Jesus Carlos Andreo [1], Carlos Henrique Bertoni Reis [4], André Luis Shinohara [1], Geraldo Marco Rosa Júnior [5,6], Daniel de Bortoli Teixeira [4], Marcelie Priscila de Oliveira Rosso [1], Daniela Vieira Buchaim [4,7] and Rogério Leone Buchaim [1,4,*]

1. Department of Biological Sciences (Anatomy), Bauru School of Dentistry, University of São Paulo (USP), Bauru, São Paulo 17012-901, Brazil; irish_knaan@hotmail.com (I.J.S.G.); karinatorrespomini@gmail.com (K.T.P.); anacarolinacb25@gmail.com (A.C.C.B.); jcandreo@usp.br (J.C.A.); andreshinohara@yahoo.com.br (A.L.S.); marcelierosso@usp.br (M.P.d.O.R.)
2. Department of Dentistry, Faculty of Health Science, Universidad Iberoamericana (UNIBE), Santo Domingo 10203, Dominican Republic
3. Mother and Teacher Pontifical Catholic University (PUCMM), Santo Domingo 10203, Dominican Republic
4. Postgraduate Program in Structural and Functional Interactions in Rehabilitation, University of Marília (UNIMAR), Marília, São Paulo 17525-902, Brazil; carloshbreis@yahoo.com.br (C.H.B.R.); daniel.dbt@hotmail.com (D.d.B.T.); danibuchaim@usp.br (D.V.B.)
5. University of the Ninth of July (UNINOVE), Bauru, São Paulo 17011-102, Brazil; geraldomrjr@yahoo.com.br
6. University of the Sacred Heart (USC), Bauru, São Paulo 17011-160, Brazil
7. Medical School, University Center of Adamantina (UniFAI), Adamantina, São Paulo 17800-000, Brazil
* Correspondence: rogerio@fob.usp.br

Received: 13 December 2019; Accepted: 27 January 2020; Published: 4 February 2020

Abstract: To assess the effects of chronic alcoholism on the repair of bone defects associated with xenograft. Forty male rats were distributed in: control group (CG, $n = 20$) and experimental group (EG, $n = 20$), which received 25% ethanol ad libitum after a period of adaptation. After 90 days of liquid diet, the rats were submitted to 5.0-mm bilateral craniotomy on the parietal bones, subdividing into groups: CCG (control group that received only water with liquid diet and the defect was filled with blood clot), BCG (control group that received only water with liquid diet and the defect was filled with biomaterial), CEG (alcoholic group that received only ethanol solution 25% v/v with liquid diet and the defect was filled with blood clot), and BEG (alcoholic group that received only ethanol solution 25% v/v with liquid diet and the defect was filled with biomaterial). In the analysis of body mass, the drunk animals presented the lowest averages in relation to non-drunk animals during the experimental period. Histomorphologically all groups presented bone formation restricted to the defect margins at 60 days, with bone islets adjacent to the BCG biomaterial particles. CEG showed significant difference compared to BEG only at 40 days (17.42 ± 2.78 vs. 9.59 ± 4.59, respectively). In the birefringence analysis, in early periods all groups showed red-orange birefringence turning greenish-yellow at the end of the experiment. The results provided that, regardless of clinical condition, i.e., alcoholic or non-alcoholic, in the final period of the experiment, the process of bone defect recomposition was similar with the use of xenograft or only clot.

Keywords: bone repair; biomaterial; alcoholism; alcohol

1. Introduction

The concept of alcoholism peaked in the eighteenth century, shortly after the growing production and marketing of distilled alcohol, resulting from the industrial revolution. Ethanol is a water-soluble organic solvent with the ability to penetrate all compartments of the human body, and its systemic effects produce changes in the central nervous system, muscular system, liver disease, chronic pancreatitis, cardiovascular disease, lung disease, among others [1].

Ethanol impairs bone formation by inhibiting osteoblast proliferation. In addition, alcohol induces oxidative stress and participates in the regulation of osteoclast differentiation, resulting in increased signaling of RANKL-RANK, kB nuclear factor activating receptor ligand in bone cells, and increased osteoclastogenesis [2].

In addition, ethanol changes the levels of cytokines responsible for regulating bone metabolism, such as increased tumor necrosis factor alpha (TNF-alpha) and interleukin-6 (IL-6). These, in turn, cause the suppression of osteoblast synthesis and consequently the deposition of osteoid matrix [3].

Stimulation of ethanol on osteocytes causes increased secretion of sclerostin protein, which binds to cellular receptors in order to antagonize the action of BMP in the Wnt signaling cascade. The morphology and osteocyte apoptosis number are also altered with alcohol consumption, which sends signs of osteoclast recruitment, increasing bone resorption [4].

This change in bone remodeling may be expressed by bone loss and serum osteocalcin levels, a marker of bone formation. Moreover, it compromises collagen gene expression, also non-collagenous matrix proteins, and significantly reduces the levels of carboxy-terminal procollagen I propeptide [5]. These changes are also related to the duration and amount of alcohol exposure [6].

Alcohol consumption causes harmful effects on bone density [7] and is related to osteoporosis, due to imbalance in bone repair [8], altering bone microarchitecture, decreasing trabecular and cortical bone thickness, and consequently increasing the risk of fracture [9].

There are many scientific reports about xenografts in the bone repair process, whose results evidence that it is a viable alternative to autologous grafts [10]. There is currently a social and commercial incentive for alcohol consumption worldwide, which may lead to a larger number of patients using ethyl derivatives in clinical practice. Given this context, associated with the lack of experimental studies in the literature that relate alcoholism to biomaterial used in this innovative research and translational projection, it was decided to analyze a low-cost bone substitute used in the areas of dentistry and medicine with alcoholism.

Therefore, the aim of this study was to evaluate the effects of chronic alcoholism on the repair of bone defects in rats filled or not with a bovine bone matrix.

2. Materials and Methods

2.1. Biomaterial—Bone Graft Substitutes

The Bonefill biomaterial (Bionnovation Biomedical A.B., Bauru, São Paulo, Brazil) evaluated in this study is produced by decalcification of the cortical portion of bovine femur, totally denatured and sterilized by 25 kGy gamma radiation. It is commercially available in average particle size between 0.6–1.5 mm in diameter and 0.5 gr package. This dental product is registered in Brazilian Ministry of Health by The National Sanitary Surveillance Agency (ANVISA10392710012).

2.2. Experimental Design

Forty adult 60-day-old male Wistar rats (*Rattus norvegicus*) were used, weighing approximately 250 g. The animals were kept in conventional cages containing 4 animals/box. The macroenvironment presented artificial timer-managed lighting, which controlled the 12/12-h light/dark cycle, with 220 lux brightness, 55% humidity, exhaust fan and air conditioning, maintaining an average temperature of 22 °C. All animals had free access to standard rat chow (Nuvilab™, Nuvital, Colombo, Paraná, Brazil) and their daily consumption was not measured (Table 1). All experimental procedures on the animals were conducted after approval by the Institutional Review Board on Animal Studies of Bauru School of Dentistry, University of São Paulo (Protocol: CEEPA-023/2012).

Table 1. Nutritional composition of the chow used in the experiment for all animals. Ad libitum feeding of chow.

Parameter	Chow *
Humidity (%)	2.30
Brute protein (%)	21.96
Ether (fatty) extract (%)	4.61
Mineral residue (%)	8.36
Brute fiber (%)	4.04
Nitrogen-free extract (%)	48.73
Calcium (%)	1.32
Phosphorus (%)	0.82
Brute energy (Kcal/kg)	3913

* Nuvilab CR1. Nuvital, Colombo, Paraná, Brazil.

The animals were randomly separated into two large groups according to the type of liquid diet (drinking water or ethanol): CG, $n = 20$ (control group - non-alcoholic), which received a standard rat chow and water ad libitum; and EG, $n = 20$ (experimental group–alcoholic), which received a standard rat chow and ethanol ad libitum. After the surgical procedure, the groups were subdivided according to the type of treatment (clot or biomaterial) and the clinical condition (alcoholic or non-alcoholic), as follows: CCG (control group that received only water with liquid diet and the defect was filled with blood clot), BCG (control group that received only water with liquid diet and the defect was filled with biomaterial), CEG (alcoholic group that received only ethanol solution 25% v/v with liquid diet and the defect was filled with blood clot), and BEG (alcoholic group that received only ethanol solution 25% v/v with liquid diet and the defect was filled with biomaterial) (Figure 1A1).

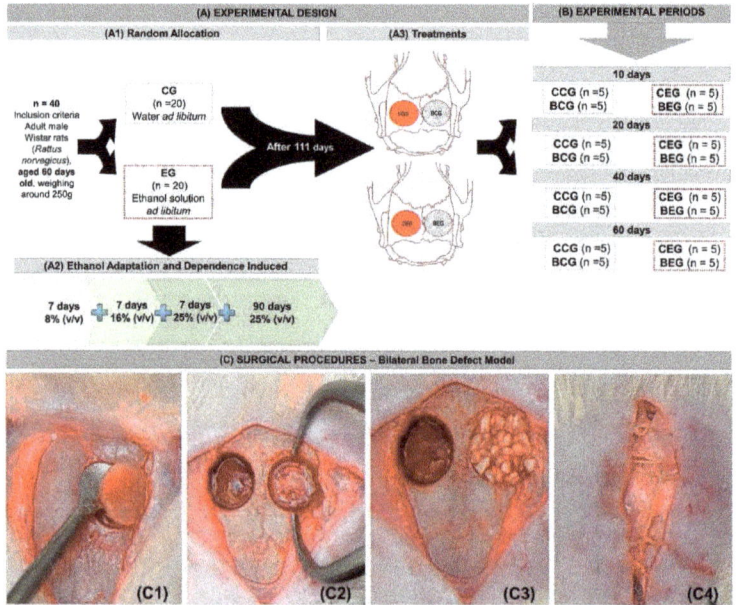

Figure 1. (**A**) Experimental design. (**A1**) Random Allocation—Forty adult male Wistar rats, (*Rattus norvegicus*), aged 60 days old, weighing around 250 g were divided into two broad groups: CG—Control group (*n* = 20)—that received only water with liquid diet and EG—Experimental group (*n* = 20)—that received ethanol solution 25% (*v/v*) with liquid diet after adaptation period. (**A2**) Ethanol Adaptation and Dependence Induced—Animals were gradually drunk at progressive concentrations of ethanol solution (8–16–25% *v/v*). After 21 days of alcohol adaptation, the animals remained at 25% (*v/v*) until the surgical procedure for 90 days. (**A3**) Treatments—After surgical procedures—bilateral bone defect model in the parietals, four subgroups were preformatted according to treatment (blood clot vs. biomaterial) and clinical conditions (alcoholic vs. non-alcoholic): Animals that received only water with liquid diet: CCG (right parietal bone defect was filled with blood clot) and BCG (left parietal bone defect was filled with biomaterial); Animals that received ethanol solution 25% (*v/v*): CEG (left parietal bone defect was filled with blood clot); BEG (right parietal bone defect was filled with biomaterial). B) Experimental Periods—at 10, 20, 40 and 60 days the skulls of 5 animals/group were collected, totalizing 5 defects/period of each subgroup CCG, BCG, CEG and BEG. (**C**) Surgical Procedures–Bilateral Bone Defect Model–(**C1**) 5-mm left parietal osteotomy; (**C2**) Two bone defects in parietal bones; (**C3**) One defect filled with biomaterial and the contralateral only with blood clot; (**C4**) Periosteum suture with nylon 5-0.

2.3. Ethanol Adaptation and Induced Dependence—Semi-Voluntary Ethanol Administration—Alcohol-Liquid Diet

This study followed the determined chronic alcoholism model of "semi-voluntary", where alcohol was the only liquid food available. Animals in the experimental group (EG) were submitted to an alcohol adaptation model, where the only available liquid source was ethanol ad libitum. In the first 7 days, the animals received 8% (*v/v*) ethanol solution, in the second week 16% (*v/v*) and third week 25% (*v/v*). After this period of gradual adaptation, the animals remained on the 25% (*v/v*) liquid ethanol diet for further 90 days, when they underwent experimental surgery and remained on the same diet until the corresponding euthanasia period. Animal health was monitored daily (Figure 1A2,B).

2.4. Surgical Procedures

After 111 days of ethanol intake and induced dependence, all rats were weighed and subjected to intramuscular general anesthesia with ketamine at a dose of 50 mg/kg i.m. (Dopalen™, Ceva, Paulinia, São Paulo, Brazil) plus xylazine at the dose of 10 mg/kg i.m. (Anasedan™, Ceva, Paulinia, São Paulo, Brazil), with strict monitoring of anesthesia mainly in alcoholic animals [11].

After frontoparietal trichotomy and antisepsis with 2% chlorhexidine, a cranio-caudal longitudinal incision of approximately 20 mm in length was made for tissue exposure and divulsion. Circular bilateral osteotomy was performed on the parietal bones with a 5 mm diameter trephine drill (Neodent, Curitiba, Paraná, Brazil) at low speed (1500 rpm) under constant saline irrigation (Figure 1C1,C2).

The defect on the left parietal was filled with 14 mg of biomaterial (previously established in a pilot study) and the right parietal was filled only with blood Clot (Figure 1A3,C3). The periosteum was repositioned and sutured with Vicryl™ polyglactin suture (Ethicon J&J, São Paulo, Brazil) 5-0 and the integument with 4-0 silk suture (Ethicon™, J&J, São Paulo, Brazil) (Figure 1C4).

After the surgical procedure, the animals were placed under incandescent light for complete anesthetic recovery and submitted to a single intramuscular injection of enrofloxacin 2.5 mg/kg (Flotril™; Schering-Plough SA, Rio de Janeiro, Brazil) and intramuscular injections of 0.06 mg/kg dipyrone (Analgex™; Agener União, São Paulo, Brazil) for 3 days.

2.5. Collection of Specimens and Histological Procedures

Five animals from each group were euthanized using the aforementioned anesthetic overdose at the respective periods of 10, 20, 40 and 60 days (Figure 1B). The specimens were removed and fixed in 10% buffered formalin for 48 h and then demineralized in EDTA, a solution containing 4.13% Titriplex™ III (Merck KGaA, Darmstadt, Germany) and 0.44% sodium hydroxide, for a period of approximately 40 days. Then, the specimens were subjected to standard histological procedures and included in Histosec ™ (Merck KGaA, Darmstadt, Germany). Histological sections were obtained with 5 μm thickness prioritizing the defect centers for hematoxylin and eosin, Masson's trichrome and picrosirius-red staining.

2.6. Body Mass Analysis

Body mass was determined by simple weighing on a Bel Mark 3500 precision scale (BEL™ Analytical Equipment Ltda, São Paulo, Brazil), with maximum capacity of 3500 g and minimum of 200 g with the aid of a Styrofoam box for animal containment. Measurements occurred on the day of the surgical procedure (initial mass) and on the corresponding days of euthanasia (10, 20, 40, and 60 days).

2.7. Histomorphological and Histomorphometric Evaluation

The histologic sections were analyzed in the Histology Laboratory of Bauru School of Dentistry, University of São Paulo (São Paulo, Brazil) by light microscopy (Olympus model BX50, Tokyo, Japan) at approximate magnifications of ×4, ×10, ×40 and ×100. To establish a standard criteria for judgment, there was a training session with an experienced pathologist.

For histomorphological description of the bone defect area, the central region (edge-to-edge measurement) was considered with the aid of free-scale image capture system (DP Controller software (3.2.1.276 version, Olympus, Tokyo, Japan) to analyze tissue formation granulation, inflammatory infiltrate, formation of primary bone tissue and bone maturation.

A virtual overall image of the defect (Masson's trichrome, ×10) was generated to quantify the volume density (%) of newly formed bone tissue and biomaterial by AxioVision Rel. 4.8 Ink (Carl Zeiss MicroImaging GmbH, Jena, Germany). For determination of volumetric density (%), the equation $Vvi = AAi = Ai/A \times 100$ was considered, considering Vvi (volume density), AAi (area density), Ai (area filled with newly formed bone tissue or particle of biomaterial), A (total area examined) (Figure S1) [12].

Images from picrosirius-red stained sections were captured using a higher resolution digital camera Leica DFC 310FX (Leica™, Microsystems, Wetzlar, Germany) connected to a confocal laser microscope Leica DM IRBE and capture system LAS 4.0.0 (Leica™, Microsystems, Heerbrugg, Switzerland). The quality of newly formed bone in the defects was evaluated by the orientation pattern and width of the collagen fibers detected by the birefringence of polarization colors ranging from red-orange (primary disorganized bone tissue) to green-yellow (organized bone-lamellar bone tissue).

2.8. Statistical Evaluation of Data

Data on body mass, volume density (%) of newly formed bone tissue and biomaterial particle were expressed as mean ± standard deviation of the mean (SEM). All tests were performed using Statistica 10.0 software (StatSoft Inc., Tulsa, OK, USA) and the significance level was set at $p < 0.05$. The independent "t" test was used to compare the initial body mass in alcoholic and non-alcoholic groups, and the paired "t" test was applied to compare the initial and final body mass within the same group. The percentage of bone formation and biomaterial in groups in different periods was assessed by one-way ANOVA variance test (time) for independent samples, and Tukey's post hoc test, at a significance level of $p < 0.05$. To compare the percentage of bone formation in drunk vs. non-alcoholic animals (CCG vs. CEG and BCG vs. BEG) in the different periods, the t-test was applied for independent samples, and post hoc Tukey test, at a significance level of $p < 0.05$. To compare the percentage of bone formation in animals of the same group for different treatments (CCG vs. BCG and CEG vs. BEG), at different periods, the paired t-test and post hoc Tukey test were applied, at a significance level of $p < 0.05$.

3. Results

3.1. Effects of Ethanol Administration on Behavioral and Clinical Profiles of Rats

Regarding the general clinical profile, all animals presented good physical condition throughout the experiment, with no signs of morbidity and no mortality rate. There was no infection in either group, nor in the surgical area. However, some animals in the experimental group-alcohol group showed changes in their behavioral profile, especially regarding the parameters of agitation, aggressiveness and exploratory activity.

In the evaluation of body mass, after the period of alcoholic induction (111 days), the animals of the experimental group showed lower initial mass gain (day of surgery) compared to the control group (343.82 ± 41.93 vs. 444.59 ± 45.20, weight in grams, respectively). At 20 days, the mean mass of EG group showed no significant difference, but between 40–60 days there was a slight increase in the means from 5.9% to 6.4%, relative to day 0. In non-alcoholic animals, CG, the increase between 20–60 days was 7.9% to 14.1%, relative to day 0 (Figure 2).

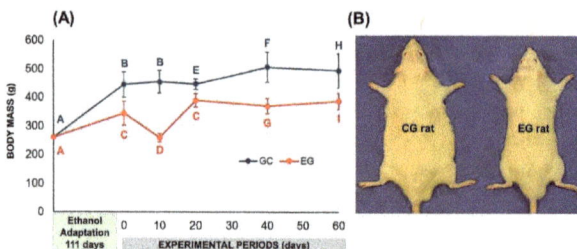

Figure 2. (A) Graphic representation of body mass (g) during ethanol adaptation-induced dependence (111 days) and experimental periods of control (CG-water ad libitum) and experimental (EG-ethanol solution ad libitum) groups. (B) Comparison between body masses of CG vs. EG showing the negative effect of alcohol. Different letters $p < 0.05$ (independent t-test and paired t-test showed interaction between group and period).

3.2. Histological Evaluation

At 10 days, all experimental groups presented bone formation at the defect margins (Figure 3A–D). In groups treated with blood clot, CCG and CEG, there was predominance of richly vascularized granulation tissue filling the entire surgical area. However, BCG and BEG were shown to be reaction tissue surrounding the biomaterial particles (Figure 4A,B and Figure 5).

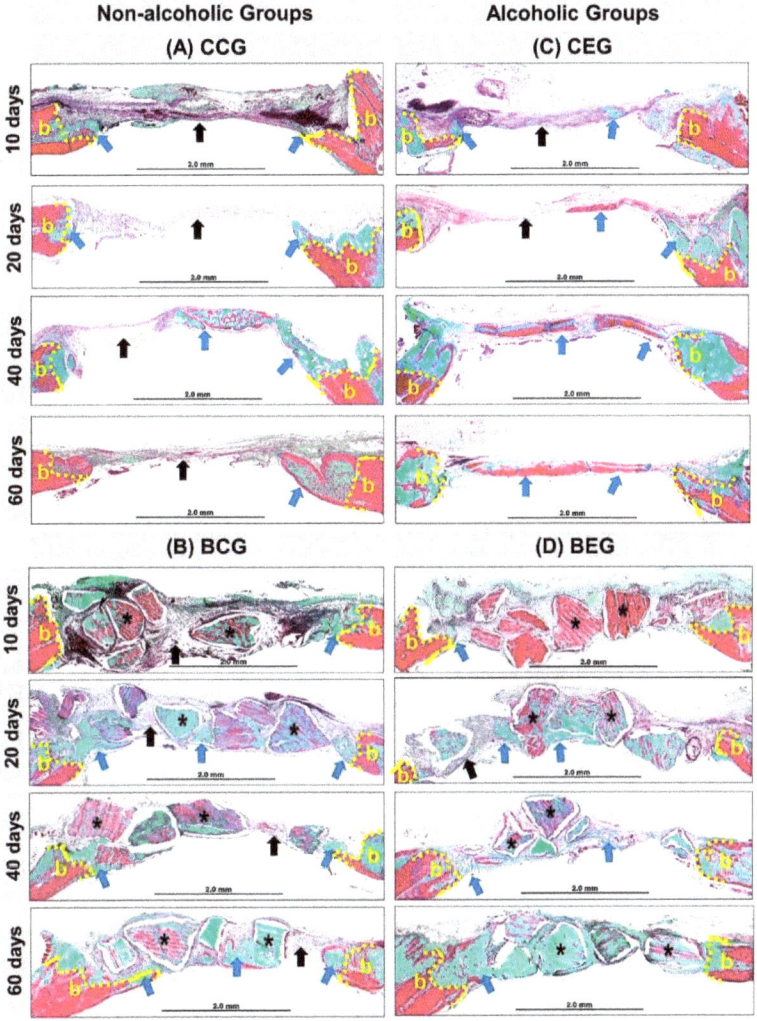

Figure 3. Panoramic histological views in skull defects created in the animals (A–D). Non-alcoholic (A,B) and alcoholic (C,D) treated with blood clot or biomaterial at different experimental periods, 10, 20, 40 and 60 days. (A,B) Non-alcoholic Groups: showed new bone formation (blue arrows) from the defect border (b), with partial closure by fibrous connective tissue in GCC (black arrows) or by particles of biomaterial surrounded by fibrous connective tissue in BCG (asterisk). (C,D) Alcoholic groups: new bone formation was observed from the border and on the dura-mater surface, with bone islets on the defect center in CEG and bone islets adjacent to the biomaterial particles in BEG. In both groups, at 60 days, lamellar tissue transition to compact tissue was observed (Masson's Trichrome; original magnification ×4; bar = 2 mm).

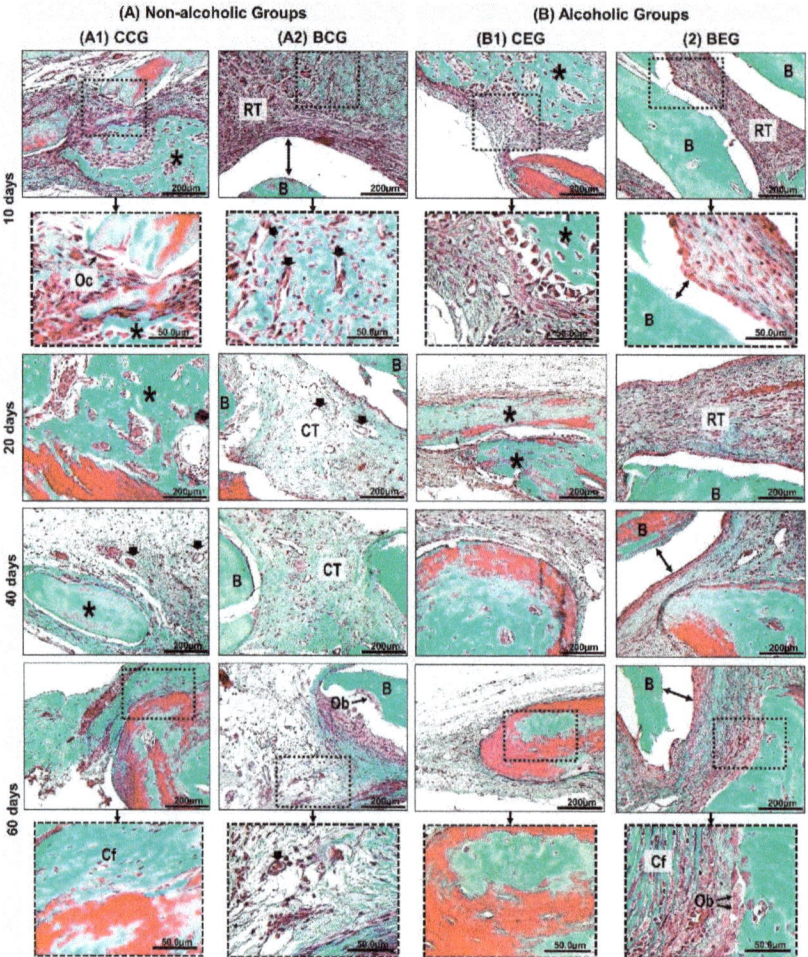

Figure 4. Details of evolution of bone healing of cranial defects created in the Non-Alcoholic (**A**) and Alcoholic (**B**) animals treated with blood clot or biomaterial. (**A1**) CCG (Non-alcoholic Group; defects filled with blood clot): at 10 days, defects showed trabecular bone formation (asterisk), with the presence of osteoclastic cells (Oc) on the edge of the remaining bone tissue. 20–40 days, the new bone formed showed bone maturation phase surrounded by blood capillaries (black arrow). At 60 days, collagen fibers were arranged in a more regular manner (Cf). (**A2**) BCG (Non-alcoholic Group: defects filled with biomaterial): at 10 days, defects showed tissue reaction (RT) surrounding the particles of the biomaterial (B); artifacts in histologic sections (double arrow–gap between biomaterial and tissue). Between 20–60 days, connective tissue (CT) presented scarce inflammatory cells with thin and thick collagen fibers, which were parallelly arranged at the end of the experimental period. (**B1**) CEG (Alcoholic Group; defects filled with blood clot): in the early periods, the defects presented inflammatory cells, decreasing at 40 days. The bone tissue formed at 60 days was predominantly compact and mature. (**B2**) BEG (Alcoholic Group; defects filled with biomaterial): 10–20 days, sections showed discrete bone formation, and biomaterial particles permeated by reaction tissue. In the later periods, collagen fibers were organized in parallel, and osteoblastic cells (Ob) forming a single cell line adjacent to the matrix. Masson's Trichrome; original magnification ×40; bar = 100 µm; and Insets, magnified images ×100; bar = 50 µm.

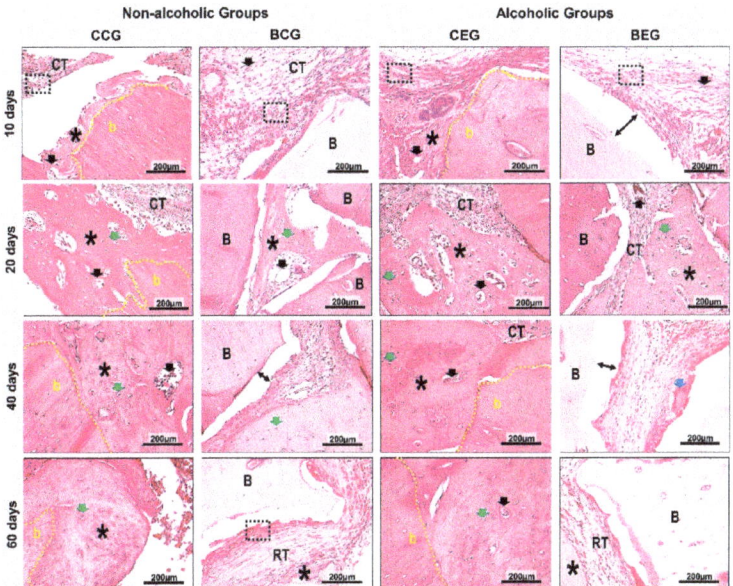

Figure 5. Histological details with HE staining of evolution of cranial defect bone healing created in the Non-Alcoholic and Alcoholic animals treated with blood clot or biomaterial. At 10-20 days, CCG and CEG showed the presence of richly vascularized connective tissue (CT) (black arrow) and new bone tissue (asterisk) at the defect margin (b) with trabecular arrangement. All groups presented inflammatory cells (inside the black lined area), more evident in BCG and BEG, permeating the particles (B). At 40 days there was typical lamellar arrangement, interspersed with osteocytes (green arrow), and in BCG and BEG inflammatory cells and multinucleated giant cells (blue arrow). At 60 days, bone tissue was mature and compact in defects filled with blood clot, there was decreased inflammatory reaction in BCG and BEG groups and regularly organized collagen fibers. Artifacts in histologic sections (double arrow–gap between biomaterial and tissue). HE; original magnification x 40; bar = 100 µm.

At 20 days, in the CCG and CEG groups, immature trabecular bone formation was in the transition phase of bone maturation, obtaining a denser lamellar arrangement at 40 days. In BCG, the reaction tissue was in resolution phase, with sparse inflammatory infiltrate, unlike that observed in the alcohol-treated / biomaterial-treated animals, BEG (Figure 4A,B and Figure 5).

At the end of the experimental period, at 60 days, all groups presented complete closure of the surgical area by fibrous connective tissue and / or particles of biomaterial. In defects treated with blood clot, in CCG and CEG, the new bone formation remained restricted to the defect margins and over the dura mater with a more evident bone maturation pattern, but with smaller thickness than the remaining bone. In the BCG and BEG groups, the particles were encased in evenly arranged, thicker collagen fibers (Figure 3A–D, Figure 4A,B and Figure 5).

3.3. Histomorphometric Evaluation

In relation to biomaterial volume density (%) in the BCG and BEG groups, a tendency of decrease between 10 (mean of 30.4%) to 60 days (mean of 18.98%) was observed (compare data in the Table 2). However, no statistical differences between periods in each group (ANOVA, $p > 0.2$) or groups per period ("t" test, $p > 0.6$) were observed.

Table 2. Mean ± standard deviation of volume density of comparison of biomaterial in the different experimental groups.

Group	10 Days	20 Days	40 Days	60 Days	One-Way ANOVA (p)
BCG	32.69 ± 11.71 [a]	21.60 ± 5.16 [a]	28.03 ± 10.64 [a]	18.43 ± 11.86 [a]	0.160
BEG	28.24 ± 11.60 [a]	22.93 ± 11.46 [a]	25.58 ± 4.00 [a]	19.54 ± 11.06 [a]	0.601
Unpaired t-test (p)	0.5875	0.8190	0.6426	0.8829	

Same lowercase letters indicate that there was no statistically significant difference. Significant differences $p < 0.05$.

Regarding bone formation volume density (Table 3), a significant increase was observed only in the defects created in alcoholic rats filled with blood clot, CEG group ($p = 0.007$), between 10 (mean of 6.67%) and 40 days (mean of 17.42%), as well as between 10 and 60 days (mean of 18.29%). In the same animals, the contralateral defects filled with biomaterials, BEG group, the bone formation volume density was 0.43 times smaller than CEG. Regardless of clinical condition (Table 4), no statistical differences in the bone formation were observed between CCG vs. CEG ($p > 0.1$) and BCG vs. BEG ($p > 0.3$).

Table 3. Mean ± standard deviation of volume density of new bone formation. Comparison among periods within the same group was evaluated by one-way ANOVA (column, 10 vs. 20 vs. 40 vs. 60 days). Comparison between defects treated with biomaterial vs. clot per condition (non-alcoholic and alcoholic) (line, CCG vs. BCG and CEG vs. BEG).

Period (Days)	Volume Density of New Bone Formation (%)					
	Non-Alcoholic Rat (n = 5/Period)			Alcoholic Rat (n = 5/Period)		
	CCG	BCG	Paired t-Test (p)	CEG	BEG	Paired t-Test (p)
10	5.30 ± 3.08 [a]	7.54 ± 6.56 [a]	0.326	6.67 ± 3,09 [a]	6.98 ± 5,97 [a]	0.852
20	8.41 ± 5.17 [a]	13.79 ± 11.14 [a]	0.370	12.33 ± 1,89 [a,b]	7.96 ± 4,40 [a]	0.156
40	15.50 ± 7.14 [a]	12.97 ± 7.07 [a]	0.610	17.42 ± 2,78 [b]	9.59 ± 4,59 [a]	0.018
60	14.51 ± 7.69 [a]	13.34 ± 12.45 [a]	0.865	18.29 ± 7,89 [b,*]	12.85 ± 7,94 [a,*]	0.122
One way ANOVA (p)	0.05	0.618		0.007	0.466	

Different letters ([a] ≠ [b]) indicate a statistically significant difference ($p < 0.05$), except for asterisks (*) that do not show significant difference (CEG vs. BEG; $p = 0.122$).

Table 4. Mean ± standard deviation of volume density of new bone formation Comparison between same treatment per animal condition (non-alcoholic and alcoholic). Unpaired t-test (line, CCG vs. CEG and BCG vs. BEG).

Period (Days)	Volume Density of New Bone Formation (%)					
	Clot			Biomaterial		
	CCG (n = 5/Period)	CEG (n = 5/Period)	Unpaired t-Test (p)	BCG (n = 5/Period)	BEG (n = 5/Period)	Unpaired t-Test (p)
10	5.30 ± 3.08 [a]	6.67 ± 3.09 [a]	0.528	7.54 ± 6.56 [a]	6.98 ± 5.97 [a]	0.899
20	8.41 ± 5.17 [a]	12.33 ± 1.89 [a]	0.151	13.79 ± 11.14 [a]	7.96 ± 4.40 [a]	0.307
40	15.50 ± 7.14 [a]	17.42 ± 2.78 [a]	0.591	12.97 ± 7.07 [a]	9.59 ± 4.59 [a]	0.678
60	14.51 ± 7.69 [a]	18.29 ± 7.89 [a]	0.464	13.34 ± 12.45 [a]	12.85 ± 7.94 [a]	0.942

Same lowercase letters indicate that there was no statistically significant difference. Significant differences $p < 0.05$.

3.4. Influence of Clinical Condition (Alcoholic Versus non-Alcoholic) and/or Type of Treatment (Clot vs. Biomaterial) on Collagen Content During the Bone Repair Process

Between 10–20 days, in the analysis of collagen matrix birefringence in polarized microscopy, all groups showed predominance of red-orange birefringence in the defect margins related to immature bone formation. In the BCG and BEG groups, the biomaterial presented red-orange birefringence surrounded by thin and disorganized collagen fibers (Figure 6).

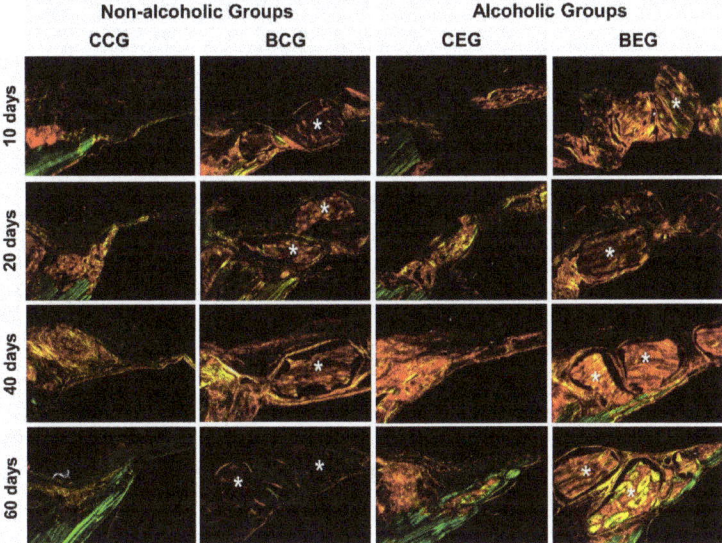

Figure 6. Photomicrographies of birefringent fibers stained with Picrosirius red under polarized light at 10, 20, 40 and 60 days of repair. In the initial periods, non-alcoholic groups, CCG and BCG showed red-orange collagen fibers becoming yellow-green at the end of the experiment. Alcoholic groups demonstrated evidently disorganized bone collagenous matrix, with red-orange birefringence around the grafted biomaterial (asterisk) and adjacent to the defect edges, CCG; BCG, at 60 days. The xenogeneic graft (BCG; BEG) presented red-orange birefringence. Picrosirius red staining, original magnification ×5. Scale bar = 50 µm.

In the final periods, 40–60 days, in non-alcoholic groups (CCG and BCG), the bone collagenous matrix was more organized with green-yellow birefringence than the alcoholic animals (Figure 6).

4. Discussion

The growing increase in chronic alcohol consumption in the last decades has encouraged the development of numerous researches in the medical and dental areas to alleviate the deleterious effects of ethanol on bone loss due to osteoporotic conditions or with difficulty in consolidating extensive bone defects [13,14].

However, there are still few studies in the scientific literature evaluating its impact on the repair and osseointegration process of biomaterials. Thus, the results of this in vivo study showed that the biomaterial served as a scaffold for bone cells, a biological event that can attenuate the harmful effects of ethanol on the bone repair process.

Experimentally several bone defect sites are tested, but in rats the most commonly used is the critical defect in calvaria. The experimental animal model used in this study involving 5-mm bilateral craniotomies in the parietal bones has been used in scientific research because of its ability to produce paired analyzes in a standardized manner and allow the evaluation of bone substitute materials in the reconstruction of critical size bone defects without involvement of the sagittal suture [15].

In addition, the bones of the calvaria and face have intramembranous ossification, a typical location of dental defects [16]. It is also possible to take into consideration the constitution of the biomaterial tested, being particulate and of medium size (0.6–1.5 mm), more commonly indicated in smaller defects because, in large orthopedic defects, typical of long bones, it is preferable to the use of block grafts.

Initial studies have graphically described a U-shaped curve relating ethanol consumption to various chronic diseases, but the literature still remains controversial regarding bone tissue effects [17]. Thus, we adopted in this experiment the 25% (v/v) alcohol dosage, as we based on preliminary analyzes by De Souza et al. [18], Buchaim et al. [19] and [20] who reported destructive effects on bone with the use of 20% and 25% (v/v) ethanol without inducing animal death, contrary to the 5% dosage that had a protective effect [21].

In addition, previous studies by our research group evaluated the plasma concentrations of three ethanol dosages and their effects on bone repair, which showed that 25% alcoholization had pharmacologically relevant plasma concentration (540 mg/dL) to cause alterations bones that compromised the morphofunctional restoration of lost tissue [19].

In this study, we evaluated the effects of alcohol on body mass and new bone formation by descriptive and histomorphometric analysis by Masson's trichrome, and collagen fiber birefringence analysis by picrosirius-red staining in order to observe the alignment of the bone collagen and fiber structure at 10, 20, 40, and 60 days after injury.

The results of the body mass analysis showed lower averages in alcoholic than in non-alcoholic animals during the whole experimental period [20]. According to the literature consulted, the extensive use of alcohol can lead to dysfunctions in nutrient metabolism, causing changes such as decreased digestion and absorption, as alcohol has an influence on the stomach [22] and intestines [23] that can increase nutrient excretion, and consequently the risk of malnutrition [24,25]. Therefore, prolonged alcohol consumption may result in the lower body mass gained in animal experiments compared to non-alcoholic animals [26].

In the behavioral evaluation, some alcoholic rats initially presented aggressiveness and irritability, which are associated with hyperexcitability of the central nervous system, triggered by physiological dependence of alcohol [27,28].

All panoramic histological images of this experiment showed no integumentary invasion in the surgical bed, and the formation of fibrous connective tissue observed on the defect center and/or adjacent to the particles originated from the injured margins. Proper repositioning of the periosteum acted as a mechanical barrier preventing the collapse of extracranial tissues and consequently the migration of competing cells to osteoblasts [29,30].

Histomorphologically, defects filled with blood clot, CCG and CEG, presented the same pattern of bone repair, forming new bone tissue at the defect margins and extending centripetally, but with complete closure by fibrous connective tissue. This finding agrees with previous authors who suggest that it is a critical bone defect when spontaneous bone regeneration does not occur during the experimental period, requiring reconstruction of these defects by grafting [31].

In the initial periods, all experimental groups presented inflammatory infiltrate, a local defensive process against tissue aggression [32,33]. Between 20–40 days, the newly formed bone tissue was in transition from fine to compact trabeculae as observed by the study of Rocha et al. [34].

In the same period, BCG and BEG showed persistence of material implanted in the receptor bed with some dispersed inflammatory cells, without formation of macrophage aggregates and multinucleated giant cells, characteristic of chronic granuloma inflammatory process [35]. Thus, it is believed that the biomaterial proved to be biocompatible, i.e., the organism recognized the particles as part of its structure and not as an aggressor to its microenvironment [36,37].

At 60 days, the new bone formation remained restricted to the defect margins in all groups, with compact lamellar arrangement [38]. In BCG, bone islets adjacent to the particles were observed, which is consistent with studies that reported to be characteristic of osteoconductive biomaterial by providing a scaffold for osteoblastic cells, facilitating the deposition of new bone on its surface [39].

At the end of the experimental period, the inflammatory process present in the BCG and BEG groups was in the resolution phase, being more evident in non-alcoholic animals (BCG). These results lead to the perception of increase in serum levels of proinflammatory cytokines, derived from ethanol-induced liver disorders, which may have contributed to the persistence of the reaction tissue [40].

Histomorphometrically, in the analysis of the influence of time on bone formation, CEG showed a gradual increase in averages up to 40 and 60 days with statistically significant difference (Table 3, column). All other groups presented higher average in the final period of the experiment (60 days) in relation to the initial period (10 days), but without significant difference. Previous investigations have reported the physiological events that occur after the accommodation of biomaterial particles in the surgical bed, which alters the microenvironment and consequently delays the new bone formation [41].

Regarding the percentage of biomaterial in the BCG and BEG groups (Table 2, line), there was a decrease in final periods, but without significant difference. The delayed degradation of particles, as evidenced by their presence at 60 days, even in non-alcoholic animals, may be related to the intrinsic characteristics of the biomaterial [42]. This finding is corroborated by the study of Desterro et al. [43], who stated that non-sintered bovine apatites (<1000 °C) with organic matrix have lower dissolution rates, which directly impacts the biodegradation time.

Regarding the analysis of interference of the clinical condition, alcoholic and non-alcoholic, for each treatment, there was no statistical difference in all experimental groups (Table 4). In animals that used the biomaterial filling the bone defect, the results show that the formation of new bone was similar between the groups (BCG and BEG), regardless of the clinical condition (alcoholic or non-alcoholic), possibly for its osteoconductive property [39].

However, in comparing the percentage of bone formation according to treatment, clot or biomaterial (Table 3), CEG showed significant difference compared to BEG at 40 days (17.42 ± 2.78 vs. 9.59 ± 4.59, respectively). This fact may be correlated to the impact of ethanol on activities of the immune system, leading to changes in the phagocytic activity of polymorphonuclear cells. Thus, the persistence of particles may also have contributed to the delayed new formation in later periods compared to blood clot filling [44].

In the histochemical analysis of collagen fibers by picrosirius-red, all experimental groups initially presented red-orange birefringence, characteristic staining of formation of thin and disorganized collagen fibers turning greenish-yellow over the periods [45].

The biomaterial particles showed birefringence close to the newly formed bone tissue, precluding the measurement of specific fibers from bone repair. The study by Desterro [43] proved the presence of residual organic material in the particles of this biomaterial by X-ray diffraction analysis (XRD diffractogram of Bonefill^{TM}), justifying the markup in this analysis.

Based on the experimental model employed, it can be concluded that, regardless of the clinical condition, alcoholic or non-alcoholic, in the final period of the experiment, the process of bone defect recomposition was similar with the use of xenograft or only clot. The use of biomaterial can provide a scaffold that guides bone growth, especially in larger defects.

For prospective research in the field of tissue engineering, it is suggested to associate with biomaterials, plasma-derived biodegradable polymers such as PRP, PRF and fibrin sealants in order to make graft material moldable in the surgical bed, facilitate insertion and agglutination, and prevent its dispersal [46]. In addition, adjuvant and noninvasive methods are also recommended to accelerate and improve the regeneration process such as hyperbaric chamber, pulsed ultrasound (LIPUS), and laser photobiomodulation therapy [46–49].

5. Limitations

Knowing that ethanol causes β-catenin signaling pathway dysregulation by increasing/decreasing the activity or expression of its protein constituents, suggests that future studies employ molecular and biochemical analyses in order to detail the effects of ethanol on filled bone defects with biomaterial.

Supplementary Materials: The following are available online at http://www.mdpi.com/1996-1944/13/3/695/s1, Figure S1: Representative image of the methodology used to quantify the area of newly formed bone, graft particles, by Axio Vision software.

Author Contributions: Conceptualization—I.J.S.G. and R.L.B.; Data collection—I.J.S.G., K.T.P.; Formal analysis—I.J.S.G., K.T.P., D.d.B.T., A.C.C.B. and A.C.C.B.; Methodology—I.J.S.G., K.T.P., D.V.B., A.L.S., G.M.R.J., M.P.d.O.R.; C.H.B.R.; Resources—J.C.A. and R.L.B.; Supervision—R.L.B.; Visualization—I.J.S.G., K.T.P., D.V.B., A.C.C.B. and R.L.B.; Writing of original draft—I.J.S.G., K.T.P. and R.L.B.; Writing, review, and editing—I.J.S.G. and R.L.B. All authors have read and agreed to the published version of the manuscript.

Funding: This study was financed in part by the Coordenação de Aperfeiçoamento de Pessoal de Nível Superior—Brasil (CAPES)—Finance Code 001.

Conflicts of Interest: The authors declare no conflict of interest.

References

1. González-Reimers, E.; Santolaria-Fernández, F.; Martín-González, M.; Fernández-Rodríguez, C.; Quintero-Platt, G. Alcoholism: A systemic proinflammatory condition. *World J. Gastroenterol.* **2014**, *20*, 14660–14671. [CrossRef] [PubMed]
2. Ronis, M.; Mercer, K.; Chen, J. Effects of Nutrition and Alcohol Consumption on Bone Loss. *Bone* **2008**, *23*, 1–7. [CrossRef] [PubMed]
3. Maurel, D.B.; Boisseau, N.; Benhamou, C.L.; Jaffre, C. Alcohol and bone: Review of dose effects and mechanisms. *Osteoporos. Int.* **2012**, *23*, 1–16. [CrossRef] [PubMed]
4. Maaurel, D.B.; Jaffre, C.; Rochefort, G.Y.; Aveline, P.C.; Boisseau, N.; Uzbekov, R.; Gosset, D.; Pichon, C.; Fazzalari, N.L.; Pallu, S.; et al. Low bone accrual is associated with osteocyte apoptosis in alcohol-induced osteopenia. *Bone* **2011**, *49*, 543–552. [CrossRef]
5. Shetty, S.; Kapoor, N.; Bondu, J.; Thomas, N.; Paul, T. Bone turnover markers: Emerging tool in the management of osteoporosis. *Indian J. Endocrinol. Metab.* **2016**, *20*, 846.
6. Johnson, T.L.; Gaddini, G.; Branscum, A.J.; Olson, D.A.; Caroline-Westerlind, K.; Turner, R.T.; Iwaniec, U.T. Effects of chronic heavy alcohol consumption and endurance exercise on cancellous and cortical bone microarchitecture in adult male rats. *Alcohol. Clin. Exp. Res.* **2014**, *38*, 1365–1372. [CrossRef]
7. Mikosch, P. Alcohol and bone. *Wien. Med. Wochenschr.* **2014**, *164*, 15–24. [CrossRef]
8. Chakkalakal, D.A. Alcohol-induced bone loss and deficient bone repair. *Alcohol. Clin. Exp. Res.* **2005**, *29*, 2077–2090. [CrossRef]
9. Ventura, A.S.; Winter, M.R.; Heeren, T.C.; Sullivan, M.M.; Walley, A.Y.; Holick, M.F.; Patts, G.J.; Meli, S.M.; Samet, J.H.; Saitz, R. Lifetime and recent alcohol use and bone mineral density in adults with HIV infection and substance dependence. *Medicine (United States)* **2017**, *96*, e6759. [CrossRef]
10. Rosso, M.P.D.O.; Buchaim, D.V.; Pomini, K.T.; Botteon, B.D.C.; Reis, C.H.B.; Pilon, J.P.G.; Duarte Júnior, G.; Buchaim, R.L. Photobiomodulation Therapy (PBMT) Applied in Bone Reconstructive Surgery Using Bovine Bone Grafts: A Systematic Review. *Materials (Basel)* **2019**, *12*, 4051. [CrossRef]
11. Flintoff, K. Oh rats! A guide to rat anaesthesia for veterinary nurses and technicians. *N. Z. Vet. Nurse* **2014**, *20*, 22–27.
12. Weibel, E.R. Stereological Principles for Morphometry in Electron Microscopic Cytology. *Int. Rev. Cytol.* **1969**, *26*, 235–302. [PubMed]
13. Lauing, K.; Roper, P.; Nauer, R.; Callaci, J. Acute alcohol exposure impairs fracture healing and deregulates β-catenin signaling in the fracture callus. *Alcohol. Clin. Exp. Res.* **2012**, *36*, 2095–2103. [CrossRef] [PubMed]
14. Dguzeh, U.; Haddad, N.C.; Smith, K.T.S.; Johnson, J.O.; Doye, A.A.; Gwathmey, J.K.; Haddad, G.E. Alcoholism: A multi-systemic cellular insult to organs. *Int. J. Environ. Res. Public Health* **2018**, *15*, 1083. [CrossRef] [PubMed]

15. Vajgel, A.; Mardas, N.; Farias, B.C.; Petrie, A.; Cimões, R.; Donos, N. A systematic review on the critical size defect model. *Clin. Oral Implant Res.* **2014**, *25*, 879–893. [CrossRef] [PubMed]
16. Wang, W.; Yeung, K.W.K. Bone grafts and biomaterials substitutes for bone defect repair: A review. *Bioact. Mater.* **2017**, *2*, 224–247. [CrossRef]
17. Wood, A.M.; Kaptoge, S.; Butterworth, A.; Willeit, P.; Warnakula, S.; Bolton, T.; Paige, E.; Paul, D.S.; Sweeting, M.; Burgess, S.; et al. Risk thresholds for alcohol consumption: Combined analysis of individual-participant data for 599912 current drinkers in 83 prospective studies. *Lancet* **2018**, *391*, 1513–1523. [CrossRef]
18. De Souza, D.M.; Ricardo, L.H.; Prado, M.D.A.; Prado, F.D.A.; Da Rocha, R.F. The effect of alcohol consumption on periodontal bone support in experimental periodontitis in rats. *J. Appl. Oral. Sci.* **2006**, *14*, 443–447. [CrossRef]
19. Buchaim, R.L.; Buchaim, D.V.; Andreo, J.C.; Roque, D.D.; Roque, J.S.; de Castro Rodrigues, A. Effects of three alcoholic diets on the bone repair in the tibia of rats. *Cienc. Odontol. Bras.* **2009**, *12*, 17–23.
20. Pomini, K.; Cestari, M.; German, I.; Rosso, M.; Gonçalves, J.; Buchaim, D.; Pereira, M.; Andreo, J.C.; Rosa Júnior, G.M.; Della Coletta, B.; et al. Influence of experimental alcoholism on the repair process of bone defects filled with beta-tricalcium phosphate. *Drug Alcohol Depend.* **2019**, *197*, 315–325. [CrossRef]
21. Liberman, D.N.; Pilau, R.M.; Gaio, E.J.; Orlandini, L.F.; Rösing, C.K. Low concentration alcohol intake may inhibit spontaneous alveolar bone loss in Wistar rats. *Arch. Oral. Biol.* **2011**, *56*, 109–113. [CrossRef] [PubMed]
22. Lieber, C.S. Relationships between nutrition, alcohol use and liver disease. *Alcohol Res. Health* **2003**, *27*, 220–231. [PubMed]
23. Feinman, L.; Lieber, C.S. Nutrition and diet in alcoholism. In *Modern Nutrition in Health and Disease*, 9th ed.; Shils, M.E., Olson, J.A., Shike, M., Ross, A.C., Eds.; Williams & Wilkins: Baltimore, MD, USA, 1998; pp. 1523–1542.
24. Velvizhi, S.; Nagalashmi, I.; Essa, M.M.; Dakshayani, K.B.; Subramanian, P. Effects of α-ketoglutarate on lipid peroxidation and antioxidant status during chronic ethanol administration in Wistar rats. *Pol. J. Pharmacol.* **2002**, *54*, 231–236. [PubMed]
25. Thomson, A.D.; Pratt, O.E. Interaction of nutrients and alcohol: Absorption, transport, utilization and metabolism. In *Nutrition and Alcohol*; Watson, R.R., Watzl, B., Eds.; CRC Press: Boca Raton, FL, USA, 1992; pp. 75–99.
26. Aruna, K.; Rukkumani, P.; Varma, S.P.; Menon, V.P. Therapeutic role of Cuminum on ethanol and thermally oxidized sunflower oil induced toxicity. *Phytother. Res.* **2005**, *19*, 416–421. [CrossRef] [PubMed]
27. Kovács, G.L.; Toldy, E. Basal and isoproterenol-stimulated cyclic-adenosine monophosphate levels in mouse hippocampus and lymphocytes during alcohol tolerance and withdrawal. *Alcohol Alcohol.* **2003**, *38*, 11–17. [CrossRef] [PubMed]
28. Kimbrough, A.; de Guglielmo, G.; Kononoff, J.; Kallupi, M.; Zorrilla, E.; George, O. CRF1 receptor-dependent increases in irritability-like behavior during abstinence from chronic intermittent ethanol vapor exposure. *Alcohol. Clin. Exp. Res.* **2017**, *41*, 1886–1895. [CrossRef]
29. Neagu, T.P.; Ţigliş, M.; Cocoloş, I.; Jecan, C.R. The relationship between periosteum and fracture healing. *Rom. J. Morphol. Embryol.* **2016**, *57*, 1215–1220.
30. Spicer, P.; Kretlow, J.; Young, S.; Jansen, J.; Kasper, F.; Mikos, A. Evaluation of Bone Regeneration Using the Rat Critical Size Calvarial Defect. *Nat Protoc.* **2012**, *7*, 1918–1929. [CrossRef]
31. Gosain, A.; Song, L.; Yu, P.; Mehrara, B.; Maeda, C.; Gold, L.; Longaker, M. Osteogenesis in cranial defects: Reassessment of the concept of critical size and the expression of TGF-beta isoforms. *Plast. Reconstr. Surg.* **2000**, *106*, 360–371. [CrossRef]
32. Browne, S.; Pandit, A. Biomaterial-Mediated Modification of the Local Inflammatory Environment. *Front. Bioeng. Biotechnol.* **2015**, *3*, 67. [CrossRef]
33. Chen, Z.; Klein, T.; Murray, R.Z.; Crawford, R.; Chang, J.; Wu, C.; Xiao, Y. Osteoimmunomodulation for the development of advanced bone biomaterials. *Mater. Today* **2016**, *19*, 304–321. [CrossRef]
34. Rocha, C.A.; Cestari, T.M.; Vidotti, H.A.; De Assis, G.F.; Garlet, G.P.; Taga, R. Sintered anorganic bone graft increases autocrine expression of VEGF, MMP-2 and MMP-9 during repair of critical-size bone defects. *J. Mol. Histol.* **2014**, *45*, 447–461. [CrossRef] [PubMed]

35. Ma, M.; Liu, W.; Hill, P.; Bratlie, K.; Siegwart, D.; Chin, J.; Park, M.; Guerreiro, J.; Anderson, D.G. Development of Cationic Polymer Coatings to Regulate Foreign Body Responses. *PLoS ONE* **2017**, *32*, 736–740. [CrossRef] [PubMed]
36. Chaikof, E.L.; Matthew, H.; Kohn, J.; Mikos, A.G.; Prestwich, G.D.; Yip, C.M. Biomaterials and scaffolds in reparative medicine. *Ann. N. Y. Acad. Sci.* **2002**, *961*, 96–105. [CrossRef] [PubMed]
37. Wubneh, A.; Tsekoura, E.K.; Ayranci, C.; Uludağ, H. Current state of fabrication technologies and materials for bone tissue engineering. *Acta Biomater.* **2018**, *80*, 1–30. [CrossRef]
38. Zhang, N.; Ma, L.; Liu, X.; Jiang, X.; Yu, Z.; Zhao, D.; Zhang, L.; Zhang, C.; Huang, F. In vitro and in vivo evaluation of xenogeneic bone putty with the carrier of hydrogel derived from demineralized bone matrix. *Cell Tissue Bank.* **2018**, *19*, 591–601. [CrossRef]
39. Winkler, T.; Sass, F.A.; Duda, G.N.; Schmidt-Bleek, K. A review of biomaterials in bone defect healing, remaining shortcomings and future opportunities for bone tissue engineering. *Bone Jt. Res.* **2018**, *7*, 232–243. [CrossRef]
40. Hanak, C.; Benoit, J.; Fabry, L.; Hein, M.; Verbanck, P.; de Witte, P.; Walter, H.; Dexter, D.T.; Ward, R.J. Changes in pro-inflammatory markers in detoxifying chronic alcohol abusers, divided by lesch typology, reflect cognitive dysfunction. *Alcohol Alcohol.* **2017**, *52*, 529–534. [CrossRef]
41. Shiu, H.; Goss, B.; Lutton, C.; Crawford, R.; Xiao, Y. Formation of Blood Clot on Biomaterial Implants Influences Bone Healing. *Tissue Eng. Part B Rev.* **2014**, *20*, 697–712. [CrossRef]
42. Wang, X.; Luo, Y.; Yang, Y.; Zheng, B.; Yan, F.; Wei, F.; Friis, T.E.; Crawford, R.W.; Xiao, Y. Alteration of clot architecture using bone substitute biomaterials (beta-tricalcium phosphate) significantly delays the early bone healing process. *J. Mater. Chem. B* **2018**, *6*, 8204–8213. [CrossRef]
43. De Paula do Desterro, F.; Sader, M.S.; de Almeida Soares, G.; Vidigal, G.M.; Vidigal, G.M. Can inorganic bovine bone grafts present distinct properties? *Braz. Dent. J.* **2014**, *25*, 282–288. [CrossRef] [PubMed]
44. Hamidabadi, H.; Shafaroudi, M.; Seifi, M.; Bojnordi, M.; Behruzi, M.; Gholipourmalekabadi, M.; Shafaroudi, A.; Rezaei, N. Repair of Critical-Sized Rat Calvarial Defects With Three-Dimensional Hydroxyapatite-Gelatin Scaffolds and Bone Marrow Stromal Stem Cells. *Med. Arch.* **2018**, *72*, 88. [CrossRef] [PubMed]
45. Biguetti, C.; Cavalla, F.; Tim, C.; Saraiva, P.; Orcini, W.; de Andrade Holgado, L.; Rennó, A.; Matsumoto, M. Bioactive glass-ceramic bone repair associated or not with autogenous bone: A study of organic bone matrix organization in a rabbit critical-sized calvarial model. *Clin. Oral Investig.* **2019**, *23*, 413–421. [CrossRef] [PubMed]
46. Pomini, K.T.; Buchaim, D.V.; Andreo, J.C.; de Oliveira Rosso, M.P.; Della Coletta, B.B.; German, Í.J.S.; Biguetti, A.C.C.; Shinohara, A.L.; Rosa Júnior, G.M.; Cosin Shindo, J.V.T.; et al. Fibrin Sealant Derived from Human Plasma as a Scaffold for Bone Grafts Associated with Photobiomodulation Therapy. *Int. J. Mol. Sci.* **2019**, *20*, 1761. [CrossRef] [PubMed]
47. De Oliveira Gonçalves, J.; Buchaim, D.; de Souza Bueno, C.; Pomini, K.; Barraviera, B.; Júnior, R.; Andreo, J.; de Castro Rodrigues, A.; Cestari, T.; Buchaim, R.L. Effects of low-level laser therapy on autogenous bone graft stabilized with a new heterologous fibrin sealant. *J. Photochem. Photobiol. B* **2016**, *162*, 663–668. [CrossRef]
48. Pomini, K.T.; Andreo, J.C.; De Rodrigues, A.C.; De Gonçalves, J.B.O.; Daré, L.R.; German, I.J.S.; Rosa, G.M.; Buchaim, R.L. Effect of low-intensity pulsed ultrasound on bone regeneration biochemical and radiologic analyses. *J. Ultrasound Med.* **2014**, *33*, 713–717. [CrossRef]
49. Escudero, J.S.B.; Perez, M.G.B.; de Oliveira Rosso, M.P.; Buchaim, D.V.; Pomini, K.T.; Campos, L.M.G.; Audi, M.; Buchaim, R.L. Photobiomodulation therapy (PBMT) in bone repair: A systematic review. *Injury* **2019**, *50*, 1853–1867. [CrossRef]

© 2020 by the authors. Licensee MDPI, Basel, Switzerland. This article is an open access article distributed under the terms and conditions of the Creative Commons Attribution (CC BY) license (http://creativecommons.org/licenses/by/4.0/).

MDPI
St. Alban-Anlage 66
4052 Basel
Switzerland
Tel. +41 61 683 77 34
Fax +41 61 302 89 18
www.mdpi.com

Materials Editorial Office
E-mail: materials@mdpi.com
www.mdpi.com/journal/materials

www.ingramcontent.com/pod-product-compliance
Lightning Source LLC
LaVergne TN
LVHW070422100526
838202LV00014B/1511